FUNDAMENTALS OF VERBAL AND NONVERBAL COMMUNICATION AND THE BIOMETRIC ISSUE

NATO Security through Science Series

This Series presents the results of scientific meetings supported under the NATO Programme for Security through Science (STS).

Meetings supported by the NATO STS Programme are in security-related priority areas of Defence Against Terrorism or Countering Other Threats to Security. The types of meeting supported are generally "Advanced Study Institutes" and "Advanced Research Workshops". The NATO STS Series collects together the results of these meetings. The meetings are co-organized by scientists from NATO countries and scientists from NATO's "Partner" or "Mediterranean Dialogue" countries. The observations and recommendations made at the meetings, as well as the contents of the volumes in the Series, reflect those of participants and contributors only; they should not necessarily be regarded as reflecting NATO views or policy.

Advanced Study Institutes (ASI) are high-level tutorial courses to convey the latest developments in a subject to an advanced-level audience.

Advanced Research Workshops (ARW) are expert meetings where an intense but informal exchange of views at the frontiers of a subject aims at identifying directions for future action.

Following a transformation of the programme in 2004 the Series has been re-named and re-organised. Recent volumes on topics not related to security, which result from meetings supported under the programme earlier, may be found in the NATO Science Series.

The Series is published by IOS Press, Amsterdam, and Springer Science and Business Media, Dordrecht, in conjunction with the NATO Public Diplomacy Division.

Sub-Series

A.	Chemistry and Biology	Springer Science and Business Media
B.	Physics and Biophysics	Springer Science and Business Media
C.	Environmental Security	Springer Science and Business Media
D.	Information and Communication Security	IOS Press
E.	Human and Societal Dynamics	IOS Press

http://www.nato.int/science
http://www.springer.com
http://www.iospress.nl

Sub-Series E: Human and Societal Dynamics – Vol. 18 ISSN: 1574-5597

Fundamentals of Verbal and Nonverbal Communication and the Biometric Issue

Edited by

Anna Esposito

Department of Psychology, Second University of Naples and IIASS, Italy

Maja Bratanić

Faculty of Transport and Traffic Sciences, University of Zagreb, Croatia

Eric Keller

IMM, University of Lausanne, Switzerland

and

Maria Marinaro

Department of Physics, University of Salerno and IIASS, Italy

IOS Press

Amsterdam • Berlin • Oxford • Tokyo • Washington, DC

Published in cooperation with NATO Public Diplomacy Division

Proceedings of the NATO Advanced Study Institute on The Fundamentals of Verbal and Nonverbal Communication and the Biometric Issue
Vietri sul Mare, Italy
2–12 September 2006

ISBN 978-1-58603-733-8
Library of Congress Control Number: 2007924024

Publisher
IOS Press
Nieuwe Hemweg 6B
1013 BG Amsterdam
Netherlands
fax: +31 20 687 0019
e-mail: order@iospress.nl

Distributor in the UK and Ireland
Gazelle Books Services Ltd.
White Cross Mills
Hightown
Lancaster LA1 4XS
United Kingdom
fax: +44 1524 63232
e-mail: sales@gazellebooks.co.uk

Distributor in the USA and Canada
IOS Press, Inc.
4502 Rachael Manor Drive
Fairfax, VA 22032
USA
fax: +1 703 323 3668
e-mail: iosbooks@iospress.com

PRINTED IN THE NETHERLANDS

Fundamentals of Verbal and Nonverbal Communication and the Biometric Issue
A. Esposito et al. (Eds.)
IOS Press, 2007

Preface

This volume brings together the invited papers and selected participants' contributions presented at the International NATO-ASI Summer School on "*Fundamentals of Verbal and Nonverbal Communication and the Biometrical Issue*", held in Vietri sul Mare, Italy, September 2–12, 2006.

The School was jointly organized by the Faculty of Science and the Faculty of Psychology of the SECOND UNIVERSITY OF NAPLES, Caserta, Italy, the INTERNATIONAL INSTITUTE for ADVANCED SCIENTIFIC STUDIES "Eduardo R. Caianiello" (IIASS), Vietri sul Mare, Italy, the ETTORE MAJORANA FOUNDATION and CENTRE FOR SCIENTIFIC CULTURE (EMFCSC), Erice, Italy, and the Department of Physics, UNIVERSITY OF SALERNO, Italy. The School was a NATO event, and although it was mainly sponsored by the NATO Programme SECURITY THROUGH SCIENCE, it also received contributions from the INTERNATIONAL SPEECH COMMUNICATION SOCIETY (ISCA) and the INTERNATIONAL SOCIETY OF PHONETIC SCIENCES (ISPhS), as well as from the abovementioned organizing Institutions.

The main theme of the school was the fundamental features of verbal and nonverbal communication and their relationships with the identification of a person, his/her socio-cultural background and personal traits. The problem of understanding human behaviour in terms of personal traits, and the possibility of an algorithmic implementation that exploits personal traits to identify a person unambiguously, are among the great challenges of modern science and technology. On the one hand, there is the theoretical question of what makes each individual unique among all others that share similar traits, and what makes a culture unique among various cultures. On the other hand, there is the technological need to be able to protect people from individual disturbance and dangerous behaviour that could damage an entire community.

As regards to the problem of understanding human behaviour, one of the most interesting research areas is that related to human interaction and face-to-face communication. It is in this context that knowledge is shared and personal traits acquire their significance. In the past decade, a number of different research communities within the psychological and computational sciences have tried to characterize human behaviour in face-to-face communication through several features that describe relationships between facial expressions and prosodic/voice quality; differences between formal and informal communication modes; cultural differences and individual and socio-cultural variations; stable personality traits and their degree of expressiveness and emphasis, as well as the individuation of emotional and psychological states of the interlocutors. There has been substantial progress in these different communities and surprising convergence, and the growing interest of researchers in understanding the essential unity of the field makes the current intellectual climate an ideal one for organizing a Summer School devoted to the study of verbal and nonverbal aspects of face-to-face communication and of how they could be used to characterize individual behaviour.

The basic intention of the event was to provide broad coverage of the major developments in the area of biometrics as well as the recent research on verbal and nonverbal features exploited in face-to-face communication. The focus of the lectures and the discussions was primarily on deepening the connections between the emerging field of technology devoted to the identification of individuals using biological traits (such as voice, face, fingerprints, and iris recognition) and the fundamentals of verbal and nonverbal communication which includes facial expressions, tones of voice, gestures, eye contact, spatial arrangements, patterns of touch, expressive movement, cultural differences, and other "nonverbal" acts. The main objective of the organizers was to bring together some of the leading experts from both fields and, by presenting recent advances in the two disciplines, provide an opportunity for cross-fertilization of ideas and for mapping out territory for future research and possible cooperation. The lectures and discussions clearly revealed that research in biometrics could profit from a deeper connection with the field of verbal and nonverbal communication, where personal traits are analyzed in the context of human interaction and the communication Gestalt.

Several key aspects were considered, such as the integration of algorithms and procedures for the recognition of emotional states, gesture, speech and facial expressions, in anticipation of the implementation of other useful applications such as intelligent avatars and interactive dialog systems.

Features of verbal and nonverbal communication were studied in detail and their links to mathematics and statistics were made clear with the aim of identifying useful models for biometric applications.

Recent advances in biometrics application were presented, and the features they exploit were described. Students departed from the Summer School having gained not only a detailed understanding of many of the recent tools and algorithms utilized in biometrics but also an appreciation for the importance of a multidisciplinary approach to the problem through the analysis and study of face-to-face interactions.

The contributors to this volume are leading authorities in their respective fields. We are grateful to them for accepting our invitation and making the school such a worthwhile event through their participation.

The contributions in the book are divided into four sections according to a thematic classification, even though all the sections are closely connected and all provide fundamental insights for cross-fertilization of different disciplines.

The first section, GESTURES and NONVERBAL BEHAVIOUR, deals with the theoretical and practical issue of assigning a role to gestural expressions in the realization of communicative actions. It includes the contributions of some leading experts in gestures such as Adam KENDON and David MCNEILL, the papers of Stefanie SHATTUCK-HUFNAGEL et al., Anna ESPOSITO and Maria MARINARO, Nicla ROSSINI, Anna ESPOSITO et al., and Sari KARJALAINEN on the search for relationships between gestures and speech, as well as two research works on the importance of verbal and nonverbal features for successful communication, discussed by Maja BRATANIĆ, and Krzysztof KORŻYK.

The second section, NONVERBAL SPEECH, is devoted to underlining the importance of prosody, intonation, and nonverbal speech utterances in conveying key aspects of a message in face-to-face interactions. It includes the contributions of key experts in the field, such as Nick CAMPBELL, Eric KELLER, and Ruth BAHR as research papers, and related applications proposed by Klara VICSI, Ioana VASILESCU and Martine ADDA-DECKER, Vojtěch STEJSKAL et al., Ke LI et al., Elina SAVINO, and

Iker LUENGO et al. Further, this section includes algorithms for textual fingerprints and for web-based text retrieval by Carl VOGEL, Fausto IACCHELLI et al., and Stefano SQUARTINI et al.

The third section, FACIAL EXPRESSIONS, introduces the concept of facial signs in communication. It also reports on advanced applications for the recognition of facial expressions and facial emotional states. The section starts with a theoretical paper by Neda PINTARIĆ on pragmemes and pragmaphrasemes and goes on to suggest advanced techniques and algorithms for the recognition of faces and facial expressions in the papers by Praveen KAKUMANU and Nikolaos BOURBAKIS, Paola CAMPADELLI et al., Marcos FAUNDEZ-ZANUY, and Marco GRASSI.

The fourth section, CONVERSATIONAL AGENTS, deals with psychological, pedagogical and technological issues related to the implementation of intelligent avatars and interactive dialog systems that exploit verbal and nonverbal communication features. The section contains outstanding papers by Dominic MASSARO, Gerard BAILLY et al., David HOUSE and Björn GRANSTRÖM, Christopher PETERS et al., Anton NIJHOLT et al., and Bui TRUNG et al.

The editors would like to thank the NATO Programme SECURITY THROUGH SCIENCE for its support in the realization and publication of this edition, and in particular the NATO Representative Professor Ragnhild SOHLBERG for taking part in the meeting and for her enthusiasm and appreciation for the proposed lectures. Our deep gratitude goes to Professors Isabel TRANCOSO and Jean-Francois BONASTRE of ISCA, for making it possible for several students to participate through support from ISCA. Great appreciation goes to the dean of the Faculty of Science at the Second University of Naples, Professor Nicola MELONE, for his interest and support for the event, and to Professor Luigi Maria RICCIARDI, Chairman of the Graduate Program on Computational and Information Science, University of Naples Federico II, for his involvement and encouragement. The help of Professors Alida LABELLA and Giovanna NIGRO, respectively dean of the Faculty and director of the Department of Psychology at the Second University of Naples, is also acknowledged with gratitude.

Special appreciation goes to Michele DONNARUMMA, Antonio NATALE, and Tina Marcella NAPPI of IIASS, whose help in the organization of the School was invaluable.

Finally, we are most grateful to all the contributors to this volume and all the participants in the 2006 Vietri Summer School for their cooperation, interest, enthusiasm and lively interactions, making it not only a scientifically stimulating gathering but also a memorable personal experience.

This book is dedicated to those who struggle for peace and love, since peace and love are what keep us persevering in our research work.

The EDITORS:
Anna ESPOSITO, Maja BRATANIĆ, Eric KELLER, Maria MARINARO

Based on keynote presentations at the NATO Advanced Study Institute on Fundamentals of Verbal and Nonverbal Communication and the Biometrical Issue (the 11th *Eduardo R. Caianiello* International School on Neural Nets).

Vietri sul Mare, Italy
2–12 September 2006

The support and the sponsorship of:

- NATO Programme Security Through Science
- Second University of Naples,
  Faculty of Psychology and Faculty of Science (Italy)
- International Institute for Advanced Scientific Studies
  "E.R. Caianiello" (IIASS), Italy
- International Speech Communication Association (ISCA)
- The International Society of Phonetic Sciences (ISPHS)
- Universita' di Salerno,
  Dipartimento di Scienze Fisiche E.R. Caianiello (Italy)
- Regione Campania (Italy)
- Provincia di Salerno (Italy)

is gratefully acknowledged.

International Advisory and Organizing Committee

Maja Bratanić, Faculty of Transport and Traffic Sciences, University of Zagreb, Croatia
Anna Esposito, Department of Psychology, Second University of Naples and IIASS, Italy
Eric Keller, IMM, University of Lausanne, Switzerland
Maria Marinaro, Department of Physics, University of Salerno and IIASS, Italy

Local Organizing Committee

Anna Esposito, Second University of Naples and IIASS, Italy;
Alida Labella, Second University of Naples, Italy;
Maria Marinaro, Salerno University and IIASS, Italy;
Nicola Melone, Second University of Naples, Italy;
Antonio Natale, Salerno University and IIASS, Italy;
Giovanna Nigro, Second University of Naples, Italy;
Francesco Piazza, Università Politecnica delle Marche, Italy;
Luigi Maria Ricciardi, Università di Napoli "Federico II", Italy;
Silvia Scarpetta, Salerno University, Italy;

International Scientific Committee

Guido Aversano, CNRS-LTCI, Paris, France;
Gérard Bailly, ICP, GRENOBLE, France;
Ruth Bahr, University of South Florida, USA;
Jean-Francois Bonastre, Universitè d'Avignon, France;
Nikolaos Bourbakis, ITRI, Wright State University, Dayton, OH, USA;
Maja Bratanić, University of Zagreb, Croatia;
Paola Campadelli, Università di Milano, Italy;
Nick Campbell, ATR Human Information Science Labs, Kyoto, Japan;
Gerard Chollet, CNRS-LTCI, Paris, France;
Muzeyyen Ciyiltepe, Gulhane Askeri Tip Academisi, Ankara, Turkey,
Francesca D'Olimpio, Second University of Naples, Italy;
Anna Esposito, Second University of Naples, and IIASS, Italy;
Aly El-Bahrawy, Faculty of Engineering, Cairo, Egypt;
Marcos Faundez-Zanuy, Escola Universitaria de Mataro, Spain;
Dilek Fidan, Ankara Universitesi, Turkey;
Antonio Castro Fonseca, Universidade de Coimbra, Coimbra, Portugal;
Björn Granstrom, Royal Institute of Technology, KTH, Sweden;
David House, Royal Institute of Technology, KTH, Sweden;
Eric Keller, Université de Lausanne, Switzerland;
Adam Kendon, University of Pennsylvania, USA;

Alida Labella, Second University of Naples, Italy;
Maria Marinaro, Salerno University and IIASS, Italy;
Dominic Massaro, University of California – Santa Cruz, USA;
David McNeill, University, Chicago, IL, USA;
Nicola Melone, Second University of Naples, Italy;
Antonio Natale, Salerno University and IIASS, Italy;
Anton Nijholt, University of Twente, The Netherlands;
Giovanna Nigro, Second University of Naples, Italy;
Catherine Pelachaud, Universite de Paris 8, France;
Francesco Piazza, Università Politecnica delle Marche, Italy;
Neda Pintarić, University of Zagreb, Croatia;
José Rebelo, Universidade de Coimbra, Coimbra, Portugal;
Luigi Maria Ricciardi, Università di Napoli "Federico II", Italy;
Zsófia Ruttkay, Pazmany Peter Catholic University, Hungary;
Yoshinori Sagisaka, Waseda University, Tokyo, Japan;
Silvia Scarpetta, Salerno University, Italy;
Stefanie Shattuck-Hufnagel, MIT, Cambridge, MA, USA;
Zdenek Smékal, Brno University of Technology, Brno, Czech Republic;
Stefano Squartini, Università Politecnica delle Marche, Italy;
Vojtich Stejskal, Brno University of Technology, Brno, Czech Republic;
Isabel Trancoso, Spoken Language Systems Laboratory, Portugal;
Luigi Trojano, Second University of Naples, Italy;
Robert Vich, Academy of Sciences, Czech Republic;
Klára Vicsi, Budapest University of Technology, Budapest, Hungary;
Leticia Vicente-Rasoamalala, Alchi Prefectural University, Japan;
Carl Vogel, University of Dublin, Ireland;
Rosa Volpe, Université D'Orléans, France;

Contents

Section 1

Gestures and Nonverbal Behaviour

Fundamentals of Verbal and Nonverbal Communication and the Biometric Issue
A. Esposito et al. (Eds.)
IOS Press, 2007

Some Topics in Gesture Studies

Adam KENDON[1]
University of Naples 'Orientale', Italy
Institute for Research in Cognitive Science, University of Pennsylvania, USA

Abstract. Five topics in gesture studies are briefly discussed and references are added so that the discussion can serve as a means by which the reader can pursue them further. We begin with the question "What is a gesture?", then follow with a discussion of issues regarding the relationship between gesture and speech, what is involved in the interpretation of gestures as expressive acts, how skill in gesturing is acquired and the question of cultural differences in gesture use.

Keywords. Gesture, gesture studies, expression, speech, cultural differences

Gestures are the 'visible acts' we employ as a part of how we *say things* to one another (Kendon [1]: 7-16). We point at things to show what we are referring to and when we describe objects or actions we may use gestures to enrich, clarify or elaborate our descriptions. When we use metaphors to express abstract ideas we may use gestures as visual aids to these expressions, showing abstract processes as diagrams or movement patterns or logical relations as if they can be laid out in space. We also use gestures to show what kinds of actions we are taking with our utterances: with gestures we can, among other things, show agreement or disagreement, affirmation or denial, we can show that we are asking a question or begging another's indulgence, that we are doubtful of something or that what we are saying is hypothetical. We use gesture as a way of saying things to others when our distance from them is too great to use speech or when there is too much noise and we can't be heard. And if circumstances prevail where it is not normally possible to use speech, either because of deafness or for ritual reasons, as in some tribal societies, gestures can come to be elaborated in various ways to form a true language. Gesture also plays a very important role in how we conduct our relationships with one another: We wave in greeting, we shake hands in friendship, we embrace and kiss in showing love and affection, and with gestures we can display hostility and threat, show superiority to another, or show our subordination.

The topic of 'gesture', defined more or less as is suggested in the preceding paragraph, has been attracting much interest in recent years. Although the history of the study of gesture reaches back to classical antiquity ([1]: 17-107 provides a history), interest in gesture is nowadays more widespread than it was. Interest is no longer the province of specialists. There are several reasons for this.

---

[1] Corresponding Author: Adam Kendon, University of Naples 'Orientale' and Institute for Research in Cognitive Science, University of Pennsylvania, e-mail: adamk@dca.net.

First, studies of how gesture is organised in relation to speaking have suggested that speech and gesture are inseparable aspects of a single underlying process of utterance production (Kendon [2], [3], [1]; McNeill [4]). Yet, given that gesture, which is pantomimic and enactive, functions as a form of expression in a way that is quite different from spoken language (which uses syntax and established lexical forms), the question arises as to how a person can organise and integrate two such very different modes of expression at one and the same time. Taking gesture into consideration and recognising it as an integral part of the act of utterance production has raised important questions concerning the relationship between verbal language, thought and imagery. As David McNeill has proposed [4], [5], the study of the gestures speakers use when speaking can throw a new sort of light upon the conceptual organisation that underlies linguistic production.

Second, the close study of how interaction proceeds, made possible since the wide availability of cheap high-quality video recording, has made it clear that communication in interaction is a *multimodal* process. A full understanding of how people respond to one another in interaction requires that we take into consideration their visible bodily actions as well as their words, and this includes gesture. The work of Charles Goodwin, Jurgen Streeck, John Haviland and Christian Heath, among others, has been especially revealing here ([6], [7], [8], [9], [10], [11], [12], [13], [14], [15], [16], [17]).

Third, gesture has long attracted interest because it seems to be a 'universal' and 'natural' form of expression. Yet, at the same time, it can be shown to be governed by social conventions. Especially interesting is the fact that when speech is not available for use, whether for environmental or for physiological reasons, gesture can be used as an alternative and, in the right circumstances, it may become fashioned into a fully elaborated language ([1]: 284-306). Through the study of this process we may gain special insights into the way in which individually created forms of expression get transformed into socially shared and conventionalised forms. This seems to suggest that we may even see something of the process by which a language comes to be formed ([18], [19], [20], [21]). The great interest in sign languages that has developed since 1960 (from the work of William Stokoe [22] and Ursula Bellugi onwards [23]) has made an important contribution to this. Issues concerned with the differences and similarities between 'sign' and 'gesture' are now widely discussed (cf. [1]: 307-205; [24], [25], [26]).

Fourth, the revival in academic circles of the discussion of the problem of language origins has also contributed to the renewed interest in gesture. The idea that gestures might have constituted the first form of language was first advanced in the eighteenth century. In recent years this idea has found renewed favour with a number of scholars, for example in the work of Gordon Hewes [27], William Stokoe and colleagues, [28], and Michael Corballis [29] and their discussions have contributed to the growth of interest in gesture.

Finally, there is a trend toward a new recognition that our perception of the things around us, the way we express things, the way we think about things, the concepts we entertain are deeply conditioned by the unavoidable fact that we are *embodied* creatures. There is a growing recognition that our basic notions of space, of the nature of physical objects, of time, our sense of how we see things, depends in a deep way upon the facts of our existence in the world as corporeal beings (Anderson [30] provides a comprehensive introduction). And so it is that we are coming to see that 'language' is not an abstract system detached from practical action but that it is, rather, a system grounded in such action. Visible bodily action, as seen in gesture, is an integral part of this. For this reason, also, gesture has become an object of great interest to many scholars. So much so, indeed, that academic institutions are emerging for dealing with this. Thus an International Society for Gesture Studies has been founded and is now in its sixth year (Müller [31]). A journal devoted to gesture studies, *Gesture*, has been in publication since 2001.

In what follows I will touch briefly and lightly on a few of the topics that are current in the field of gesture studies, doing so from a personal point of view and making no attempt at being comprehensive. The bibliographical references are added to make this essay useful as a guide to the literature regarding the topics touched upon. I begin with a rather fundamental question: What is a 'gesture'? Can we establish criteria by which we may decide whether a given act is a gesture or not? I will follow with a discussion of some issues surrounding the study of the relationship between gesture and speech, touch briefly on the issue of the learning of gesture and, in the last section I will say something about cultural differences and how we might go about trying to account for them.

1. What is 'gesture'?

In our everyday dealings with one another, we tend to make distinctions in how we attend to one another's actions, treating each other's streams of action in interaction in a highly differentiated way (see Goffman [32]:210-215; and also Kendon [33], [34]: Ch. 8, [1]: Ch. 2). Thus, participants in interaction tend to show, in how they deal with one another, that for each there are actions by the other person that are regarded as part of their wilful expression, that is, part of what they intend to express to one other. This includes spoken utterances but it also includes the manual and other bodily actions that occur in co-ordination with spoken utterances or sometimes in place of them or in alternation with them. There are other actions, in contrast, that tend to be regarded as manifestations or symptoms of feelings or attitudes, rather than as wilful expressions. Blushing, spontaneous smiling or laughter, displays of embarrassment, and the like would be included here. There are yet other kinds of actions that, in the course of ordinary interaction, tend to be overlooked, or discounted, for they are regarded as purely personal and are not treated (at least, not explicitly) as being relevant to the shared focus or task of the conversation. Actions of this kind include passing adjustments to clothing, small shifts in posture, patting the hair, momentary scratchings or other kinds of self-touchings, and the like. Further, components of bodily conduct such as carriage, posture, orientation, and relative spatial positioning of the participants, which play a very important role in the way an encounter is organised (see Scheflen [35] and Kendon [34], [36]), tend not to be given explicit attention (although sometimes they are) and are not usually part of what any participant might report about an encounter to someone else, afterwards. They play a role in the 'staging' or the 'framing' of the occasion of interaction but are not a part of its 'explicit content' (or 'story-line' of the encounter, as Goffman has expressed it) .

It is the visible bodily actions that occur as an integral part of the activity treated as contributing directly to the 'explicit content' or 'story line' of the encounter that most often attract the term 'gesture'. What is interesting is that it appears that these movements can be recognised as such on the basis of their appearance alone. Elsewhere, I have attempted to list some features that actions recognised as 'gestures' tend to have ([1], [33]). I have proposed that the more of these features an action is perceived as having, the more likely it is that it will treated as if it is a wilful, gestural expression. Based on this idea, it seems, we may be able to devise a scale in terms of which we could assess movements in terms of the degree to which they exhibit the characteristics of gesture. 'Gestuality', thus, is to be seen as a matter of degree, not a category.

The features that I have proposed so far, include the following. First of all, actions regarded as 'gesture' - regardless of what parts of the body may be involved, very often have the character of excursions. That is, the action involves a departure from a base position of some kind and then a *return* to such a baseline. This is most obvious in the case of manual movements that are counted as 'gesture' but it also applies to other sorts of movements and is part of what we rely on when we distinguish a turn of the head that is a change of orientation, from a turn of the head that is part of a gesture (for example, as in a head shake).

Second, such excursions tend to manifest a phase in which the 'effort-shape' dynamics of the movement emerge as most well defined. The term 'effort-shape' is here taken from the work of Laban and Lawrence ([37]; see also Dell [38] and Bartinieff and Lewis [39]) and it refers to the manner in which a movement is made. For example, a person may raise and then lower his hand, but whether he does this rapidly or slowly, whether he adds to this raising and lowering a definite path or trajectory for the movement, whether he slows up and sustains his hand in a raised position for a moment before lowering, or not, will make all the difference in how the movement is perceived and interpreted by others.

Actions of the hands and arms that are typically deemed to be 'gestures' can often be shown to have a phase during which the hand or hands are 'merely' being moved into a certain spatial position in relation to the speaker's body and then, when they reach that position, they engage in some pattern of action that is recognised as 'expressive' in some way. Thus we can often distinguish between the 'preparation' phase of a gesture and its 'stroke' phase, the 'stroke' being the phase where a distinctive pattern of action is performed, with well articulated 'efforts' and which is judged to be the phase of action where the expression of the action, whatever it may be, is most clearly to be discerned (Kendon [3], [1]: 108-126; McNeill [4]; Kita, van Gijn and van der Hulst [40]).

I think it would be both interesting and important to pursue systematic studies of human movement perception with a view to finding out what features actions commonly deemed to be 'gestures' tend to have. It would be interesting because how we perceive and differentially attend to and so deal with each others actions is a fundamental part of the process of interaction and should be fundamental to our understanding of the domain of phenomena we are interested in. It is also of practical or methodological importance. Everybody working on 'gesture' discards some aspects of the behaviour of the people being observed as *not* being part of their concern. However, just what the criteria are that are used in deciding what is and what is not of interest for a study of 'gesture' is not usually made very explicit. Further, for those interested in the idea of using a machine to automate the processes of encoding behaviour, it will be necessary to be able to tell the machine what movements are to be counted and what are not. And for this an explicit spelling out of what features movements must *manifestly* have to be counted as 'gesture' would be very important. This is also important, of course, for those wishing to develop human-machine systems in which the machine responds to commands given by bodily actions. In such a case it will be important to ensure that the machine can recognise 'significant' human actions and not respond to those that are not 'significant'.

2. Gesture and Speech

In recent years much attention has been directed to how gesture and speech are related. Over thirty years ago I published what I think was the first attempt to examine in detail the relationship between the phrasal structure of spoken discourse and the phrasal structure of the gesturing that was concurrent with it [2]. This study was a very detailed one and the description I gave was done in relation to only about two minutes of discourse in a single speaker. I might add that in that study I also examined the patterning of head movements and postural shifts - indeed the entire kinesic profile (something rarely attempted since by me, or anyone else, as far as I know). This analysis showed how phrases of gesticulation (distinguished as units of movement which have, as a 'nucleus' a phase in which 'efforts' become most prominent, usually preceded by a preparatory phase of movement leading up to this nucleus or 'stroke', as it was later termed) could be seen as organised as a nested hierarchy of phrases that matched the nested hierarchy of speech phrases into which the spoken discourse could be analysed. In this study the speech phrases were distinguished in prosodic terms, the equivalent of *tone units* as these have been defined by David Crystal [41]. As I wrote in that paper: "Just as the flow of speech may be regarded as an hierarchically organised set of units, so we see the patterns of body motion that are associated with it as organised in a similar fashion, as if each unit of speech has its 'equivalent' in body motion." [2]: 204. Noting that the elaborate flow of gesticulatory movements analysed was co-ordinated with highly fluent speech, I concluded that "speech and movement that directly accompanies it, at least, are under the guidance of the same controlling mechanism" [2]: 206. As I put it in the title of another paper, published somewhat later, "gesture and speech are two aspects of the process of utterance." [3].

These findings were confirmed through independent observations by David McNeill who subsequently [4] went on to develop in considerable detail the argument for regarding co-speech produced gesture to be a symbolic expressive activity that is a product of a process that is the same process that governs the production of speech. To put it in another way, speech and gesture, it seems, are two components of a single underlying process through which an integrated ensemble of gesture and speech are produced.

David McNeill has given careful consideration to the manual actions of gesticulation, interpreted as expressive or symbolic actions. He has shown that they express, in various ways, aspects of the content of what the speaker is saying but he shows that this content is by no means necessarily the same as the content of the words. Nevertheless, it is intimately related to this, so that when we take both gesture and words into consideration together we may get a much enriched and enhanced view of what it is that the speaker may be saying (see also [1]: 158-198 for a discussion of how gesture can contribute to the referential meaning of an utterance).

In the discussion of the significance of the relationship between co-produced gestures and speech, a number of different theoretical views have been put forward. Perhaps with some risk of oversimplification, in recent discussions these theoretical views can be divided into two camps. On the one hand there are what I call *speech auxiliary* theories of the gesture-speech relationship. On the other hand there are what I call *partnership* theories. According to speech auxiliary theories, gestures function only as aids to the speaker as he seeks to arrive at a verbal expression. Partnership theories, on the other hand, see gesture as serving as components of the speaker's final product, the utterance being a gesture-speech *ensemble* where gesture is a part of the *final* expression.

An assumption not always made explicit in speech auxiliary theories is that, when speakers speak, it is their project to formulate *everything* in words. To the extent that they cannot do this, they are failing. Gesture, on this view, is resorted to as an *aid* to this project of complete verbal formulation. That is to say, the gestures that people produce when they speak are produced as a consequence of some failure to accomplish the task of verbalisation and the gestures serve to help in this task. Various ways in which they might do so have been proposed, including the idea that gestures are somehow an aid to the organisation of thought, as argued by Norbert Freedman [42] and Sotara Kita [43], and Bernard Rimé [44] or, as Robert Krauss has argued, that they might help in the process of finding needed words [45], [46]. Theories that look upon gesturing as an auxiliary to the process of verbal utterance tend not to regard gesture as making any significant contribution to the meaning of the speaker's utterance as finally produced, although gestures may be interesting from the point of view of what they might reveal about the mental processes involved in verbal production.

In the view of *Partnership* theories, on the other hand, speakers are not necessarily and always aiming at purely verbal formulations. Rather, it is supposed that what speakers are trying to do is to produce an *utterance* that will serve as an appropriate and adequate unit within the exchange of turns within which speakers find themselves. To produce such an utterance the speaker may use such resources as are available and can be mobilised within the frame of the pertinent moment, and this may involve the use of words, it may involve the use of gestures, or it may involve the use of both gesture and speech together in some kind of co-operative relationship. On the partnership view gestures are *partners* with speech in utterance construction ('utterance' here being regarded as the unit of expressive action that constitutes or is a component of a 'turn' in a conversation, and which may be constructed from with talk or gesture or from a combination of both modalities). The gestures a speaker may employ are to be understood as components of the utterance as *finally produced.*

This partnership view is explicitly adopted by Kendon [1], but it is also the view that is implicit in the work of McNeill [4], Calbris [47], Clark [48], Gullberg [49] and de Ruiter [50] [51], among others. These authors differ in how they understand the processes that are involved in the production of gestures and how these are co-ordinated with speech, but they all regard co-speech gestures as being produced as an integral part of the utterance production process and they all regard them as contributing directly to its referential or propositional meaning.

There are several points this view seems to imply that are worth commenting on. First of all, a speech-partnership view has suggested to some that speakers ought to be aware of their gestures, rather as they are said to be of their words. Yet it is often claimed that so-called spontaneous gestures are made unconsciously and for this reason must be beyond the full control of the speaker. In response to this there are two points to be made. First, to do something unwittingly is not the same as doing something unwilfully. What I am conscious of is largely a matter of what I happen to be paying attention to. It is this that, in consequence, I am able to recall and report about later. As with any complex skilled activity, we are rarely conscious of the details of how we achieve the results of our actions. What we are aware of, rather, are the results of our actions. Generally our attention is focused on how results of an action ensemble meet or fail to meet a current aim, not usually on the procedural details of how to carry out these actions. In much the same way, when producing an utterance, it is the meaning aimed at that is at the focus of our attention. If, after saying something, someone is asked to say again what they said, it is the *meaning* that is repeated, much less often the exact words, unless the person has been forewarned that it this is that is wanted. Even then it is not easy to recall the exact words just uttered. We do have a greater running awareness of the words we are uttering than we do of our gestures, probably because we get immediate feedback from our words, since we hear ourselves speak but do not see ourselves gesture. Yet it is perfectly possible to train oneself to be aware of one's own gestures and some people seem to be much more aware of them than others are. This would be worth further inquiry. One might expect that actors, or others who make a business of putting on performances and are concerned about the impact that they have on their audience, would often have quite a good understanding of the gestures they produce.

If one takes a speech-partnership view of gesture, one regards the gestural component of an utterance, just like the verbal component, as being produced under the guidance of a *communicative intention* on the part of the speaker. Utterances are wilful acts, shaped in the light of the speaker's aims, as well as the immediate demands of the interaction moment, and this applies as much to the gestures employed as it does to the words. However, it is one thing to shape one's utterance according to communicative demands and intentions and another thing to be successful in communicating to others (Melinger and Levelt [52] make this point in a particularly clear manner). It is certainly a matter of great interest to know how other participants pay attention to the utterances of a speaker, and what components are for them important, and how they interpret or understand them. Recent research has begun to examine the ways in which fellow participants pay attention to gesture (there are studies which have attempted to use eye-tracking technology for this purpose, see Gullberg and Holmquist [53]) and there are also a number of studies which have attempted to find out what contribution gestures appear to make to an observer's understanding of an utterance (Beattie and Shovelton [54], [55], [56], [57]; for a review of earlier work on this see Kendon [58]). Work along these lines has shown, for instance, that speakers who are recounting episodes from a cartoon use gesture to convey important information about the size and shape and spatial arrangements of objects in the cartoon. This is shown when observers of the utterance are found not to acquire such information if they are only allowed to observe the spoken component.

The issue of how co-participants pay attention to utterances, which components are important for them, and, in particular, how and when gesture makes a contribution to their understanding and the nature of their response are very interesting and important issues. However, the question as to what governs the speaker's production of gestures does not stand or fall by the results of such studies as some writers appear to have supposed. Everything I say and do in giving a lecture, for example, is governed by my communicative intentions. But this is no guarantee at all that everything I am doing and saying will be understood either in part, or even in whole, by my audience. I can try a variety of techniques - voicing variations, use various devices to structure the sequence of my arguments, or make use of gestures as I try to get my meaning across. If you do not grasp it, however, - and you certainly may very well may not grasp all of my meanings - this does not mean that what I was trying to do was not governed by my communicative intentions.

3. The problem of interpretation

It is common practice, I think, when we look at an utterance and consider its gestural component in relation to the verbal component, to *interpret* the gestures we see as if they are actions that depict something or describe something. It is interesting to consider how we arrive at the interpretations of such gestures and what is involved in the process by which they are deemed to add to the meaning of the utterance of which they are a part. Let me take a specific example. MC has been talking about his father, the owner of a grocery shop, and he is describing how the cheeses that his father used to sell were delivered to him in big crates. He has described how long the crates were in which they arrived, the shape of these crates and something of their construction. And then he says: "There used to be two cheeses in each crate." As he says "there used to be two cheeses in each crate" he holds both his hands forward, horizontally separated somewhat, each hand with the fingers spread but slightly flexed, with the palm of each hand oriented so it faces downwards (see Fig. 9.2D in [1]:166). The hands, held in this position, are moved up and down several times rapidly in movements of a very small amplitude. Now from this gesture, I maintain, most people 'see' two big round cheeses sitting side-by-side in the crate he has described. From his gesture it is understood that the cheeses were not one on top of the other nor one in front of the other. They are set side by side in the crate. But how is this understanding arrived at? If we were to be just shown the gesture and had no knowledge of the discourse of which it was a part, we would not be able to interpret it in this way. So there is something about the way in which the gestural action interacts with the discourse context which permits us this understanding and allows the gesture to contribute the additional information it does about the spatial disposition of the cheeses that is not in the speaker's words.

It would be quite wrong to say that the discourse context and the verbal component *wholly determine* our understanding of the gesture. It would not do for MC to do *any sort of gesture*. To convey information about the cheeses and how they are disposed in the crate, he is quite restricted in what he can do. And what he has done is to select from the methods of gestural representation that are available to him a type of hand shape that is used when reference is being made to a large object that has a round shape, but which is not so large that a hand cannot encompass it, at least partially. At the same time, he uses a movement feature that is used to indicate that the object referred to by the chosen hand shape is not being manipulated, is not being moved nor is it moving of its own accord, but is an object *in place*. The small amplitude back and forth movements that MC uses here are movements commonly used when the spatial position of something is being indicated. We, as fellow gesture users and we as creatures for whom the actions of others are intelligible as such, can perceive these actions of MC's gestures as actions that have to do with showing objects of a certain size and shape *in position.* Note that the neurological foundations for an understanding of this intelligibility are being laid in the recent research on 'mirror neurons'. See Stamenov and Gallese [59] and Rizzolatti and Sinigaglia [60]. What the discourse context contributes is what these object *are* - in this case they are cheeses.

I am suggesting here that when people engage in gesturing that depicts objects, show spatial relationships, indicate movement patterns and actions, and the like, they are drawing upon a widely shared *repertoire* of depictional strategies. Rather as we can, for a visual artist - a cartoonist, for example - show that he has a vocabulary of lines and forms that are used in fairly standard ways to bring off his effects, so I believe that it can be shown that the techniques used by gesturers as they depict objects and actions in relation to their discourse (and this applies as much when what is being depicted serve metaphorical functions as it does when what is being depicted is literal - as in the case I have used here) are not made up on the spot but follow certain well established rules and principles. Such gestures may be *improvised*, to be sure, but just as in musical improvisation the instrumentalist draws upon existing repertories of rhythmical forms, harmonies, note sequences, and the like so, I believe, the gesturer likewise has a repertoire of expressive forms and devices that can be drawn upon.

Already there are a number of authors who have written about techniques of representation in gesture from this point of view, including Wilhelm Wundt [61], Cornelia Müller [62], and Tim Sowa [63], [64]. I submit that this would be well worth taking further, and systematic research should be done to find out what these techniques and strategies are and how widely and consistently they are used and how they are interpreted. I think it is only because such consistent and widely shared strategies exist that people's depictive gestures are intelligible.

4. Learning to gesture

If this is right, then the next very interesting question we might raise is: *do people have to learn how to gesture in appropriate and intelligible ways?*

There have been a number of studies of gesture in very young children which seem to suggest that at very early stages in the emergence of language children will create and use gestures as alternatives to words. For a useful survey, mainly of the work carried out by Virginia Volterra and her colleagues in Rome, see [65]. Although, as spoken language becomes established, this sort of gesture use appears to decline, as the child begins to master the capacity for speaking in longer sentences, gesture employed in conjunction with speech returns. We know rather little about the further development of gesture use in children, however. After children have reached the age of two or three years, studies of the further development of linguistic expression, whether in speech or gesture, become much less frequent. I think it would be most interesting to undertake further studies of the development of gesture use in children throughout all the ages of childhood, up to and including adolescence. An early paper pertaining to this theme is by Jancovic, Devoe and Wiener [66]. Recent work by J.M.Colletta [67] has greatly expanded the study of gesture in children of later ages.

Observations by myself and Carla Cristilli, among others, collected from video-tapes made in Naples of children of various ages have suggested a number of interesting things that deserve further exploration. In children of kindergarten age we have noted that children may already be making use of some of the more highly codified gestures from the Neapolitan repertoire but that they do so in ways that suggest that have not yet mastered the contexts in which the use of these expressions is appropriate. It is somewhat reminiscent of the way in which children sometimes, in acquiring a new word, over-generalise its use. Although these children may know how to make the gestures, and although they have some notion of their contexts of use, they nevertheless may use these gestures in contexts where they would not be used by adults.

Second, we collected a number of short passages in which we noted how children of kindergarten age - between the ages of five and six years - sometimes corrected their own gesturing to align it more with adult practice. For example, a small boy telling a Donald Duck story extends his arm vertically above his head, hand with extended index finger, as he is explaining how Donald Duck goes to the top of a very tall mountain. Behind him, a teacher is aiding him in telling the story, prompting him when he has difficulty in finding words. She, also, is gesturing and here, in relation to the moment that Donald Duck goes to the top of the mountain, she also has extended her arm above her head in a way similar to the small boy. She is using an open hand shape, however. The small boy, while in the midst of his gesture, looks round at the teacher behind him and then changes his hand shape to conform to that of the teacher before he continues. In another passage we observed how a small girl was following a story being told by a teacher, making gestures as she did so, the gestures clearly being versions of those the teacher herself was using (some of these examples have now been described by Cristilli and D'Agostino [68]).

In a third set of observations children of different ages were observed telling the same story - the story of Pinocchio (this material has been described in an unpublished graduation thesis by Serafina Solimeno in 1997 of the University of Naples 'Orientale'). Comparisons were made in the gesturing observed and one thing that was very noticeable was that, in children of eleven or twelve years, their descriptive gesturing was much more precise. In describing Pinocchio's long nose, for example, younger children might make a movement forward from the nose suggestive of something elongate, but this movement was poorly defined in its form and did not end in a distinct way. Older children used much more precise gestures, delineating Pinocchio's long nose with movements which showed a high degree of shaping. It is as if, as children approach the beginning of the second decade, their control over the gestures they produce becomes more refined. They have, by this age, acquired the skill necessary to make well defined and intelligible depictive gestures, which at earlier ages they were much less able to do.

From these preliminary observations it appears that small children often pay close attention to the gestures of the adults interacting with them and use these gestures as models in relation to which they correct their own. It also appears that children improve with age in their ability to carry out descriptive gestures, becoming able to make them in ways that are better articulated and that fit better with the requirements that such actions must have if they are to be recognisable as descriptive gestures. Much further careful observation is needed, however, so that we may have better and fuller descriptions of these changes in gesture use and performance with age, as well as a better understanding of the processes that may be involved in bringing about these changes.

5. Culture and gesture

Everybody knows that there are differences from one culture to another in the nature of gesture use. Differences in gesture use are part of the stereotypes for different cultures. For instance, it has been maintained, at least since the eighteenth century, that southern

Europeans, especially Southern Italians, gesture very much more than do Northern Europeans. The reputation for elaborate gesture use among Neapolitans has been widely acknowledged for a long time, so that it came to be mentioned in guidebooks and depicted in popular prints.

There exist remarkably few studies that can be cited that have investigated these alleged differences in any systematic way, however. The most famous study that does so, which is still the best, was first published in 1941 by David Efron [69]. Efron studied gestural behaviour in Southern Italians who had immigrated to Manhattan and compared this with the gestural behaviour of East European Yiddish speaking Jewish immigrants, also living in Manhattan. He was able to show, in a systematic manner, that there are indeed substantial differences in gestural behaviour between these two groups. He then went on to examine the gesturing of descendants of these groups who had become assimilated to American culture and showed that there was much less difference between them in this regard. His work was taken as a kind of definitive demonstration that there may be differences between ethnic groups in regard to gestural behaviour, but that these differences are *cultural*.

This point is a commonplace today, but in Efron's day this demonstration was necessary as part of a refutation of certain racial anthropological theories prevalent at the time he worked. His work was taken by writers such as Weston La Barre [70] and Ray Birdwhistell [71] as being very important for the point of view that they were developing, which was that all aspects of communication conduct, not just language, are culturally patterned. Efron's work was deemed of crucial importance for the demonstration of Edward Sapir's famous dictum that "the unwritten code of gestured messages and responses, is the anonymous work of an elaborate social tradition." He maintained that this unwritten code of gestured messages cannot be regarded as "simple organic responses. On the contrary, it is as finely certain and artificial, as definitely a creation of social tradition, as language or religion or industrial technology." [72]:556.

It was this idea, specifically as it was expressed by Sapir, that bodily behaviour, insofar as it functioned communicatively, was culturally patterned as language is patterned, that was the original inspiration for Birdwhistell's attempt to develop a *kinesics* which would be a sort of parallel to *linguistics* (See Howard [73]: 316-317). Kinesics, as Birdwhistell tried to develop it, never fulfilled the promise that was claimed for it (Kendon and Sigman [74]), yet the idea that gestural expression is culturally patterned and not something that is "universal" came to be widely accepted, and the concept of kinesics was historically very important in bringing about the systematic analysis of bodily action from a communicative point of view.

It was in this era - that of the middle decades of the last century, in particular - that saw the publication of a number of vocabularies or dictionaries of gestures. In part these had been inspired by Ekman and Friesen's [75] concept of the 'emblem' which proposed that vocabularies of gestures that were conventionalised in form and that had great stability in meaning could be found in different communication communities. It was widely accepted that this sort of gesture is a thoroughly cultural product and a number of publications have demonstrated that cultures differ both in the gestural forms employed and in how these forms are used. Desmond Morris' study of 1979 [76] is often cited in this connection (see [77] for Kendon's extended discussion of this book).

The sharp division that has so often been maintained in recent years between 'emblems' and 'illustrators' or 'gesticulation', as the so-called spontaneous non-codified gesturing typically associated with speech is sometimes referred to, is not sustainable - at least if it is taken to imply that such a sharp division is reflected in the gestural behaviour of individuals. In fact, at least as far as I can see from my own observations in Naples, in terms of how so-called emblems are used, they are so thoroughly integrated with whatever gesturing a person may be engaged in that it is not useful to maintain this separation. What we observe in the flow of gesturing going along with speech among Neapolitans is the employment of forms of expression that vary in the degree to which they are well-articulated and vary in the degree to which they appear as easily recognisable or conventionalised forms, but the way these forms of expression function in relation to the spoken component of the utterances of which they are a part does not seem to differ according to the degree to which they appear to be conventionalised. They operate alongside forms of gestural expression fashioned through improvisation, as integral components of the process of utterance construction.

All gesturing is shaped by social processes. Nevertheless it remains that in any community a repertoire of more or less stable forms tends to become established. The various vocabularies of gestures that have been published in the past are witness to this. What is not well understood is what determines the emergence of these stable forms. What are the conditions under which a vocabulary of such gestural forms becomes established? To account for this studies are needed that look at the contexts of interaction in which such gestures are used, as well as taking into consideration such factors as the social and environmental ecology of the community in question, and we have to do this historically, since communication practices establish themselves as social traditions and may develop a degree of autonomy, as of course has been the case with spoken languages. A few investigations can be mentioned that begin to tackle some of these issues. Sherzer [78] has published an elegant analysis of the contexts of use of the thumbs up gesture, as this is used in urban settings in Brazil and Heather Brookes [79] has published a similar analysis of a gesture in common use among black urban youth in Johannesburg, South Africa. Brookes [80] [81] has added to this work with some further studies of other gestures in the repertoire of the youthful black male society in Johannesburg. This work exemplifies the sort of systematic observational work on gesture that is very much needed.

In my own work, I have offered some cultural ecological observations regarding the city of Naples, Italy, in an effort to provide a context in terms of which the prominent place that gesture plays in everyday interaction in that city (and related areas) may be understood (see [82], [83], [1]: 349-354). In regard to the culture of the Neapolitan area (or, more broadly, the urban culture of coastal Campania) there is a very long and continuous history of habitation, urban environmental structure and modes of behaviour in public places (to use Goffman's phrase), that stretches back very many centuries and which, so I have argued, can be seen as creating conditions that, in several different ways, would favour the elaboration of the gestural modality in everyday interaction. Note that, as a form of expression, gesture is silent and is a form of *visible action*. In consequence it can be useful for communication over distances too great for speech exchanges, it can be used in noisy environments, but it is also useful, if performed in the right manner, as a mode of communication that does not attract immediate attention. On the other hand, if performed in a conspicuous manner it can be a means by which a person may make himself vivid and interesting, it can serve as a visual elaboration of figures of spoken expression, it can be recruited to permit side exchanges in circumstances where there are numerous simultaneously overlapping behaviour settings. It thus has a number of different properties which, in several different ways, can suit it for use in the traditional Neapolitan environment.

One may suppose that, within any given culture there is what might be called a "communication economy" (an idea derived from Dell Hymes [84]). That is, that there is system of interrelated communicative modalities which are balanced in relation to one another in different ways according to the various communicative circumstances typical in the culture. Depending upon these circumstances, in some communication economies the gestural modality may be more enhanced than it is in others. I think that there is an adaptive relationship between the structure of a communication economy and the structure and circumstances of interaction and communication within a culture and that it is in these terms that we will come to an understanding as to why, in traditional Naples, let us say, gesture in many different ways played so prominent a role where in some other cultures, such as those in the north of Europe, it played a less prominent role and came to be elaborated in rather different ways.

But in regard to gesture and culture, we need far more studies. We need studies that compare gesture use according to social situation and types of discourse (very little along these lines exist), we need to look at social class differences, at the role of literacy in the society and educational practices. Thus, why is it that it so widely observed that members of the governing classes use gesture so much less than members do the governed classes? For example, in Italy during periods of Spanish domination, Spanish officials were particularly noted for their almost complete immobility in their public appearances (see Burke [85]). At the time when Britain was ascendant as a world power, upper class Britons, those in a position of authority, became famous for their "stiff upper lip" and their general disinclination to engage in anything that might smack of 'French' or 'Italiante' or 'Continental' bodily expression. With regard to literacy, it is notable that societies, such as that of Naples, at least until the end of the nineteenth century, relied very little on written communication (refer here to the ordinary people, not to the upper reaches of educated society). It may be that in societies that tend to be mainly "oral" in their culture gesture is more widely used than in societies where literate modes of communication are more common (Hibbitts [86] provides a very interesting discussion). Factors such as these must be considered, if we are to develop an understanding of what it is that shapes the extent and nature of gesture use within a given society.

A final point may now be mentioned. It is also possible that there are aspects of gesture use and organisation that are related to the strictly linguistic features of the language used. There are a few studies that suggest this. Studies by David McNeill and Susan Duncan [87] comparing Spanish, Mandarin Chinese and English speakers, by Kita and Özyürek [88] comparing Turkish, Japanese and English speakers, suggest that the use of gesture when speakers are describing an event they have observed may be influenced by such factors as how topic is structured in spoken discourse or the way in which verbs of motion may or may not incorporate manner of action. Recent work by Marianne Gullberg [89] has suggested that where Dutch and French differ in how verbs used to describe object placement incorporate aspects of the character of the object being handled, so Dutch and French speakers will differ in how they display these aspects gesturally. Keiko Yoshioka [90] reports observations that suggest that differences in how Dutch and Japanese mark, in spoken discourse, reference to items previously referred in the same discourse, may influence whether or not gestures are used in marking them. Work along these lines is of great interest. Much further work is needed however. In particular, it will be necessary to take into consideration the rhetorical nature of the speaker's talk. Who the audience is, what the speaker can assume about the interlocutor's prior knowledge of what is being talked about, the nature of the discourse style or register being employed, among many other factors, all have consequences for whether and how gesture is used in association with speech. Only when we have a better understanding of these matters will we be able to arrive at a balanced interpretation of such findings. Nevertheless, the idea that differences in the structure and expressive capacities of spoken languages could have consequences for how users of these languages employ gestures remains a very interesting one.

References

[1] Kendon, A. (2004). *Gesture: Visible Action as Utterance.* Cambridge: Cambridge University Press.
[2] Kendon, A. (1972). Some relationships between body motion and speech. An analysis of an example. In A. Siegman & B. Pope (Eds.), *Studies in Dyadic Communication.* (pp. 177-210). Elmsford, New York: Pergamon Press.
[3] Kendon, A. (1980). Gesticulation and speech: two aspects of the process of utterance. In M. R. Key (Ed.), *The Relationship of Verbal and Nonverbal Communication.* (pp. 207-227). The Hague: Mouton and Co.
[4] McNeill, D. (1992). *Hand and Mind.* Chicago: Chicago University Press.
[5] McNeill, D. (2005). *Gesture and Thought.* Chicago: Chicago University Press.
[6] Goodwin, C. (1986). Gesture as a resource for the organization of mutual orientation. *Semiotica, 62*(1/2), 29-49.
[7] Goodwin, C. (2000). Pointing as situated practice. In S. Kita (Ed.), *Pointing: Where Language, Culture and Cognition Meet.* (pp. 217-241). Mahwah, NJ: Lawrence Erlbaum.
[8] Goodwin, C. & Goodwin, M. H. (1992). Context, activity and participation. In P. Auer & A. di Luzio (Eds.), *The Contextualization of Language.* Amsterdam: Benjamins.
[9] Haviland, J. B. (1993). Anchoring, Iconicity and Orientation in Guugu Yimithirr Pointing Gestures. *Journal of Linguistic Anthropology, 3*(1), 3-45.
[10] Haviland, J. B. (2000). Pointing, gesture spaces and mental maps. In D. McNeill (Ed.), *Language and Gesture.* (pp. 13-46). Cambridge: Cambridge University Press.
[11] Haviland, J. B. (2003). How to point in Zinacantán. In S. Kita (Ed.), *Pointing: Where language, culture and cognition meet.* (pp. 139-169). Mahwah, NJ: Lawrence Erlbaum Associates.
[12] Heath, C. (1992). Gesture's discrete tasks: Multiple relevancies in visual conduct in the contextualization of language. In P. Auer & A. di Luzio (Eds.), *The Contextualization of Language.* (pp. 102-127). Amsterdam/Philadelphia: John Benjamins.
[13] Streeck, J. (1993). Gesture as Communication I: Its Coordination with Gaze and Speech. *Communication Monographs, 60*(4), 275-299.
[14] Streeck, J. (1994). Gesture as communication II: The audience as co-author. *Research on Language and Social Interaction, 27*(3), 239-267.
[15] Streeck, J. (2002). A body and its gestures. *Gesture, 2*(1), 19-44.
[16] Streeck, J. & Hartege, U. (1992). Previews: gestures at the transition place. In P. Auer & A. d. Luzio (Eds.), *The Contextualization of Language.* (pp. 135-157). Amsterdam/Philadelphia: John Benjamins.
[17] Le Baron, C. & Streeck, J. (2000). Gestures, knowledge and the world. In D. McNeill (Ed.), *Language and Gesture.* (pp. 118-138). Cambridge: Cambridge University Press.
[18] Yau, S.-C. (1992). *Creations gestuelle et debuts du langage: Creation de langues gestuelles chez des sourds isoles.* Paris: Editions Langages Croisés.
[19] Kegl, J., Senghas, A., & Coppola, M. (1999). Creation through contact: Sign language emergence and sign language change in Nicauragua. In M. DeGraff (Ed.), *Language Creation and Language Change. Creolization, Diachrony and Development.* (pp. 179–237). Cambridge, Mass.: MIT Press.
[20] Polich, L. (2005). *The emergence of the deaf community in Nicaragua: With sign language you can learn so much.* Washington, D.C: Gallaudet University Press.
[21] Sandler, W., Meir, I., Padden, C., & Aronoff, M. (2005). The emergence of grammar: Systematic structure in a new language. *Proceeding of the National Academy of Sciences, 102*(7), 2661-2665.
[22] Stokoe, W. C. (1960). Sign Language Structure: An outline of the visual communication systems of the American deaf. *Studies in Linguistics Occasional Papers, 8*, 1-78.
[23] Klima, E. A. & Bellugi, U. (1979). *The Signs of Language.* Cambridge, MA: Harvard University Press.

[24] Emmorey, K. (1999). Do signers gesture?. In L. H. Messing & R. Campbell (Eds.), *Gesture, Sppech and Sign*. (pp. 132-159). Oxford: Oxford University Press.
[25] Liddell, S. K. (2000). Blended spaces and deixis in sign language discourse. In D. McNeill (Ed.), *Language and Gesture*. (pp. 331-357). Cambridge: Cambridge University Press.
[26] Liddell, S. K. (2003). *Grammar, Gesture and Meaning in American Sign Language*. Cambridge: Cambridge University Press.
[27] Hewes, G. W. (1973). Primate communication and the gestural origins of language. *Current Anthropology*, *14*, 5-24.
[28] Armstrong, D. F., Stokoe, W. C., & Wilcox, S. E. (1995). *Gesture and the Nature of Language*. Cambridge: Cambridge University Press.
[29] Corballis, M. C. (2002). *From hand to mouth : the origins of language*. Princeton: Princeton University Press.
[30] Anderson, M. L. (2003). Embodied cognition: a field guide. *Artificial Intellgence*, *149*, 91-130.
[31] Müller, C. (2002). A brief history of the origins of the International Society for Gesture Studies (ISGS). *Gesture*, *2*(1), 127-132.
[32] Goffman, E. (1974). *Frame analysis*. Cambridge, MA: Harvard University Press.
[33] Kendon, A. (1978). Differential perception and attentional frame: two problems for investigation. *24*, 305-315.
[34] Kendon, A. (1990). *Conducting Interaction: Patterns of Behavior in Focused Encoounters*. Cambridge: Cambridge University Press.
[35] Scheflen, A. E. (1964). The significance of posture in communication systems. *Psychiatry: Journal of Interpersonal Relations*, *27*, 316-331.
[36] Kendon, A. (1972). The role of visible behaviour in the organization of social interaction. In M. von Cranach & I. Vine (Eds.), *Social Commnication and Movement: Studies of Interaction in Man and Chimpanzee*. (pp. 29-74). London: Academic Press, in cooperation with the European Association of Experimental Social Psychology.
[37] Laban, R. & Lawrence, F. C. (1947). *Effort*. London: Macdonald and Evans.
[38] Dell, C. (1970). *A Primer for Movement Description using Effort-Shape and Supplementary Concepts*. New York: Dance Notation Bureau.
[39] Bartinieff, I. & Lewis, D. (1980). *Body Movement: Coping with the Environment*. New York: Gordon and Breach Science Publishers.
[40] Kita, S., van Gijn, I., & van der Hulst, H. (1998). Movement phases in signs and co-speech gestures and their transcription by human coders. In I. Wachsmuth & M. Frölich (Eds.), *Gesture and Sign Language in Human Computer Interaction. Lecture Notes in Artificial Intelligence*. (Vol. 1371, pp. 23–35). Berlin: Springer Verlag.
[41] Crystal, D. (1969). *Prosodic Systems and Intonation in English*. Cambridge: Cambridge University Press.
[42] Freedman, N. (1977). Hands, words and mind: On the structuralization of body movements during discourse and the capacity for verbal representation. In N. Freedman & S. l. Grand (Eds.), *Communicative Structures and Psychic Structures: A Psychoanalytic Approach*. (pp. 109-132). New York and London: Plenum Press.
[43] Rimé, B. & Schiaratura, L. (1991). Gesture and speech. In R. S. Feldman & B. Rimè (Eds.), *Fundamentals of Nonverbal Behavior*. (pp. 239-281). Cambridge: Cambridge University Press.
[44] Kita, S. (2000). How representational gestures help speaking. In D. McNeill (Ed.), *Language and Gesture*. (pp. 162-185). Cambridge: Cambridge University Press.
[45] Krauss, R., Chen, Y., & Chawla, P. (1996). Nonverbal behavior and nonverbal communication: What do conversational hand gestures tell us?. In M. Zanna (Ed.), *Advances in Experimental Social Psychology: Volume 28*. (pp. 389–450). New York: Academic Press.
[46] Krauss, R. M., Chen, Y., & Gottesman, R. F. (2000). Lexical gestures and lexical access: a process model. In D. McNeill (Ed.), *Language and Gesture*. (pp. 261-283). Cambridge: Cambridge University Press.
[47] Calbris, G. (1990). *Semiotics of French Gesture*. Bloomington: Indiana University Press.
[48] Clark, H. H. (1996). *Using Language*. Cambridge: Cambridge University Press.
[49] Gullberg, M. (1998). *Gesture as a Communication Strategy in Second Language Discourse: A Study of Learners of French and Swedish*. Lund: Lund University Press.
[50] de Ruiter, J. P. (2000). The production of gesture and speech. In D. McNeill (Ed.), *Language and Gesture*. (pp. 284-311). Cambridge: Cambridge University Press.
[51] de Ruiter, J. P. (In Press). Postcards from the mind. *Gesture*.
[52] Melinger, A. & Levelt, W. M. (2004). Gesture and the communicative intention of the speaker. *Gesture*, *4*(2), 119-141.

[53] Gullberg, M. & Holmqvist, K. (2006). What speakers do and what listeners look at. Visual attention to gestures in human interaction live and on video. *Pragmatics and Cognition*, *14*(1), 53-82.
[54] Beattie, . G. & Shovelton, . H. (1999). Mapping the range of information contained in the iconic hand gestures that accompany spontaneous speech. *Journal of Language and Social Psychology*, *18*, 438–462.
[55] Beattie, G. & Shovelton, H. (1998). The communicational significance of the iconic hand gestures which accompany spontaneous speech. An experimental and critical appraisal. In S. Santi, I. Guaïtella, C. Cavé, & G. Konopczynski (Eds.), *Oralité et gestualité: Communication multimodale, interaction.* (pp. 371-385). Paris: L'Harmattan.
[56] Beattie, G. & Shovelton, H. (1999). Do iconic hand gestures really contribute anything to the semantic information conveyed by speech? An experimental investigation. *Semiotica*, *123*(1/2), 1-30.
[57] Beattie, G. W. & Shovelton, H. (2001). An experimental investigation of the role of different types of iconic gesture in communication: A semantic feature approach. *Gesture*, *1*(2), 129-149.
[58] Kendon, A. (1994). Do Gestures Communicate? A Review. *Research on Language and Social Interaction*, *27*(3), 175-200.
[59] Stamenov, M. I. & Gallese, V. (Eds.). (2002). *Mirror Neurons and the Evolution of Brain and Language.* Amsterdam and Philadelphia: John Benjamins.
[60] Rizzolatti, G. & Sinigaglia, C. (2006). *So quel che fai: il cervello che agisce e i neuroni specchio.* Milan: Raffaello Cortina Editore.
[61] Wundt, W. (1973). *The Language of Gestures* (J. S. Thayer, C. M. Greenleaf, & M. D. Silberman, Trans.). The Hague: Mouton.
[62] Müller, C. (1998). *Redebegleitende Gesten: Kulturgeschichte, Theorie, Sprachvergleich.* Berlin: Arno Spitz.
[63] Sowa, T. (2006). *Understanding coverbal iconic gestures in shape descriptions.* Amsterdam: AKA Press.
[64] Sowa, T. & Wachsmuth, I. (2005). *A model for the representation and processing of shape in coverbal iconic gestures.* Paper presented at the Proceedings of KogWis05. The German Cognitive Science Conference 2005, Basel.
[65] Volterra, V., Caselli, M. C., Capirci, O., & Pizzuto, E. (2005). Gesture and the emergence and development of language. In M. Tomasello & D. I. Slobin (Eds.), *Beyond Nature-Nurture. Essays in Honor of Elizabeth Bates.* (pp. 3-78). Mahwah, NJ: Lawrence Erlbaum Associates.
[66] Jancovic, M. A., Devoe, S., & Wiener, M. (1976). Age-related changes in hand and arm movements as nonverbal communication: some conceptualizations and an empirical exploration. *Child Development*, *46*(4), 922-928.
[67] Colletta, J.-M. (2004). *Le développement de la parole che l'enfant âgé de 6 à11 ans: corps, langage et cognition.* Sprimont, Belgium: Pierre Mardarga.
[68] Cristilli, C. & D'Agostino, A. (2005). A proposito di rapporti fra codici e di strategie universali della comunicazione: linguaggio verbale e linguaggi gestuale nell'interazione adulto-bambino. *Atti del IV Congresso della Associazione Italiana di Linguistica Applicata (AltLA), Modena 19-20 febbraio 2004.* (pp. 114-143). Perugia: Guerra Edizioni.
[69] Efron, D. (1972). *Gesture, Race and Culture.* The Hague: Mouton [Originally published as *Gesture and Environment* in 1941 by Kings Crown Press, New York].
[70] La Barre, W. (1947). The cultural basis of emotions and gestures. *Journal of Personality*, *16*, 49-68.
[71] Birdwhistell, R. L. (1970). *Kinesics and Context: Essays in Body Motion Communication.* Philadelphia: University of Pennsylvania Press.
[72] Sapir, E. (1951). The unconscious patterning of behavior in society. In D. G. Mandelbaum (Ed.), *Selected Writings of Edward Sapir in Language, Culture and Personality.* (pp. 554-559). Berkeley: University of California Press.
[73] Howard, J. (1984). *Margaret Mead: A Life.* New York: Simon and Schuster
[74] Kendon, A. & Sigman, S. J. (1996). Ray L. Birdwhistell (1918-1994). *Semiotica*, *112*(3/4), 231-261.
[75] Ekman, P. & Friesen, W. (1969). The repertoire of non-verbal behavior: Categories, origins, usage and coding. *Semiotica*, *1*(1), 49-98.
[76] Morris, D., Collett, P., Marsh, P., & O'Shaughnessy, M. (1979). *Gestures: Their Origins and Distribution.* London: Jonathan Cape.
[77] Kendon, A. (1981). Geography of gesture. *Semiotica*, *37*(1/2), 129-163.
[78] Sherzer, J. (1991). The Brazilian thumbs-up gesture. *Journal of Linguistic Anthropology*, *1*(2), 189-197.
[79] Brookes, H. (2001). O clever 'He's streetwise.' When gestures become quotable: The case of the clever gesture. *Gesture*, *1*, 167–184.
[80] Brookes, H. (2004). A repertoire of South African quotable gestures. *Journal of Linguistic Anthropology*, *14*(2), 186-224.

[81] Brookes, H. (2005). What gestures do: Some communicative functions of quotable gestures in conversations among Black urban South Africans. *Journal of Pragmatics*, *37*, 2044-2085.
[82] Kendon, A. (2004). Contrasts in gesticulation: A Neapolitan and a British speaker compared. In C. Mueller & R. Posner (Eds.), *Semantics and Pragmatics of Gesture: Proceedings of the Berlin Conference, April 1998.* Berlin: Weidler Buchverlag.
[83] Kendon, A. (2000). Andrea de Jorio and his work on gesture. Editor's Introduction. *Gesture in Naples and Gesture in Classical Antiquity. A translation of Andrea de Jorio's La mimica degli antichi investigata nel gestire napoletano, and with an introduction and notes by Adam Kendon.* (pp. xix-cvii). Bloomington: Indiana University Press.
[84] Hymes, D. (1974). *Foundations in Sociolinguistics. An Ethnographic Approach.* Philadelphia: University of Pennsylvania Press.
[85] Burke, P. (1992). The language of gesture in early modern Italy. In J. Bremmer & H. Roodenburg (Eds.), *A Cultural History of Gesture.* (pp. 71-83). Ithaca, New York: Cornell University Press.
[86] Hibbitts, B. J. (1992). "Coming to our senses": Communication and legal expression in performance cultures. *Emory Law Journal*, *41*(4), 873-960.
[87] McNeill, D. & Duncan, S. (2000). Growth points in thinking for speaking. In D. McNeill (Ed.), *Language and Gesture.* (pp. 141-161). Cambridge: Cambridge University Press.
[88] Kita, S. & Ozyürek, A. (2003). What does cross-linguistic variation in semantic coordination of speech and gesture reveal? Evidence for an interface representation of spatial thinking and speaking. *Journal of Memory and Language*, *48*, 16-32.
[89] Gullberg, M. (Forthcoming). Language-specific encoding of placement events in gestures. In . E. Pederson & J. Bohenmeyer (Eds.), *Event Representations in Language and Cognition.* Cambridge: Cambridge University Press.
[90] Yoshioka, K. (2005). *Linguistic and gestural introduction and tracking of referents in L1 and L2 discourse.* Groningen: GRODI: Groningen Dissertations in Linguistics.

A. Esposito et al. (Eds.)
IOS Press, 2007

Gesture and Thought

David MCNEILL[1]
University of Chicago

Abstract. Both a synopsis and extension of *Gesture and Thought* (the book), the present essay explores how gestures and language work together in a dialectic. In this analysis the 'purpose' of gesture is to fuel and propel thought and speech. A case study illustrates the dependence of verbal thought on context and how it functions. Problems for computational modeling, the presence and absence of gesture 'morphemes, and speculation on how an imagery-language dialectic evolved are also provided.

Keywords. Gesture, growth points, imagery-language dialectic, dynamic dimension of language, psychological predication

1. Dialectic

Gesture and Thought, a book with the same title as this essay [1], explores the unification of language and gesture and what it implies: a new conception of language or rather, as I describe below, an old conception, long forgotten, that re-emerges as new: language as an imagery-language dialectic. In this dialectic gestures provide the imagery, and the dialectic itself fuels speech and thought. Gesture is an integral component of language in this conception, not merely an accompaniment or ornament. Such gestures are synchronous and co-expressive with speech, not redundant, and not signs, salutes, or emblems. They are frequent—about 90% of spoken utterances in descriptive discourse are accompanied by them [2]. They occur in similar form across many cultures (we have observed speakers from more than 20, including so-called 'high-gesture' cultures). They synchronize with speech at points where speech and gesture embody shared underlying meanings in discourse, and possess communicative dynamism, points of maximal discursive force at the moment of speaking [3]. A host of phenomena reveal the tight bond—to the point of fusion—of speech-gesture combination at these points.[2]

---

[1] Corresponding Author: David McNeill, University of Chicago, Department of Psychology, Chicago, e-mail: dmcneill@uchicago.edu.

[2] Among them: 1) The disruption of speech flow caused by delayed auditory feedback does not interfere with speech-gesture synchrony: the cross-modal unit remains intact ([4], first DAF experiment). 2) The onset of a gesture stroke inoculates against clinical stuttering. The onset of stuttering, once a stroke has begun, causes immediate cessation of the stroke [5]. In both cases, stuttering and gesture stroke are incompatible. 3) Gestures and speech spontaneously exchange semantic complexity in memory—information presented in gesture may be recalled in speech but not in gesture [6] and information in speech recalled in gesture but not in speech [7]. 4) Congenitally blind speakers perform gestures even to a known blind listener [8]. That is, so strong is the speech-gesture bond, that speakers with no experience of gesture, speaking to listeners known to have no perception of gesture, perform gestures (presumably unwittingly) with the flow of speech.

This fusion creates conditions for a dialectic. The dialectic is of unlike cognitive modes. Speech and gesture contrast semiotically—a gesture is global, synthetic, instantaneous, and not specified by conventions of form; a linguistic form is analytic, combinatoric, linear, and defined by socially-constituted rules. Their synchrony sets up a combination of opposites expressing the same idea unit. Such a combination is unstable and seeks resolution, and this fuels thinking for speaking. Such is the process of utterance formulation in a dialctic (contrasting with such standard psycholinguistic models as, e.g., [9]). A dialectic implies:

- A conflict or opposition of some kind, and
- Resolution of the conflict through further change or development.

The concept of an imagery-language dialectic extends (with gesture) a concept initiated by Vygotsky, in the 1930s [10].

This new conception also recaptures an insight lost for almost a century, that language requires two simultaneous modes of thought—what Saussure, in recently discovered notes composed around 1910 [11], termed the 'double essence' of language (although he expressed this without reference to gestures). Wundt [12], writing at almost exactly the same time, had a similar insight in this famous passage:

> "From a psychological point of view, the sentence is both a simultaneous and a sequential structure. It is simultaneous because at each moment it is present in consciousness as a totality even though the individual subordinate elements may occasionally disappear from it. It is sequential because the configuration changes from moment to moment in its cognitive condition as individual constituents move into the focus of attention and out again one after another" (p. 21).[3]

Gesture and Thought focuses on the real-time actualization of thought and language, regarding language multimodally and in context—its dynamic dimension. On the dynamic dimension, language appears to be a process, not an object. On the crosscutting static dimension, it looks to be an object but not a process. In fact, both dimensions must be considered, as both are indispensable to a full theoretical explication of utterances

2. Imagery

'Imagery' in this discussion is defined as a *symbolic form that lacks duality of patterning*—to use Hockett's term for one of his 15 design features of language [13]. It is a symbol whose form is determined by its meaning, not shaped by contrasts with other forms or by conventional standards of form qua form. In addition. imagery in gesture is actional as well as visuospatial. It is also non-photographic. The form of the gesture is driven by its meaning as much as by simulation.

2.1. Metaphor expands imagery

Thanks to metaphor, imagery is not limited to concrete references. Metaphoricity is a fundamental property of gesture [14]; without it the dialectic is restricted to a narrow band of concrete description. Metaphor extends imagery to encompass abstract meaning. A famous example is the 'conduit' metaphor, which appears in such purely verbal uses as "there was a lot in that book"; the image being that meaning is a

---

[3] I am grateful to Zenzi Griffin for alerting me to this passage.

substance and the book is a container [15, 16]. A gestural conduit metaphor is the Palm Up Open Hand (PUOH), described by Cornelia Müller [17], where the palm 'holds' or 'contains' some 'discursive substance'. Via PUOH, totally abstract content can, as imagery, engage co-expressive linguistic material dialectically, and thus initiate conceptual change. An example is a colloquium speaker saying "you [=a computer program] just wander around idly ..." The metaphor of a wanderer was accompanied by a simultaneous gestural metaphor, in which the right hand was rotating and the left hand was motionless, with these meanings: [4]

RH = the state of the program now
LH = the state of the program just before

The palms were facing inward, which suggests a flow of time away from speaker, who was the zero point for the deixis of the scene. The gestures were otherwise like the PUOH, presenting discursive objects that were in space and behaving in a wandering way. The gesture realizes the abstract notion of a computer operation as an object; speech at the same time refers to "you" and "wandering", two further objectifications. Thus gesture and verbal metaphors were in creative tension. A detailed illustration of dialectic, together with an analysis of how the form of the utterance resolves the dialectic, is given in Section 4.

3. The specific gesture type

The semiotic combinations that create an imagery-language dialectic are summarized in what I once termed Kendon's Continuum [4], named after [19]:

Spontaneous Gesticulation (Mode 1) → Language-slotted (Mode 2) → Pantomime (Mode 2) → Emblems (Mode 2) → Signs

Mode 1=unwitting gestures, Mode 2=gestures intended as symbols (due to S. Duncan).

As one goes from gesticulation to sign language:

- The obligatory presence of speech declines.
- Language like properties increase.
- Socially regulated signs replace spontaneously generated form-meaning pairs.

The simultaneous presence of gesticulation and speech pits unlanguage and language at the same moment. Gesticulation is the type of gesture we analyze in depth; language-slotted gestures are also gesticulations but have a different timing relation—replace speech rather than synchronize with it ("he goes [gesture]", the gesture timed to coincide with a vacant grammatical slot); pantomime is dumb-show and occurs without speech at all; an emblem is a culturally established morpheme (or semi-morpheme, because it does not usually have syntagmatic potential), such as the "OK" sign and others, and can occur with or without speech; and sign languages are full socially-constituted languages and combine with speech only with distortion (American Sign Language and others).

Even though 'gesticulation' (hereafter, 'gesture') is only one point on the Continuum it dominates gesture output in storytelling, living space description, academic discourse (including prepared lectures) and conversation—commonly 99% if

[4] This example is discussed in [18]

not all gestures (the propensity to combine conventionalized emblems or 'quotable' gestures [20] with gesticulation varies across cultures; in the genres listed, among North American speakers at least, it is essentially absent). As the Mode 1/Mode 2 distinction indicates, gesticulation alone is unwitting, not intended as symbols. They are integrated with linguistic content that appears to the speaker, unaware of gesture, to be the main object of semiosis.[5]

4. The growth point

The smallest unit of the imagery-language dialectic is posited to be a 'growth point,' so named because it is theoretically the initial unit of thinking for speaking out of which a dynamic process of organization emerges. A growth point combines imagery and linguistic categorial content, and the theory is that such a combination initiates cognitive events. A growth point is an empirically recoverable idea unit, inferred from speech-gesture synchrony and co-expressiveness.

4.1. A case study

An example recorded in an experiment (offered in part because of its ordinariness) is a description by one speaker of a classic Tweety and Sylvester escapade, which went in part as follows: "and Tweety Bird runs and gets a bowling ba[ll and drops **it down** the drainpipe]." Speech was accompanied by a gesture in which the two hands thrust downward at chest level, the palms curved and angled inward and downward, as if curved over the top of a large spherical object. At the left bracket, the hands started to move up from the speaker's lap to prepare for the downward thrust. Then the hands, at the very end of "drops," held briefly in the curved palm-down position, frozen in midair (the first underlining). Next was the gesture stroke—the downward thrust itself—timed exactly with "it down" (boldface). Movement proper ceased in the middle of "down," the hands again freezing in midair until the word was finished (the second underlining). Finally, the hands returned to rest (end of second underlining up to the right bracket). The two holds reveal that the downward thrust was targeted precisely at the "it down" fragment: the downward thrust was withheld until the speech fragment could begin and was maintained, despite a lack of movement, until the fragment was completed. Significantly, even though the gesture depicted downward thrusting, the stroke bypassed the very verb that describes this motion, "drops," the preparation continuing right through it and holding at the end—an explanation for this seeming overshoot is provided later.

The growth point was thus the fragment, "it down," plus the image of a downward thrust. Both sides of the growth point are essential, and are opposed dialectically in that the linguistic components have combinatoric potential and categorize the image; the imagery component embodies these categories in an instantaneous whole; the different modes are simultaneously active (for the speaker and the listener, who is trying to recreate the growth point). That one idea exists in two such different modes is the motive force for the utterance and its linked meaning formation.

---

[5] The Continuum grew into four Continua in [21].

4.2. Unpacking

The growth point is resolved by unpacking it into a more stable form, with a grammatical construction being the most stable outcome possible. Intuitions of good form (called 'intuitions-1,' the individual's direct perceptual experience of the static structure language) arise and are the *stop orders* for the dialectic. Once the speaker sensed a well-formed construction, she resolved the conflict by distributing the imagery and categorial content of the growth point into its prepared slots, and this stopped the dialectic process (how this might work is illustrated below). In this way, the dynamic intersects the static, as expected by Saussure's double essence insight—intersects it in fact in several ways: in the growth point, in the unpacking, and in the stop order. It is not that unpacking invariably reaches a full grammatical construction. It proceeds until some threshold of stability is reached, which may often be less than a complete outcome; or it may just break off if stability proves unattainable in the time spans attainable at socially realistic speech rates, e.g., because of an inappropriate construction attempt. Thus pauses and grammatical approximations, rife in daily discourse, can be explained as products of the dialectic resolution and the speaker's efforts toward it within limited time spans.

In this example, nonetheless, the growth point smoothly unpacked into a construction, the causative "someone drops (=causes to move by dropping) something down some landmark object." Intuitions-1 of the caused-motion construction arose and became the stop order, the construction plausibly resolving the dialectic by providing slots for the growth point image and its categorial content.

Subj	V	**Obj**	**Obl**
↕	↕	↕	↕
Ø (Tweety)	drops	**it (b-ball)**	**down**

(boldface for the slots that gathered the pieces of the growth point; the Tweety subject and the verb "drops" are explained below).

4.3. Context and fields of oppositions

Context is a second source of dynamism. Theoretically, a growth point is a psychological predicate in Vygotsky's [9] sense, a significant contrast within a specific context (cf. the concept of communicative dynamism [3]. While context reflects the physical, social and linguistic environment, it is also a mental construction; the speaker constructs this representation of context, in order to make the intended contrast meaningful within it. The growth point is thus not fixed and implies the context from which it is differentiated. Finding this context in actual data is an essential part of validating the growth point empirically. The mental construction of the context is modeled as a field of oppositions; what the speaker creates is a field of oppositions to make the psychological predicate differentiable within it. This is a model in which meaning is a *relationship* between a point of contrast and the background or field of oppositions from which it is being differentiated, not an accumulated 'substance'.

4.3.1. The catchment

A further concept, the catchment, provides an empirical route for finding this field of oppositions. A catchment comprises multiple gestures with recurring form features, and reveals the discourse segment to which the growth point belongs. More than one catchment can be simultaneously active for the same growth point. The full complement of catchments can suggest the oppositions from which the growth point is being differentiated.

To identify the catchment in the "it down' case, we look for other gestures in which the hands are shaped and/or move similarly to the target gesture, and see if these gestures comprise a family with thematic continuity. We find such a family; in the speaker's rendition, similar two-handed gestures had to do with the bowling ball in the role of an antagonistic force, contra-Sylvester.

4.4. A full analysis

We can thus further specify the "it down" growth point: it was a psychological predicate differentiating the bowling ball *as* this antagonistic force. Various antagonistic forces against Sylvester were the field of oppositions; the differentiated version was this force in the form of the bowling ball moving downward. The growth point and this context provide a richer picture of the speaker's idea unit than a purely referential reading of the phrase, "drops it down the drainpipe," suggests:

Ways of Thwarting Sylvester: Bowling Ball Downwards

Also, we can now explain the timing of the gesture: the downward thrust coincided exactly with the linguistic categorial content with which it formed a growth point, idea unit, or psychological predicate. It skipped the verb "drops," despite the fact that this verb described the bowling ball's motion down, precisely because the verb does not describe the bowling ball in its role as an antagonistic force; it describes what Tweety did, not the bowling ball, and thus could not have categorized the image with the intended meaning. The speaker's core idea was not dropping but the idea of the bowling ball moving down as an antagonistic force. Hence, the details of how gesture and speech combined, including timing, can be explained as aspects of the speaker's construction of the psychological predicate in the context, which is to say her thought process in context. (Other psychological predicates in the same catchment also conveyed the antagonistic force theme, specifying its effects on the unfortunate Sylvester: how he became a kind of living bowling ball, rolled down a street, into a bowling alley, and knocked over all the pins. Each of these can be analyzed in turn as psychological predicates differentiating further contrasts within the Antagonistic field of oppositions.)

The growth point was unpacked into a caused-motion construction, as noted, and we can analyze this and explain where the remaining pieces of the utterance, Ø (Tweety) and "drops", came from as well. Unpacking is more than just finding a construction in which to house a growth point; it includes the differentiation of further meanings each with their own contexts, and integrating them with the growth point so that the construction, including its semantic frame, can resolve it. The unpacking took place in a second catchment, also active during the speaker's representation of the bowling ball episode. The immediately preceding utterance was, "he tries going [**up**] the in**sid**][e of t**he drai**npipe]," which segued directly into our target utterance. The three gestures (in bold) were made the same way, with one hand rising upward, the first finger extended. Although this may include pointing, the gestures occurred with the theme of Sylvester acting as a force of his own (for many speakers an extended first-finger gesture conveys compression: Sylvester, inside the pipe, squeezes his plump body down to about half size). So, for this speaker, the utterance and the target utterance comprised a paradigm of opposed forces. Opposed forces was her way of construing the episode: not merely the bowling ball and Sylvester colliding, but Sylvester, a force moving up, versus the bowling ball, a force moving down—each force with its own gesture imagery. The bowling ball moreover was not the original antagonistic force; the sentence was "(Tweety) drops it down," which starts out with Tweety in the subject slot as the force. The speaker understood from the cartoon that she had to make the bowling ball into this force. The verb "drops" plus the caused-motion construction neatly achieved the shift from Tweety to the bowling ball. This is the growth point account how the verb and the Tweety subject made their way into the utterance.

The whole target utterance was thus the product of two contexts: 1) the growth point in the context of the bowling ball as an antagonistic force: this was the core idea unit; and 2) caused-motion with "drops" and Tweety as subject: the further meanings in the paradigm of opposed forces that resolved the imagery-language dialectic, and shifted the antagonistic force to the bowling ball. The target utterance, although a single grammatical construction, grew out of two distinct contexts and gained oppositional meaning from each.

The linguistic side of a growth point is not necessarily grammatical. The "it down" growth point is not grammatical but nonetheless formed a growth point with the downward image in the context of thwarting Sylvester. Nor is it necessarily a verb (a popular psycholinguistic hypothesis that the verb is the starting point, is contradicted by the preparation phase passing straight through "drops"). The growth point can be any co-expressive linguistic category(ies) that enables the intended point of contrast to be differentiated within a field of oppositions built in part to make the contrast possible. Unpacking then must find a construction to resolve the growth point into a stable pattern.

Metaphoricity is present also. The downward moving bowling ball existed as something else, as an abstract idea of an antagonistic force. The importance of the metaphor is to enable the abstract, non-imaged meaning of an antagonistic force to become an image and to take part, as an image, in an imagery-language dialectic. In this way metaphoricity was an essential part of the growth point (not only in this case but in numerous others, perhaps all). This bowling ball metaphor was an impromptu creation but other gesture metaphors are culturally established but play the same role of enabling imagery-language dialectics with abstract unimageable meanings. An illustration is the 'palm up open hand', in which the hand(s) appear to present a discursive object. The metaphor is recognizable as the so-called 'conduit' metaphor, an image of the general metaphor culture (but not universal), in which an abstract idea is presented as if it were a substance in the hand or a container (cf. verbal examples like "the movie had a lot of meaning," where the movie is a container, or "she handed him that idea," where an idea is on the hand).[6]

---

[6]The gesture includes iconicity obviously, but also, in the placement of the hands in the upper central space, deixis indicating an upper space locus; and following Kevin Tuite's [22] proposal that in every gesture there is a rhythmical pulse, something like a beat indicating that content has significance beyond its immediate setting, in the wider discourse, for example. Thus, one gesture includes all semantic components, and this is

4.5. Summary

The growth point is thus a theory of the cognitive core of utterances; what thought units are like as they begin; their incorporation of context; how they evolve dialectically and how imagery intersects linguistic form to create a surface utterance. However, it is a limited model. It says nothing of how growth points are activated. This includes lexical activation, as in the case study, where part of the categorial core—the "it"—was triggered by the ball reference (and word "ball") in the preceding clause. Models of lexical retrieval may apply but it is also possible that such models are inadequate to explain this kind of feedforward (since it is not actually 'feedforward'—the word "ball" was not just shipped ahead to become the next GP; rather it triggered a whole new precise idea in the speaker's mind, where the ball took on the role of antagonist). Also, we see in the "it down" case study that tracking the scope of recent co-references is assumed in the model but not explained—the "it" indicates co-reference vis-à-vis the earlier "ball" but there is no mechanism for this at present. It may be that some of the missing ingredients are matters of new elaborations (how the GP was initiated at the first mention of "ball" in the preceding clause for example, how "it" indexes the co-reference of the bowling ball), but others belong to another realm altogether—the proper modeling of speaker's purposes, for example, including the seemingly correct intuition that local purposes are created by the process of verbal thought as much as guiding them, beacon-like. Some of the lexical activation problems may be solved only once this further mystery is plumbed.

I apply this theoretical framework over a range of situations—discourse and gesture in different languages (Turkish, Spanish, Mandarin, as well as English); gestures of children; the Whorfian hypothesis compared with Slobin's thinking for speaking [23, 24]; linguistic impairments (aphasia; right-hemisphere damage, which impairs discourse cohesion; and the split-brain state, all of which were described in [4] but are now integrated into a new neurogestural model) [1].

5. Gestures and inhabitance

A further point places this entire discussion on a different plane and in so doing provides an answer to the question: what becomes of an imagery-language dialectic when gestures do *not* appear? We get a deeper understanding of the imagery-language dialectic by introducing the concept of a 'material carrier'. A material carrier is the embodiment of meaning in a concrete enactment or material experience. A material carrier appears to enhance the symbolization's representational power. The concept implies that the gesture, *the actual motion of the gesture itself*, is a dimension of meaning. Such is possible if the gesture *is* the very image; not an 'expression' or 'representation' of it, but is *it*. The gesture *itself* is a component of the dialectic. From this viewpoint, a gesture is an image in its most developed—that is, most materially, naturally embodied—form. The absence of a gesture is the converse, an image in its least material form. The greater the felt departure of being from the immediate context, the more likely its materialization in a gesture, because of its contribution to being.

---

not a unique case. Multiple components is a reason for rejecting the idea of gesture types and thinking instead of dimensions—metaphoricity, iconicity, deixis, emphasis (i.e., beats), etc.—on which gestures load to differing degrees.

Thus, gestures are more or less elaborated depending on the importance of material realization to being. Absence of gesture is then explained in the dialectic framework: it is the predictable result of a minimal departure from context; with repetitive or denatured contexts imagery fades and, Cheshire Cat-like, only the leer of purely word-thought remains. Merleau-Ponty [25] expressed a similar view of language in *The Phenomenology of Perception*: "The link between the word and its living meaning is not an external accompaniment to intellectual processes, the meaning inhabits the word … What then does language express, if it does not express thoughts? It presents or rather it *is* the subject's taking up of a position in the world of his meanings" (p. 193). The "it down" growth point was this speaker's taking up of a position in the world of her cartoon narration, her momentary state of being, materialized in the image of the bowling ball as an antagonistic force.

6. The social/mental interface

A further dimension, as Vygotsky [9] famously argued, brings out that human thought is fundamentally social in character, even in the absence of an active interlocutor. Such implies that growth points are intrinsically social; hence a dialectic with conventionalized social gestures fits right in (cf. [19]). The growth point indeed does not describe a mind-in-isolation. Social context effects were present even in the case study—that a gesture occurred at all presumed a listener, and the gesture was presented to the listener in central space. Any social minimalism reflects the limits of the circumstances, not a restriction of the concept itself, and in fact work in my lab in recent years, especially by my PhD students, has revealed social-interactive contexts of the growth point.

Özyürek [26] showed that changing the number and the spatial loci of listeners has an effect on the speaker's gestural imagery. Thus, among the shaping factors in a field of oppositions was the speaker's social interactive context. Plugging this result into the growth point model, we infer that an imagery-language dialectic can be altered by changes of the social context.

And dialogues result in individuals inhabiting *similar growth points*. One can find two-party growth points, gestures from one person synchronizing with a second person's speech, and vice versa—someone's speech accompanied by another person's gestures (experiments by [27], [28], and Duncan pers. comm., respectively). Conversations are dynamically affected by the participant's gestures, even decisively altering direction when a conflict arises over the meanings metaphorized in the shared gesture space [29]. Such conflicts produce diverging imagery-language dialectics, which speakers attempt to realign.

On the other hand, when one speaker attempts to insert a false scene into a narration a joint GP is often impossible, as shown in the immediate breakdown of the interchange with the listener (the listener's confusion is the 'lie-detector'; research by Franklin). Finally, turn taking exchanges and interactions in group meetings can be explained in terms of 'mind-merging', in which turn-exchange signals synchronize GPs between outgoing and incoming speakers [30].

7. Problems with modeling[7]

As we have seen, gestures can be conceptualized as objects of cognitive inhabitance and as agents of social interaction. Inhabitance seems utterly beyond current modeling. An agent of interaction may be modelable, but here we focus on what is called the global property of gestures: can it be modeled? A main sticking point for a model of the GP appears to be its character as a minimal dialectic unit. One aspect of this is the global character of gesture imagery. Being global seemingly cannot be modeled, but read on.

Global refers to the fact that the determination of meaning in a gesticulation proceeds in a downward direction. The meanings of the 'parts' of the gesture are determined by the meaning of the whole. In fact, parts come into being only in the meaning landscape of the whole; they have no independent existence (so, for example, the palms facing down mean agenthood, but the individual fingers meaning nothing; or in a different case, the first finger extended means compression but the palm means nothing—the parts depend in both cases on the global significances of their gestures). This semiotic model contrasts to the upward determination of meanings in sentences. *Synthetic* refers to the fact that a single gesticulation concentrates into one symbolic form distinct meanings that might be spread across the entire surface of the accompanying sentence.

The problem is that the use of features in computational models appears to force the process of gesture creation to be combinatoric, thus losing the opposition of semiotic modes essential to the dialectic (global imagery vs. combinatoric language). Features would be combinations of forms and meanings like: the hands a) facing down (→force downward), b) shaped around and over an imaginary sphere (→bowling ball), and c) moving downward jointly (→direction of bowling ball and force). In a model, such form-meaning pairs combine to create a gesture with the intended significance.

To be global, however, the process wants to work from the overall meaning downward. Even if we force a model to proceed in this direction, it appears that form features need to have their own meanings in order for a global meaning to find them—but do they? Here are some thoughts:

- The specific form features of the gesture are constrained by mechanical factors—where the hands already are, their current orientation, etc, which need not have anything to do with current significance.
- Suppose that significances trickle down into a configuration that already exists and the speaker then improvises something that we, on analysis, decide means 'spherical', 'downward', and 'effort'—what does she need to do for this?
- She needs to perform *an action that embodies these meanings*. Does this imply combining form-meaning features? *Or is it enough to 'act'?* Is the action of propelling a bowling ball downward sufficient to generate a gesture with the significances we are after?
- The idea of *coordinative structures* (the Haskins-related action model) seems to apply, with the addition of a thought-language-hand link (accessing and steering coordinative structures using significances). Coordinative structures are not themselves significant forms; they are "flexible patterns of cooperation

[7] I am grateful to the GP to the Max group at ZiF, the University of Bielefeld—Sue Duncan, Timo Sowa and Stefan Kopp—for freewheeling discussions of the material in this section.

among a set of articulators to accomplish some functional goal" (anonymous Yale linguistics handout found by Google).

Using coordinative structures. The goal is to exploit the inherent flexibility of coordinative structures in such as way that significances activate and shape them. Do coordinative structures so managed avoid the combination problem or are they just a fig leaf? The question is: does the idea of the bowling ball as an antagonistic force moving downwards automatically take care of features such as size (largish), placement (upper), direction (down), and motive force (agenthood)? As I understand coordinative structures, they work like tuned springs. They start off from some initial state and tamp down as they approximate the target: an object or an image. If the attractor can be a real object, with a thought-language-hand link, as IW reveals, it can also be a *significance* (e.g., the idea of a bowling ball being thrust downward and its metaphoric meaning of an 'antagonistic force').

So, the resolution: ideas or significances are attractors of coordinative structures; the coordinative structures zero in on these attractors; the properties of the *attractor* bring out features in the coordinate structures interactively: so features are outcomes, not initial conditions, with significances that derive from the action as a whole, and this is the global property. There is no lexicon of feature-meaning pairs ('facing down →force downward' and the like). The features arise during the action itself. Once a gesture has been created it is usually true that we can identify features of form that carry meanings, but these are the outcomes of the gesture, not the source. Each coordinative structure is an 'action primitive', but the critical difference from a feature is that coordinative structures do not have significances. Cornelia Müller's implicit actions in gesture (drawing, outlining, sculpting, grasping, etc.) reemerge as packages of coordinative structure, or patterns of patterns adapted to objects, actions or shapes [17], now adapted to ideas, as kinds of metaphors at the origin of an imagery-language dialectic.

I can't judge the computational feasibility of this resolution, but it does seem to provide a way to generate global imagery with significances that descend from wholes to parts non-compositionally. It is conceivable, at least worth mentioning, that a hybrid analog-digital machine could correctly model the growth point. The analog device itself could be simulated digitally, of course, but should simulate such properties as three-dimensional space, limited but varying granularity, differentiation of spatial blocks, orientations, etc. These properties establish the coordinative structures targeting significances, as described.

But not modeled: not modeled by coordinative structures is the growth point itself. Coordinative structures explain the global property, essential to a dialectic, but not the differentiation of psychological predicates; growth; inseparability from context; co-presence of imagery and linguistic categorization; the co-expressiveness of imagery and language; internal tension and motivation; or change/unpacking. In short, the 'essential duality' of language [10] of which the growth point is a minimal unit, seems at present impossible to model by a computational system.

8. Language origins: 'the ultimate answer'

An important new source of observations is the case of IW, a man who suffered, as a young adult, sudden and complete deafferentation from the neck down [31]. IW relearned movement control by utilizing vision and cognition, and he now, decades later, controls motion in this way to perfection. He also performs gestures with speech synchrony and co-expressiveness and does so even without vision, a condition where nongesture instrumental actions are impossible for him. In other words, actions for IW organized by language and thought have properties beyond those of goal-directed actions. His case suggests a partial dissociation in the brain of the organization of gesture from the organization of instrumental action, and the existence of a dedicated thought-language-hand link that would be the common heritage of all humankind.

8.1. Evolution of the thought-language-hand link

We accordingly end with an attempt to provide 'the ultimate answer' to the question of an imagery-language dialectic—why it exists at all—with a theory of language evolution that focuses on this thought-language-hand link. I develop a hypothesis that the origin of language crucially depended at one point on gestures (I do not mean that the first form of language was gestural: I intend something quite different, as I will explain below). Without gestures, according to this hypothesis, the brain circuits required for language could not have evolved in the way they apparently have. In common with much recent speculation, the theory presupposes the recently discovered 'mirror neurons', but adds something theoretical. This is 'Mead's loop' (named after the philosopher, George Herbert Mead [32], who wrote that "Gestures become significant symbols when they implicitly arouse in an individual making them the same response which they explicitly arouse in other individuals."), which I propose supplements the mirror neuron circuit.

8.1.1. Mirror neurons and Mead's loop

According to Mead's loop, what was selected in human evolution is a capacity, not present in other primate brains, for the mirror neuron circuit to respond to one's *own gestures* as if they belonged to someone else (thus gesture is activated as part of social interaction, producing among other things the social dependence of gestures when the addressee is invisible—speaking on the phone, a blind person talking to another blind person—but not speaking into a tape recorder). Crucially, Mead's loop brings the meanings of gestures into an area of the brain where actions are orchestrated. It provides a way for significances other than the significances of actions themselves to co-opt the action orchestration machinery of Broca's area, and explains how and under what conditions the IW-revealed thought-language-hand link could have evolved. A creature who possessed such a capacity, however minimally, would have had advantages in child rearing, for example, being better able to scaffold and error correct (plausible vectors for the origin of language being mothers and infant children; an origin of language *acquisition* as well as of language).

8.1.2. But not gesture-first

Contrary to a theory from Condillac that has been enthusiastically revisited in recent years [e.g. 33, 34], that the initial form of language was gesture, I am advocating that evolution selected an ability to *combine* speech and gesture. Speech and gesture would have evolved *together*, with no gesture-first stage. The plausibility of this hypothesis is enhanced by William Hopkins' observation that chimpanzees show hand dominance for gestures only when the movements co-occur with vocalization (Hopkins pers. comm.). The last common ancestor may therefore already have had a vocalization-gesture link. The postulated thought-language-hand link could build on this precursor during its own selection via Mead's Loop, but gesture-first would had to have circumvented it. Further, if there had been a gesture-first step it would not have led to human language but to pantomime (pantomime could have its own evolution, landing at a different point on Kendon's Continuum, reflected today in different timing re speech—alternating rather than simultaneous). But speech could not have evolved without simultaneous gesture, and gestures could not have evolved without a never-ending duet with speech [35].

References

[1] McNeill, David 2005. *Gesture and thought.* Chicago: University of Chicago Press.

[2] Nobe, Shuichi 2000. Where do *most* spontaneous representational gestures actually occur with respect to speech? In D. McNeill (ed.), *Language and gesture*, pp. 186-198. Cambridge: Cambridge University Press.

[3] Firbas, Jan. 1971. On the concept of communicative dynamism in the theory of functional sentence perspective. *Philologica Pragensia* 8: 135-144.

[4] McNeill, David. 1992. *Hand and mind: What gestures reveal about thought.* Chicago: University of Chicago Press.

[5] Mayberry, Rachel & Jaques, Joselynne. 2000. Gesture production during stuttered speech: insights into the nature of gesture-speech integration. In D. McNeill (ed.). *Language and gesture*, pp. 199-214. Cambridge: Cambridge University Press.

[6] Cassell, Justine, McNeill, David, & McCullough, Karl-Erik. 1999. Speech-gesture mismatches: evidence for one underlying representation of linguistic and nonlinguistic information. *Pragmatics & Cognition* 7: 1-34.

[7] Kelly, Spencer D., Barr, Dale J., Church, R. Breckinridge, & Lynch, Katheryn. 1999. Offering a hand to pragmatic understanding: the role of speech and gesture in comprehension and memory. *Journal of Memory and Language* 40: 577-592.

[8] Iverson, Jana M. & Goldin-Meadow, Susan. 1997. What's communication got to do with it? gesture in congenitally blind children. *Developmental Psychology* 33: 453-467.

[9] Levelt, Willem J. M. 1989. *Speaking: From Intention to Articulation.* Cambridge: MIT Press/Bradford Books.

[10] Vygotsky, Lev S. 1987. *Thought and language.* Edited and translated by E. Hanfmann and G. Vakar (revised and edited by A. Kozulin). Cambridge: MIT Press.

[11] Saussure, Ferdinand de. 2002. *Écrits de linguistique général* (compiled and edited by S. Bouquet and R. Engler). Paris: Gallimard.

[12] Wundt, Wilhelm. 1970. The psychology of the sentence. In Arthur Blumenthal (ed. and trans.), *Language and P\psychology: Historical aspects of psycholinguistics*, pp. 20-33. New York: John Wiley & Sons Ltd.

[13] Hockett, Charles F. 1960. The origin of speech. *Scientific American*, 203: 88-96.

[14] Ishino, Mika. 2001. Conceptual metaphors and metonymies of metaphoric gestures of anger in discourse of native speakers of Japanese. In M. Andronis, C. Ball, H. Elston & S. Neuvel (eds.), *CLS 37: The main session*, pp. 259-273. Chicago: Chicago Linguistic Society.

[15] Reddy, Michael J. 1979. The conduit metaphor: a case of frame conflict in our language about language. In A. Ortony (ed.), *Metaphor and thought*, pp. 284-297. Cambridge: Cambridge University Press.

[16] Lakoff, George & Johnson, Mark. 1980. *Metaphors we live by.* Chicago: University of Chicago Press.

[17] Müller, Cornelia 2004. The palm-up-open-hand. A case of a gesture family? In C. Müller & R. Posner (eds.), *The semantics and pragmatics of everyday gestures*, pp. 233-256. Berlin: Weidler Verlag.

[18] Parrill, Fey & Sweetser, Eve 2004. What we mean by meaning: Conceptual Integration in gesture analysis and transcription. *Gesture* 4: 197-219.

[19] Kendon, Adam 1988. How gestures can become like words. In F. Poyatos (Ed.), *Cross-cultural perspectives in nonverbal communication*, pp. 131-141. Toronto: Hogrefe.

[20] Kendon, Adam 2004. *Gesture: Visible action as utterance*. Cambridge: Cambridge University Press.

[21] McNeill, David 2000 Introduction. In D. McNeill (ed.), *Language and gesture*, pp. 1-10. Cambridge: Cambridge University Press.

[22] Tuite, Kevin. 1993. The production of gesture. *Semiotica* 93: 83-105.

[23] Slobin, Dan I. 1987. Thinking for speaking. In J. Aske, N. Beery, L. Michaelis, & H. Filip (eds.), *Proceedings of the thirteenth annual meeting of the Berkeley Linguistic Society*, pp. 435-445. Berkeley: Berkeley Linguistic Society.

[24] McNeill, David, & Duncan, Susan D. 2000. Growth points in thinking for speaking. In D. McNeill (ed.), *Language and gesture*, pp. 141-161. Cambridge: Cambridge University Press.

[25] Merleau-Ponty, Maurice. 1962. *Phenomenology of perception* (C. Smith, trans.). London: Routledge.

[26] Özyürek, Asli. 2000. The influence of addressee location on spatial language and representational gestures of direction. In D. McNeill (ed.), *Language and gesture*, pp. 64-83. Cambridge: Cambridge University Press.

[27] Kimbara, Irene 2002. On gestural mimicry. Unpublished ms., University of Chicago, Department of Linguistics.

[28] Furuyama, Nobuhiro 2000. Gestural interaction between the instructor and the learner in origami instruction. In D. McNeill (ed.), *Language and gesture*, pp. 99-117. Cambridge: Cambridge University Press.

[29] McNeill, David 2003. Pointing and morality in Chicago. In S. Kita (Ed.), *Pointing: Where language, culture, and cognition meet*, pp. 293-306. Mahwah, NJ: Erlbaum.

[30] McNeill, David 2006. Gesture, gaze, and ground. In S. Renals & S.S. Benglo (Eds.), *MLMI 2005. LNCS 3869*, pp. 1-14.

[31] Cole, Jonathan. 1995. *Pride and a daily marathon*. Cambridge, MA: MIT Press.

[32] Mead, George Herbert. 1974. *Mind, self, and society from the standpoint of a social behaviorist* (C. W. Morris ed. and introduction). Chicago: University of Chicago Press.

[33] Arbib, Michael 2005. From monkey-like action recognition to human language: An evolutionary framework for neurolinguistics. *Behavioral and Brain Sciences* 28: 105-124.

[34} Corballis, Michael C. 2002. *From hand to mouth: The origins of language*. Princeton, NJ: Princeton University Press.

[35] McNeill, David, Bertenthal, Bennett, Cole, Jonathan, & Gallagher, Shaun 2005. Gesture-first, but no gestures? Commentary on Michael Arbib "From monkey-like action recognition to human language: An evolutionary framework for neurolinguistics. *Behavioral and Brain Sciences* 28: 138-139.

Fundamentals of Verbal and Nonverbal Communication and the Biometric Issue
A. Esposito et al. (Eds.)
IOS Press, 2007

A Method for Studying the Time Alignment of Gestures and Prosody in American English: 'Hits' and Pitch Accents in Academic-Lecture-Style Speech

Stefanie SHATTUCK-HUFNAGEL[1a], Yelena YASINNIK[a], Nanette VEILLEUX[b]
Margaret RENWICK[c]
[a]*Speech Group, Research Laboratory of Electronics, Massachusetts Institute of Technology, Cambridge, MA, USA*
[b]*Department of Computer Science and Mathematics, Simmons College, Boston, MA, USA*
[c]*Department of Linguistics, Cornell University, Ithaca, NY, USA*

Abstract. This work describes a method for investigating the timing relationship between spoken accents cued by intonation (phrase-level pitch accents) and gestural accents cued by abrupt cessation of movement ('hits'), to test the hypothesis that the two kinds of events are planned by speakers to occur simultaneously. Challenges for this kind of study include i) defining the set of gestural and spoken events to be included, ii) labelling sometimes-ambiguous events such as spoken pitch accents, boundaries of accented syllables and gestural end points, and iii) providing clear criteria for what will count as alignment between events in the speech and gesture streams. Application of this method will permit a detailed test the hypothesis that prosodic planning provides a framework for the computation of a production plan for both phonological/phonetic encoding of words and segments (Keating and Shattuck-Hufnagel [2002]) and speech-accompanying gestures.

Keywords: Gestures, pitch accent, hits, prosodic hierarchy, gestural hierarchy

Introduction

Spoken utterances are commonly accompanied by movements of the hands, head, face and torso, and researchers have long been intrigued by the possibility that there is a relationship between these two streams of events. In particular, Kendon [14-16] has proposed that the gestures that accompany speech are organized into a hierarchy of constituents similar to those proposed to govern the prosody (intonation and timing) of spoken utterances. Kendon's work drew on his analysis of alignment between the gestural organization he observed in spontaneous speech interactions and the prosodic constituent called the tone group, identified by the British school of intonation (e.g. [8]). In the decades since Kendon first made this proposal, our understanding of the

[1] Corresponding Author: Stefanie Shattuck-Hufnagel, Speech Group, Research Laboratory of Electronics, Massachusetts Institute of Technology, Cambridge, MA, USA, e-mail: stef@speech.mit.edu.

phonology of intonational constituent structure has deepened, with the theoretical work of Pierrehumbert [26], Beckman and Pierrehumbert [5], Ladd [19] and others. These investigators suggest that there are several levels of intonationally-defined constituents, including the higher-level Full Intonational Phrase which is made up of one lower-level Intermediate Intonational Phrase or more. Other investigators have suggested that these intonational phrases are part of a more extensive hierarchy of prosodic constituents that ranges from the mora and/or syllable to the utterance ([12, 24, 28] and others). Although this view of prosodic constituent structure is not universally accepted (see Gussenhoven and colleagues' references [11]), it has nevertheless been widely adopted. However, the status of a number of intermediate-level constituents in the hierarchy is still unclear, and it is important to obtain empirical data to test the predictions of various competing theories of the phonology of prosody. The purpose of this paper is to begin this process by developing a method for testing and expanding on Kendon's hypothesis that prosodic and gestural elements align in time. We see such studies as laying the groundwork for a detailed examination of the more general hypothesis that a) speakers plan in a united way the vocal tract movements and other bodily movements that contribute to a communicative act (see McNeill [23], this volume, and its references), b) this planning process centrally involves the generation of prosodic structures (Keating and Shattuck-Hufnagel [13] and others), and, more speculatively, c) prosodic structures provide the organizing framework for both the spoken and the gestured aspects of the communicative act.

Prosodic structure in the sense that has been explored by recent theorists includes two major aspects: constituent structure (e.g. utterance, intonational phrase, intermediate intonational phrase, prosodic word etc.) and phrase-level prominences (pitch accents). In this paper we approach the issue of speech/gesture timing via the prominence aspect of these theories, posing the question of whether and how prosodic prominences in speech are marked by coordinated events in the gestural stream. A number of investigators have addressed this question, but results have been equivocal, in part because different studies have worked with different kinds of prominence and different aspects of gestures, making their apparently conflicting results somewhat difficult to compare. One difficulty arises from use of the term 'stress' to describe the type of spoken prominence that is under examination, because this word is used differently by different investigators. For example, in some theories, stress is held to be a single dimension which reaches higher or lower levels across the syllables of an utterance (Chomsky and Halle [7]). In such theories, the relative level of prominence of different syllables can be conveyed by different aspects of the signal, such as duration, f0 and intensity, but these variations signal differences in the degree of a single parameter, stress, whether this means a relatively strong syllable compared to other syllables within the word, or a relatively strong word compared to other words in the phrase. In contrast, Beckman and Edwards [3] propose a hierarchy of different types of prominence, analogous to the hierarchy of prosodic constituents. Their prominence hierarchy extends from non-head syllables of feet (reduced vowel syllables) and non-head segments of moras, through a range of full-vowel syllables, including unaccented lexically-stressed syllables and pre-nuclear-accented syllables to the nuclear-pitch-accented syllable that form the head of an intonational phrase. Moreover, they propose a different primary cue for each level of prominence, such as duration for lexical stress and F0 for pitch accent. Within the category of word-level or lexical-level stress, Chomsky and Halle [7] propose an unbounded number of categories, Bolinger [6] proposed only a single category of Full Vowel (a view that finds perceptual support in work by Fear, Cutler and Butterfield [10]), and more recently, several investigators have provided evidence for a voice-quality distinction between two levels of word-level prominence: main-word-stress and secondary stress [25, 30]. Kendon's [15] proposal for a timing relationship between gestural strokes and phrase-level intonational prominences refers to the 'tonic syllable', described by Crystal [8] and other members of the British School of intonation, which corresponds to the nuclear pitch accent defined by Pierrehumbert and colleagues as the final pitch accent of an intonational phrase.

This brief history of the various views that have been taken of 'stress' highlights the importance of defining the types of spoken syllables (i.e. the nature of the prominence) for which one is examining the gestural correlates. Similar issues require a careful and specific definition of the types of gestural movements being studied. For example, McClave [21] studied the timing of gestural beats (a term that is still in search of a good definition, but is broadly interpreted as a series of repeated gestures by the same articulator occurring in rapid succession) with lexically stressed syllables (as we've seen, another term in need of careful definition, but probably referring here to the main-stress syllables of words) in a sample of English speech. McClave found no consistent alignment between these two sets of events in time. In contrast, other investigators have informally noted a correspondence between gestures and prominences in the speech signal, describing their observations using a range of different terminology. Most recently, Loehr [20] has carried out a careful analysis of the alignment between pitch accents and gestural 'strokes', i.e. the meaning-bearing portion of a gesture (see Kendon [16] for discussion), and found a significant tendency for these two events to overlap in time. The present study aims to help resolve these apparent conflicts by carefully defining both the type of spoken prosodic prominence and the type of gestural movement under investigation, and by focussing on a particular type of gestural movement which lends itself to questions of temporal alignment with speech: accentual movements with sudden end points that we define as 'hits' [27].

The present study also takes advantage of several other developments which, like the evolution in phonological theories of contrastive prosodic categories over the past three decades, make it possible to study the temporal alignment of gestures and their accompanying speech in increasingly powerful ways. One of these is the emergence of Stevens' [31-34] theory of the significance of acoustic phonetic landmarks. In Stevens' view, acoustic landmarks are the perceptually salient regions of abrupt acoustic change in the speech signal that are associated with articulatory closures, narrowings and widenings that correspond to consonants, glides and vowels, and they play a critical role in the initial steps of the speech perception process. This has led to the development of an associated system for landmark labelling of speech waveforms, which provides a convenient tool for marking critical points of abrupt change in the signal associated with certain aspects of syllable structure, without requiring agonizingly subjective decisions about the detailed phonetic nature of the positional allophone that is realized by the landmarks. Landmark labels also make it possible to examine the question of whether some aspects of speech-accompanying gesture are aligned with particular points in the syllables they accompany, such as the onset of an accented syllable, its onset consonant(s) or its vowel.

Another useful advance was the development of the ToBI system for prosodic labelling [2, 4]. This transcription system provides a widely (although not universally) accepted standard for labelling prosodic constituent structure and prominence patterns. Yet another advance, the development of widely accepted gestural categories (e.g. McNeill's [22] iconic, metaphoric, deictic and beat-like gestures) and particularly of descriptions of the substructure of individual gestures (e.g. preparation, stroke and relaxation, with optional holds after each of these structural elements---see Kendon [16] pp. 112-113 for a summary) make it plausible to label the timing of various aspects of the speech-accompanying gestures while postponing for later study the complex question of their relationship to the speaker's communicative intent. Finally, the availability of a number of convenient tools for speech and video analysis make this a propitious moment to undertake a detailed study of this kind; these tools include the Praat system for displaying and labelling speech signals (http://www.fon.hum.uva.nl/praat/), iMovie's convenient tools for examining video files frame-by-frame, which have made the acquisition and analysis of substantial amounts of matched audio/video data a reasonable proposition, given a few laptops and an internet connection, and generously provided labelling tools, e.g. Anvil and Elan (http://www.dfki.de/~kipp/anvil/download.html, http://www.mpi.nl/tools/elan.html).

We view the present study, and the labelled database on which it is based, as the beginning of a broader investigation of how the speaker organizes the morphosyntactic structure of a spoken utterance along with its prosodic structure and the structure of its accompanying gestures. The question of which aspects of gesture align in time with which aspects of the speech signal is just one of the many questions to be asked about the relationship between these different aspects of the communicative signal. But it is a critical one, because if certain aspects of the speech and gestural streams are consistently timed to occur together, it will be necessary for models of speech production planning to provide a mechanism to account for this relationship. Because phrase-level pitch accents are not determined by information that might be stored in the lexicon (such as the location of main word-level stress), but must be planned afresh for each new utterance, the timing of such prosodic events with respect to speech-accompanying gesture is of particular interest for models of on-line utterance planning at the prosodic level. More generally, this study (like Kendon's 1972/1980 proposal) raises the question of how the planning of morpho-syntactic, prosodic and gestural structures are inter-related, an issue that is addressed in part by McNeil's proposal (e.g. this volume) for a common representation that governs both speech and gesture. In sum, this paper presents a framework for analysis of the temporal relationship among gestural, prosodic and morphosyntactic constituents of spoken communicative acts. It has been developed in order to explore the degree and nature of these alignment patterns, as a way of evaluating and refining competing hypotheses about how these different aspects of an utterance production plan are generated.

1. Methods

The technique for studying speech/gesture alignment described in this paper was developed on a set of 4 video samples of academic lectures, which were analysed and labelled over a 3-year period. During the evolution of this technique, we addressed three main problems: 1) defining the types of gestures and spoken prominences included in the study, 2) developing a consistent way of dealing with ambiguity in both domains, and 3) adopting explicit criteria for categorizing the alignment between speech and gesture events.

2. Materials

The speech samples for this study were drawn from a series of commercially-available academic lectures obtained from The Teaching Company (www.teachingcompany.com) They were produced by four male speakers who are university professors selected by the company for their effectiveness as lecturers in their fields. The samples were 3-6 minutes long, but did not contain a full 3-6 minutes of fluent speech because of pauses, disfluencies, self-corrections etc.

2.1 Segmental, Prosodic and Gestural Labelling of the Speech Samples

Each of the 4 samples was labelled separately for its segmental landmarks, its spoken prosody and its gestural hits. The sound and video files were labelled independently, to avoid any possibility that the gestures visible on the video could influence the listener's judgment of spoken prominence or vice versa. Earlier studies have suggested, for example, that listeners who view a speaker raising an eyebrow during a spoken utterance are more likely to judge the co-occurring syllable to be prominent (see e.g. Krahmer et al. [18]).

Segmental landmarks were labelled by 4 undergraduates who were participants in MIT's Undergraduate Research Opportunities Program and had spent the previous 2 months acquiring extensive training and experience in labelling the phonetic landmarks defined by Stevens' [31-34] theory of cues to phonological feature contrasts and their role in lexical access during perception. Stevens' theory specifies that during the process of speech recognition, listeners first detect the abrupt spectral changes in the speech signal that correspond to closures and releases of consonants, signal maxima for vowels and signal minima for glides (such as /r l w y/ in English). Hand labels for these acoustic-phonetic landmarks provide a convenient tool for aligning both prosodic events (such as pitch accents) and gestural events (such as the end points of 'hits') with segmental events related to morphosyntactic structure (such as words and syllables). Selected samples of the hand-labelled segmental landmarks were hand-checked by the authors for accuracy and determined to be acceptable.

Although the majority of landmarks predicted from the lexical representations of words are realized in connected speech signal (Shattuck-Hufnagel et al., in prep.[29]), some are not realized (e.g. the release of /t/ and closure of /b/ in *notebook*), and others are realized weakly or ambiguously (as when a stop closure results in a gradual rather than abrupt reduction in amplitude, or the release noise is of very low amplitude). The landmark labelling system provides a mechanism for marking landmarks that are either questionable or missing from the signal altogether. This is an important aspect of the study of speech/gesture alignments, because it indicates clearly where the boundaries of the accented syllable are clear, vs. where they are difficult to ascertain. This distinction will be important for distinguishing between models in which certain aspects of gestures occur precisely at certain points in the speech signal, such as pitch-accented syllables, vs. models in which the gestures simply occur in the general vicinity of those points.

Prosodic groupings (into intonational phrases) and prosodic prominences (pitch accents) were labelled by 2 undergraduate participants in the UROP program using the ToBI system for prosodic transcription [http://www.ling.ohio-state.edu/~tobi/, http://anita.simmons.edu/~tobi/, http://ocw.mit.edu/OcwWeb/Electrical-Engineering-and-Computer-Science/6-911January--IAP--2006/CourseHome/index.htm]. The ToBI labelling system prescribes that labellers provide the sparsest sequence of tone targets that can account for three aspects of an utterance: the critical aspects of the F0 contour (e.g. peaks, valleys and plateaus), the listener's intuitions about the relative prominence of syllables, and the listener's intuitions about the grouping of words into intonational phrases. The system provides two basic target levels in the F0 range (High and Low), and three classes of targets (boundary tones to mark the ends of phrases, pitch accents to mark prominent syllables, and phrase accents to describe the F0 contour between the last pitch accent in a phrase and the end of the phrase; pitch accents can be simple (e.g. H*, L*), or bitonal (e.g. L+H*, L*+H, etc), and there are a number of other complexities that do not concern us here.) The two labellers had received extensive training in the use of the ToBI system over several years.

In its current version the ToBI system provides extensive tools for marking ambiguity in the spoken prosody, such as *? when the labeller is uncertain whether a syllable is pitch-accented or not, and a separate labelling tier for recording alternative transcriptions (the ALT tier; see the MIT OpenCourseWare site above). Such ambiguities can arise, for example, in contexts where a falling pitch contour on two adjacent full-vowel syllables (as in *downtown*, *bisect*, *all right*) results in a pattern which can be interpreted either as a High pitch accent on the first syllable or a downstepped High pitch accent on the second syllable or in extended regions of low F0 where some syllables may sound possibly but not certainly accented with a L*. Marking accent uncertainty allows for separate analysis of the alignment of gestural hits with pitch-accented syllables that the prosody labeller is certain about, vs. with pitch-accented syllables that the prosody labeller is uncertain about.

Gestural hits were labelled by two undergraduates, one of whom originated the concept of a gestural hit [27] and both of whom had extensive experience in labelling 'hits'. For one speaker, these two labellers were assisted by a third undergraduate who was thoroughly trained by one of the original two. Hits are defined as gestures that display the particular kinesthetic quality of ending with an abrupt stop, usually (although not always) corresponding to a change in articulator shape (for hands) or direction of movement. We selected these gestures for study in part because this quality of abrupt stop provides a natural point of inquiry for alignment with the speech signal, since the end point frame in the video is just 33 ms long (vs. the typical range of 100-250 ms for a syllable and often more than that for the entire stroke phase of a gesture), and in part because these end points are relatively easy to identify in the signal, because the video image often clears up suddenly in that frame. In addition, such gestures often give the perceptual impression that the speaker intended the gesture to end at that point. This is compatible with the hypothesis suggested in Esposito et al. ([9] this volume) that these movements correspond to gestural accents, making them a natural starting point for asking about alignment with spoken prosodic accents. 'Hits' are thus distinguished from 'strokes' (which all gestures have), in the sense that not all identifiable gestures are hits; some display a more continuous quality (such as repeated circular motions with the hands) or a softer slowdown with a less-immediately-identifiable stopping point; such gestures would not be analysed as hits, although they have strokes. In this sense our study follows up on Loehr's [20] investigation of the alignment of gestural strokes with pitch accents, by asking more detailed questions about the alignment of this particular type of speech accompanying gesture and in a different speaking style, i.e. professional academic lecturing.

The video file, without audio, was first played at a normal display rate (using iMovie (www.apple.com) to display the speaker's movements) in order to determine which gestures had the perceptual characteristic of a sudden sharp end point that we define as a 'hit'. This characteristic is quite easy to determine at normal display rates although it does not emerge as clearly when the video is viewed frame-by-frame. Once a hit had been identified, the video was moved forward and back, one video frame at a time, to locate the single frame that corresponded to the end of the gesture, i.e. the hit end frame. Since the frame rate for the video recording was 30 frames per second, this method located the end of the hit within a 33 ms time frame. Questionable hits were marked with 'hit?'. For several of the speakers, the behavior of three different gestural articulators was recorded separately, i.e. hands, head and eyebrows, because all three articulators often participate in hit-like movements. This separate labelling will allow us to evaluate the possibility that the different articulators function differently in this type of gesture. Also for several speakers, the preparation phase, stroke phase and relaxation/recovery phase were marked, with optional holds at the end of each phase [17]. Because these four speech samples were labelled over a period of years and our labelling efforts are ongoing, individual speakers have been labelled in different ways. For example, some speakers have been labelled for movements of additional parts of the body, such as the torso, shoulders, chin, and additional movements of portions of the face, such as frown, blink, and gaze shift. Some of these movements appear to have a hit-like character, e.g. certain shoulder and chin movements, while others do not, either because they occur too rapidly (e.g. blinks) or too gradually (shifting the weight of the torso from one foot to the other).

The video frame that corresponds to end of the 'hit', i.e. the hit end frame, was generally identifiable using frame-by-frame viewing, because three types of change usually co-occurred. First, as noted earlier, the video became clearly focussed or at least strikingly less blurry, presumably reflecting the sudden cessation of movement by the articulator. Second, there was often a slight movement of the articulator in a new direction; either a 'bounce-like' movement in the opposite direction, or change in a right-angled direction. Finally, particularly for the hand there was often the beginning of a change in hand shape. If there was some uncertainty about which of two adjacent frames corresponded to the hit end frame, the first of the two frames was selected by convention.

Temporal alignments among the various labelled elements were obtained by creating Praat TextGrids which displayed, on separate labelling tiers, the locations in time of a) the 33-ms hit end frame for each gestural hit, b) the location of each phonetic landmark predicted from the segments of the word and labelled in the speech signal, from which the beginning and end of the pitch-accented syllables could be estimated, and c) each pitch accent, phrase tone and boundary tone labelled in the speech signal. The hit end frames were displayed on separate tiers for hand-hits, head-hits and brow-hits, where this information was available separately. Figure 1 shows a labelled speech sample with a Praat TextGrid showing landmark labels (on two tiers, for realized and missing implementations of the predicted landmarks), prosody labels (on separate tiers for tones, breaks and miscellaneous comments) and hit end frames (with head, hand and brow hits marked on separate tiers). Our criterion for alignment of a hit end frame with a pitch accented syllable was that some portion of the 33-ms time extent of the hit end frame overlapped with some portion of the time extent of the accented syllable as determined by the landmark labels.

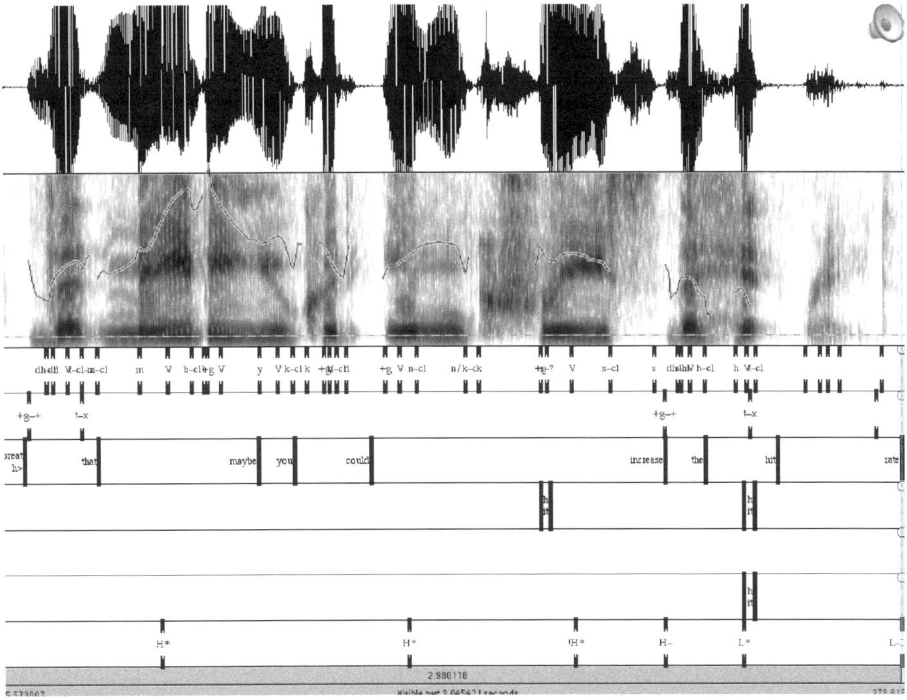

Figure 1: a labelled speech sample for Speaker B

2.2 Handling Ambiguity

A number of challenges arose in the determination of alignments among the labelled elements, and for each challenge, an analysis convention was adopted. As noted earlier, labelling systems for both the gestural hits and the prosodic pitch accents include a marker for questionable tokens, so that analyses could be carried out with and without the uncertain cases. Assigning pitch accents to syllables was not an issue, because accent labels are aligned with the accented vowel. However, determining whether a hit aligns with a pitch-accented syllable is somewhat more difficult, for several reasons. First, the syllable affiliation of some consonants of American English is not obvious to native listeners, so that if a hit end frame aligns, for example, with the closure for the word-medial flap in an utterance of *butter*, it is not certain whether it aligns with the first or second syllable of this word. To handle these cases we adopted the convention of marking such hit end frames as 'in the consonant after a pitch accented vowel', enabling us to analyse their results separately. In other cases the syllable structure of the words may be clear, but articulatory and acoustic overlap between two phonemic segments in the spoken utterance can make it difficult to determine which of several successive sets of landmarks the hit end frame is aligned with. An example of this problem might be when the consonant at the end of one word and at the beginning of the next word are identical, as in e.g. *big glove*, so that if the hit end frame aligns with the silence between the closure for the first /g/ and the release of the second one, it is unclear whether it is aligned with the unaccented syllable *big* or the accented syllable *glove*. Our convention here was to mark the alignment of the hit end frame with the accented syllable as questionable.

One final question about alignments concerns a constituent that is sometimes called the Abercrombian Foot, after Abercrombie's [1] proposal that spoken utterances are organized into rhythmic constituents that can cross word boundaries in one of two senses. First, an Abercrombian foot can combine a strong syllable with a following weak syllable a) when the weak syllable corresponds to an entire separate word, as in [*See the*][*show*], [*Give him*][*time*], or b) when the weak syllable corresponds to the initial syllable of a polysyllabic word, as in [*Make a-*][*-lignments*], [*Tell me to-*][*-morrow*]. If such rhythmic constituents are generated by speakers as part of the production plan for an utterance, then it is possible that spoken pitch accents may align somewhere in the accented Abercrombian Foot, though not always with the initial accent-bearing syllable of that foot. There is some question about precisely how to determine the Abercrombian Foot structure of an utterance, since neither a pitch accent nor a full vowel syllable appear inevitably to be heads of new Abercrombian Feet. As a result, it is not clear whether *peptide* (where the underline signifies a pitch accent) is one AbFoot or two, and similarly whether *He could go* is two AbFeet or three. Nevertheless, if hit end frames consistently align with weak syllables immediately after a strong syllable but not with weak syllables immediately before, it would provide some evidence for the hypothesis that this constituent plays a role in speech production planning. The method described here has the potential to answer this kind of detailed alignment question.

3. Discussion

This study describes a method for detailed labelling of the temporal alignments between segmental, prosodic and gestural aspects of speech acts. It provides a precise definition of the accent-lending events in the gestural and speech streams (hits and pitch accents), a set of conventions for marking clear vs. ambiguous tokens that permits the separate or combined analysis of these conditions, and an explicit quantitative criterion for what is considered to be temporal alignment between hit end frames and pitch accented syllables. During the development of this method, a number of issues arose that have potential theoretical significance, such as the syllable affiliation of a consonant that follows a main-word-stress vowel, and the status of the rhythmic constituent called the Abercrombian Foot, which may be addressed by results of future alignment studies. This method also lends itself to further exploration of Kendon's [15-16] intriguing hypothesis about the speech alignment of gestural phrases (which he defines as the preparation phase + stroke of a gesture, with any optional holds) within each gestural unit (which he defines as the events between returns to the rest position, e.g. between one episode of the hands at rest and the next episode of the hands at rest). A short quotation will give the flavor of his view of the structural, temporal and communicative interplay among speech, prosody and gesture, and the potential informativeness of close analysis of the relationships among them:

"*...when a speaker speaks, the speech is organized into a series of packages, here identified as tone units marked in terms of variations in voice, pitch level, loudness and pacing. These packages tend to correspond to units of meaning that are at a level above the lexical level, and which, for convenience may be referred to as 'idea units' (Kendon 1980). Gesture is also organized into packages of action, here termed 'gesture phrases', which coincide with and tend to be semantically coherent with the units of phrasal meaning or 'idea units' expressed in the tone units. However, the gestural expression typically takes up just a part of the idea unit the tone unit expresses. For example, it may bring out an aspect of meaning associated with the verb in the phrase..., or it may add an imagistic dimension to something referred to by a noun....The precise way in which a coincidence is achieved between a gesture phrase and that part of the tone unit to which it is related semantically appears to be variable. In our interpretation, this means that the speaker can adjust both speech and gesture to one another as if they are two separate expressive resources which can be deployed, each in relation to the other, in different ways according to how the utterance is being fashioned.*" (Kendon [16], pp. 126)

References

[1] Abercrombie, D. (1965/1971), *Syllable quantity and enclitics in English: Studies in Phonetics and Linguistics.* Oxford: Oxford University Press

[2] Beckman, M.E. and Ayers, G.E. (1994), Guidelines for ToBI labeling. Unpublished Ohio State University ms. (see www.ling.ohio-state.edu/research/phonetics/)

[3] Beckman, M.E. and Edwards, J. (1994), Articulatory evidence for differentiating stress categories. In Patricia A. Keating, ed., *Papers in Laboratory Phonology I: Between the Grammar and the Physics of Speech.* Cambridge: Cambridge University Press

[4] Beckman, M.E., Hirschberg, J. and Shattuck-Hufnagel, S. (2005), The original ToBI system and the evolution of the ToBI framework. In Jun, Sun-Ah, ed., *Prosodic Typology.* Oxford: Oxford University Press

[5] Beckman, M. E. and Pierrehumbert, J. B. (1986), Intonational structure in Japanese and English. *Phonology Yearbook 3*, pp. 255-310

[6] Bolinger, D. (1958), A theory of pitch accent in English. *Word 14*, pp. 109-149

[7] Chomsky, N. and Halle, M. (1968), *The Sound Pattern of English.* MIT Press: Cambridge, Mass.

[8] Crystal, D. (1969), *Prosodic Systems and Intonation in English.* Cambridge: Cambridge University Press

[9] Esposito A., Esposito D., Refice M., Savino M., Shattuck-Hufnagel S., A Preliminary Investigation of the Relationship between Gesture and Prosody in Italian. In A. Esposito, M. Bratanic, E. Keller, M. Marinaro (Eds.), *The Fundamentals of Verbal and Nonverbal Communication and the Biometrical Issue*, IOS press (this volume).

[10] Fear, B., Cutler, A. and Butterfield, S. (1995), The strong/weak syllable distinction in English. *Journal of the Acoustical Society of America 97*, pp. 1893-1904

[11] Gussenhoven, C. (2004), *The Phonology of Tone and Intonation.* Cambridge: Cambridge University Press.

[12] Hayes, B. (1989), The prosodic hierarchy in meter. In *Rhythm and Meter: Phonetics and Phonology 1*, ed. Paul Kiparsky and Gilbert Youmans. New York: Academic Press

[13] Keating, P.A. and Shattuck-Hufnagel, S. (2002), A prosodic view of word form encoding for speech production. *UCLA Working Papers in Phonetics 101*, pp. 112-156

[14] Kendon, A. (1972), Some relationships between body motion and speech: An analysis of an example. In Studies in Dyadic Communication, ed. Aaron Siegman and Benjamin Pope. Elmsford, New York: Pergamon Press, pp. 177-210

[15] Kendon, A. (1980), Gesticulation and speech: two aspects of the process of utterance. In *The Relationship of Verbal and Nonverbal Communication*, ed. Mary Ritchie Key. The Hague: Mouton and Co. pp. 207-227

[16] Kendon, A. (2004), *Gesture: Visible Action as Utterance.* Cambridge: Cambridge University Press

[17] Kita, S. (1993), *Language and Thought Interface: A Study of Spontaneous Gestures and Japanese Mimetics.* PhD Dissertation, Departments of Psychology and Linguistics, University of Chicago

[18] Krahmer, E., Ruttkay, Z., Swerts, M. and Wesselink, W. (2002), Pitch, eyebrows and the perception of focus. In *Proceedings of Speech Prosody 2002*, Aix-en-Provence

[19] Ladd, D. R. (1996), *Intonational Phonology.* Cambridge: Cambridge University Press

[20] Loehr, D. (2004), *Gesture and Intonation.* PhD Dissertation, Georgetown University, Washington D.C.

[21] McClave, E. (1994), Gestural beats: The rhythm hypothesis. *Journal of Psycholinguistic Research 23*, pp. 45-66

[22] McNeill, D. (1992), *Hand and Mind: What Gestures Reveal about Thought.* Chicago: University of Chicago Press

[23] McNeill, D: Gesture and Thoughts. In A. Esposito, M. Bratanic, E. Keller, M. Marinaro (Eds.), *The Fundamentals of Verbal and Nonverbal Communication and the Biometrical Issue,* IOS press, this volume.

[24] Nespor, M. and Vogel, I. (1986), *Prosodic Phonology.* Dordrecht: Foris Publications

[25] Okobi, T. (2006), *Acoustic Correlates of Word Stress in American English.* PhD Thesis, Department of Elecgtrical Engineering and Computer Science and Program in Speech and Hearing Bioscience and Technology, Massachusetts Institute of Technology

[26] Pierrehumbert, J. B. (1980), *The Phonology and Phonetics of English Intonation.* PhD thesis, Department of Linguistics, Massachusetts Institute of Technology

[27] Renwick, M., Shattuck-Hufnagel, S. and Yassinik, Y. (2004), The timing of speech-accompanying gestures with respect to prosody. *Journal of the Acoustical Society of America 115*, 2397

[28] Selkirk, E. O. (1984), *Phonology and Syntax: The Relation between Sound and Structure.* Cambridge, Mass.: MIT Press

[29] Shattuck-Hufnagel, S., Veilleux, N.M. and Stevens, K.N. (in prep), The reliability of acoustic landmarks in a sample of American English spontaneous speech.

[30] Sluijter, A. M.C. and van Heuven, V. J. (1996), Spectral balance as an acoustic correlate of linguistic stress. *Journal of the Acoustical Society of America 100*, pp. 2471-2485

[31] Stevens, K.N. (1998), *Acoustic Phonetics.* MIT Press: Cambridge, MA

[32] Stevens, K.N. (2002), Toward a model for lexical access based on acoustic landmarks and distinctive features. *Journal of the Acoustical Society of America 111,* pp. 1872-1891

[33] Stevens, K.N. (2003), Acoustical and perceptual evidence for universal phonological features. In *Proceedings of the 15th International Congress of Phonetic Sciences, Barcelona*, pp. 33-38

[34] Stevens, K.N. and Keyser, S.J. (in press), Quantal theory, enhancement and overlap. In Clements, N. and Ridouane, R., eds., *Proceedings of the Workshop on Quantal Theory*, The Sorbonne, Paris, July 2006

A. Esposito et al. (Eds.)
IOS Press, 2007

What Pauses Can Tell Us about Speech and Gesture Partnership

Anna ESPOSITO[1], Maria MARINARO
Department of Psychology, Second University of Naples (SUN), and IIASS, Italy
Department of Physics, Salerno University, and IIASS, Italy

Abstract. Considering the role that speech pauses play in communication we speculate on the possibility that holds (or gesture pauses) may serve to similar purposes supporting the view that gestures as language are an expressive resource that can take on different functions depending on the communicative demand. The data reported in the present paper seem to support this hypothesis, showing that 93% of the children and 78% of the adult speech pause variation is predictable from holds, suggesting that at the least to some extent, the function of holds may be thought to be similar to speech pauses. While speech pauses are likely to play the role of signalling mental activation processes aimed at replacing the "old spoken content" of an "utterance" with a new one, holds may signal mental activation processes aimed at replacing the "old visible bodily actions" (intimately involved in the semantic and/or pragmatic contents of the old "utterance") with new bodily actions reflecting the representational and/or propositional contribution that gestures are engaged to convey in the new "utterance".

Keywords: Speech pauses, gesture pauses, synchrony

Introduction

In daily human-to-human interaction we usually encode the messages we want to transmit in a set of actions that go beyond verbal modality. Nonverbal actions (grouped under the name of gestures) help to clarify meanings, feelings, and contexts, acting for the speaker as an expressive resource exploited in partnership with speech for appropriately shaping communicative intentions and satisfying the requirements of a particular message being transmitted.

Said in this way, it is clear that we are taking a position attributing to gestures similar semantic and pragmatic functions as in speech and rejecting the hypothesis that either gestures or speech alone might have the primary role in the communicative act [35, 39, 45, 47]. There is a considerable body of evidence supporting our position [13, 23, 24, 25, 29, 35, 41, 45, 54] but there are also data suggesting that the role of gestures is secondary to speech being only of support to the speaker's effort in encoding his/her message [27, 43, 44, 48, 49, 50, 55, 57].

The latter appears to be a reasonable position since during our everyday life interactions we are aware of generating verbal messages and of the meaning we

[1] Corresponding Author: Anna Esposito, Department of Psychology, Second University of Naples, and IIASS, Via Vivaldi, 43, 81100 Caserta, Italy, e-mail: iiass.annaesp@tin.it, anna.esposito@unina2.it.

attribute to these, thanks to the continuous auditory feedback we get on our speech, whereas, we are not endowed with a similar feedback on our gesticulation, posture, and facial expressions. Objectively, this is not the only argument in favour of the above position. As a matter of fact, most of the gesturing is made without a conscious control, since we do not pay special attention to our gesturing while speaking, and additionally, humans carry out a successful interaction also in situations where they cannot see each other (on the telephone for example, see Short et al., and Williams [60, 63]). Conversely, it is really hard to infer the meaning of a message when only gestures and no speech is provided [27, 43, 44, 57]. It seems just common sense to assert the primary role of speech in communication and since we undoubtedly gesture during visible and non-visible interactions (see to this aim the works of Rimé [56], and Rimé & Schiratura [57]) it might appear obvious, if not trivial (see Krauss et al. [26, 43, 44, 55]) to deduce that the role of gestures, if any, in communication, is just of assistance to the listener, or to the speaker, or both.

More in-depth analyses, however, shed doubts on the above position and prove that gestures and speech are partners in shaping communication and giving kinetic and temporal (visual and auditory) dimensions to our thoughts. Some hints of this may simply be experienced in our everyday life. Gestures resolve speech ambiguities and facilitate comprehension in noisy environment [35, 58, 62], they act as a language when verbal communication is impaired [29, 35, 38], and in some contexts not only they are preferred but produce more effective results than speech in communicating ideas (see the mismatches theory in Goldin-Meadow [29]). More interesting, it has been shown that gestures are used in semantic coherence with speech [35, 45] and may be coordinated with tone units and prosodic entities, such as pitch-accented syllables and boundary tones (see to this aim the works of Kendon [35], Yasinnik et al. [64], Shattuck-Hufnagel et al. [61], and Esposito et al. [14]). Moreover gestures add an imagistic dimension to the phrasal contents [35, 39, 45, 47] and are synchronized with speech pauses [6, 8, 18, 19, 20, 34]. In the light of these considerations, gestures are to be regarded as an expressive system that, in partnership with speech, provide means for giving form to our thoughts [13, 35, 39, 41, 42, 45, 46, 47].

Have we been convincing? Are the provided data able to definitely assess the role of gestures in communicative behaviours? Since the experimental data are somewhat conflictual, the question of how to integrate and evaluate the different positions on the relevance of gestures in communication is still open and the results we are going to present may be relevant in evaluating their relative merits.

In previous works (see Esposito et al. [18, 19, 20]) we adopted the theoretical framework that gestures, acting in partnership with speech, have similar semantic and pragmatic functions. Starting from these assumptions, we tried to answer the following questions about hand movements: assuming that speech and gestures are co-involved in the production of a message, is there any gestural equivalent to filled and empty pauses in speech? Assuming that we have found some equivalent gestural entities, to what degree do these synchronize with speech pauses?

As an answer to our first question, in two pilot studies we identified a gestural entity that we called hold (or gesture pause). A careful review of speech and gesture data showed that in fluent speech contexts, holds appear to be distributed similarly to speech pauses and to overlap with them, independently from the language (the gesture data were produced by Italian and American English speakers) and independently from the context (a planning and a narration context). In the present study we found further

support to our previous speculations in narrative discourse data from children and adults who participated in a similar type of elicitation. Both adults and children were native speakers of Italian. There were two goals motivating this extension of the previous research:
1) If the relationships previously found are robust, they should be independent of the age; i.e., they should also be evident in children discourse; 2) If at least some aspects of speech and gesture reflect a unified planning process, these should be similar for all human beings provided that the same expressive tools are available.

1. Speech pauses

A characteristic of spontaneous speech, as well as of other types of speech, is the presence of silent intervals (empty pauses) and vocalizations (filled pauses) that do not have a lexical meaning. Pauses seem to play a role in controlling the speech flow. Several studies have been conducted to investigate the system of rules that underlie speaker pausing strategies and their psychological bases. Research in this field has shown that pauses may play several communicative functions, such as building up tension or raising expectations in the listener about the rest of the story, assisting the listener in her/his task of understanding the speaker, signalling anxiety, emphasis, syntactic complexity, degree of spontaneity, and gender, and transmitting educational and socio-economical information [1, 4, 12, 30, 31, 40].

Studies on speech pause distribution in language production have produced evidence of a relationship between pausing and discourse structure. Empty and filled pauses are more likely to coincide with boundaries, realized as a silent interval of varying length, at clause and paragraph level [5, 28, 33, 59]. This is particularly true for narrative structures where it has been shown that pausing marks the boundaries of narrative units [9, 10, 15, 16, 17, 51, 52, 53, 59].

Several cognitive psychologists have suggested that pausing strategies reflect the complexity of neural information processing. Pauses will surface in the speech stream as the end product of a "planning" process that cannot be carried out during speech articulation and the amount and length of pausing reflects the cognitive effort related to lexical choices and semantic difficulties for generating new information [7, 9, 15, 16, 17, 30].

We can conclude from the above considerations that pauses in speech are typically a multi-determined phenomenon [3,7, 32], attributable to physical, socio-psychological, communicative, linguistic and cognitive causes. Physical pauses are normally attributed to breathing or articulatory processes (i.e. pauses due to the momentary stoppage of the breath stream caused by the constrictors of the articulatory mechanism or the closure of the glottis). Socio-psychological pauses are caused by stress or anxiety [3]. Communicative pauses are meant to permit the listener to comprehend the message or to interrupt and ask questions or make comments. Linguistic pauses are used as a mean for discourse segmentation. Finally, cognitive pauses are related to mental processes connected to the flow of speech, such as replacing the current mental structure with a new one, in order to continue the production [9, 10, 11] or difficulties in conceptualization [30].

In the light of these considerations it seems pretty logical to ask what could be, if any, the role of gesture pauses (holds henceforth) in communication and the possible functions they are assumed to play with respect to speech pauses. To this aim, in the reported data, socio-psychological, articulatory, and communicative pauses were ruled out from the analysis. The first ones were assumed not to be a relevant factor, by virtue of the particular elicitation setting (see next section for details). The second and third ones were identified during the speech analysis and eliminated from the dataset. The speech pauses considered in this work, therefore, are linguistic, cognitive and breathing pauses. On the basis of the assumption that breathing and linguistic pauses are part of the strategy the speakers adopt for grouping words into a sentence, we assumed that they are part of the planned communication process.

In this study, empty pauses are simply defined as a silence (or verbal inactivity) in the flow of speech equal to or longer than 120 milliseconds. Filled pauses are defined as the lengthening of a vowel or consonant identified perceptually (and on the spectrogram) by the transcriber or as one of the following expressions: "*uh, hum, ah, ehm, ehh, a:nd, the:, so, the:n, con:, er, e:, a:, so:*"[2].

A gesture pause (or hold) is detected when the arms and hands remained still for at least three video frames (i.e., approximately 120 ms.) in whatever position excluding the rest position. The latter is defined as the home position of the arms and hands when they are not engaged in gesticulation and typically this is at the lower periphery of the gesture space (see Mc Neill [47] p.89). The holds associated with gesture rest were included in the analysis by virtue of the particular elicitation setting (see next section for details).

Note that the absence of movement is judged perceptually by an expert human coder. Therefore the concept of hold is ultimately a perceptual one. A hold may be thought to be associated with a particular level of discourse abstraction. In producing a sentence, the speaker may employ a metaphoric gesture [47] with a hold spanning the entire utterance. However, the speaker may also engage in word search behaviour (characterized by a slight oscillatory motion centred around the original hold) without any change in hand shape (the *Butterworth gesture* cited in Mc Neill [47]).
The speaker may also add emphatic beats [47] coinciding with points of peak prosodic emphasis in the utterance. While the word search and emphatic beats may sit atop the original hold, most observers will still perceive the underlying gesture hold.

2. Materials

The video recordings on which our analysis is based are of narrations by 6 female children (9 years old ± 3 months) and 4 adults (2 males and two females, average age 28 years old ± 3 years). The children and the adults told the story of a 7-minute animated colour cartoon they had just seen. The cartoon is of a familiar type to Italians (adults and children) involving a cat and a bird. The listener in the children's case was the child's teacher together with other children also participating in the experiment, whereas in the adults' case the listener was the experimenter who also was a close friend of them. Children's recordings were made after the experimenter had spent two months with the children in order to become familiar to them and after several preparatory recordings had been made in various contexts in order for the children to get used to the camera. In the adults' case, the recordings were made by the experimenter, in a friendly environment (at the experimenter's home) after a relaxing

[2] The notation ":" indicates vowel or consonant lengthening.

dinner and justifying the story-telling elicitation as a memory recall experiment. This kept out stranger-experimenter inhibitions from the elicitation setting; i.e., factors that could result in stress and anxiety. Limiting these factors allowed us to rule out the "socio-psychological" type of pauses [3].

The cartoon had an episodic structure, each episode characterized by a "cat that tries to catch a bird and is foiled" narrative arc. Because of the cartoon's episodic structure, typically children and adults would forget entire episodes. In this case the experimenter was allowed to give suggestions in order to remember them. However, given the length and the difficulty of these kinds of analysis, in the present paper only partial data are reported. Therefore, only three episodes were analyzed for all the children, whereas we reported the analysis of seven episodes for the two female adults and of six for the male adults. None of the participants was aware that speech and gesture pauses were of interest. The video was analyzed using commercial video analysis software (VirtualDub™) that allows to view video-shots, and forward and backward movements through the shots. The speech waves, extracted from the video, were sampled at 16 kHz and digitalized at 16 bits. The audio was analyzed using Speechstation2™ from Sensimetrics. For the audio measurements the waveform, energy, spectrogram, and spectrum were considered together, in order to identify the beginnings and endings of utterances, filled and empty speech pauses and phoneme lengthening. The details of the criteria applied to identify the boundaries in the speech waveform are described in Esposito and Stevens [9]. Both the video and audio data were analyzed perceptually, the former frame-by-frame and the latter clause-by-clause, where a "*clause*" is assumed to be "*a sequence of words grouped together on semantic or functional basis*" [16].

3. Results

The results are summarized in Tables 1 and 2. Table 1 summarizes the children's data. The symbol # stands for the word "number". The percentage of speech pauses overlapping or not with holds, as well as the percentage of holds that occur only in continuous speech is indicated in parentheses and has been computed over the total number of speech pauses each subject achieved during the narration of the analyzed episodes. The sample mean of percentage of overlaps and non is reported in the last row, together with the sample Standard Deviation (SD). Also reported (in seconds) is the total duration of the narrated episodes for each subject. Table 2 summarizes the adults' data and is structured similarly to Table 1.

The *Pearson product-moment correlation coefficient* was computed as a descriptive statistic of the magnitude or the amount of information that can be inferred about speech pause frequency from the known hold frequency.

The Pearson correlation coefficient between holds and speech pauses for children was $r = 0.97$, and the proportion of the variation of speech pauses that is determined by the variation of holds (i.e. the *coefficient of determination*) was $r^2 = 0.93$, which means that 93% of the children's speech pause variation is predictable from holds. For adults $r = 0.88$, and $r^2 = 0.78$. The two groups of speakers produced a similar distribution of hold and speech pause overlaps.

Table 1: Summary of children data

Subjects.	*# Holds*	*# Spech Pauses (Empty, Filled, and Vowel Lengthening)*	*# Speech Pauses Overlapping with Holds*	*# Speech Pauses Not Overlapping with Holds*	*# Holds in Continuous Speech*
S1 (160.08s)	106	114	95(83%)	19(17%)	26 (25%)
S2 (148.76s)	102	109	91(83%)	18(17%))	28 (27%)
S3 (50.84s)	32	38	35(93%)	3(8%))	4 (13%)
S4 (110.20s)	68	90	83(92%)	7(8%)	15 (22%)
S5 (92.96s)	45	44	42(95%)	2(4%)	13 (29%)
S6 (127.36s)	69	69	43(62%)	26(38%)	28 (40%)
Total	**422**	**464**	**389 (84.8%, SD ±12.4)**	**75 (15,3%, SD ±12.2)**	**114 (26%, SD ±8.9)**

Table 2: Summary of adults' data

Subjects.	*# Holds*	*# Spech Pauses (Empty, Filled, and Vowel Lengthening)*	*# Speech Pauses Overlapping with Holds*	*# Speech Pauses Not Overlapping with Holds*	*# Holds in Continuous Speech*
S1 (male, 204.16s)	112	90	72 (**80%**)	18(**20%**)	41 (36%)
S2 (male 230.59s)	135	120	98 (**82%**)	22(**18%**)	52 (39%)
S3 (female 352.55s)	250	167	155 (**93%**)	12(**7%**)	121 (48%)
S4 (female 245.80s)	206	121	93 (**77%**)	28(**23%**)	110 (53%)
Total	**714**	**498**	**418 (83%, SD ±9.9)**	**80 (17%, SD ±6.9)**	**324 (44%, SD ±7.8)**

The frequency of overlaps between holds and speech pauses not only was remarkably high but was much the same for adults and children, clearly indicating that both children and adults tended to synchronize speech pauses with holds independent of their age. The degree of synchronization was so high that further statistic analyses to assess its significance were not necessary, if the word "*synchronization*" is interpreted more loosely to mean "*the obtaining of a desired fixed relationship among corresponding significant instants of two or more signals* [www.its.bldrdoc.gov]".

Figure 1 graphically shows the percentage of overlaps against the percentage of speech pauses that do not overlap with holds, in children (1a) and adults (1b).

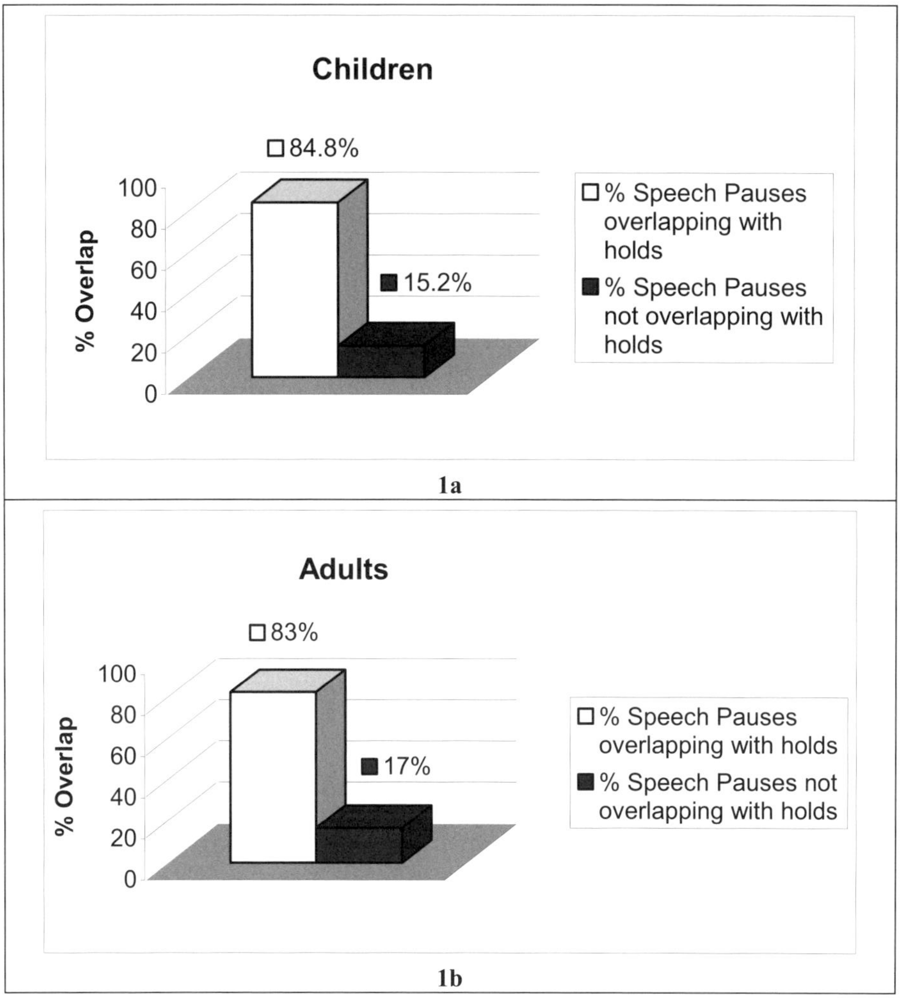

Figure 1: Percentage of overlaps and non-overlaps between speech pauses and holds in children (1a) and adults (1b)

Figure 2 displays, for each subject in each group (children and adults), the holds and speech pause rates computed as the ratios between the number of holds and/or speech pauses over the length of the subject's narrations measured in seconds. Figure 2a is for children and Figure 2b is for adults.

In order to assess if the differences in the two rates were statistically significant, a one way ANOVA test was performed for each group, with hold and speech pause rates as within-subject variables. The ANOVA results showed no significant difference for children ($F(1,10) = 1.09$, $\rho = 0.32$)) suggesting that holds and speech pauses were equally distributed along children's narrations. For adults, the differences were statistically significant ($F(1,6) = 11.38$, $\rho = 0.01$)), suggesting that adults used holds more frequently than speech pauses.

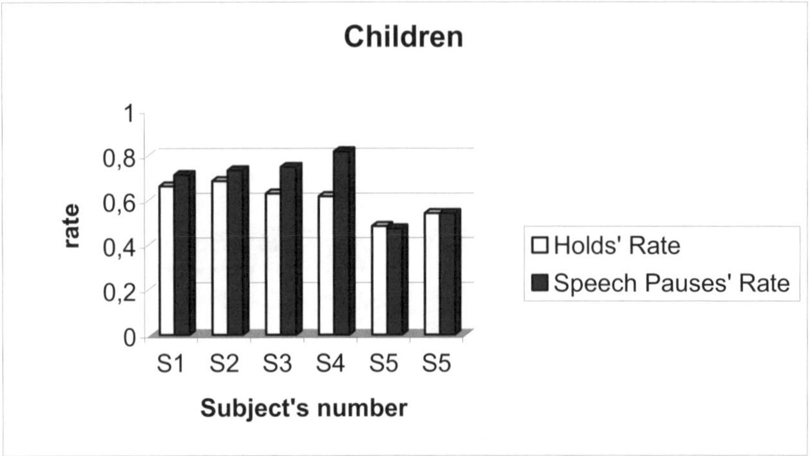

2a

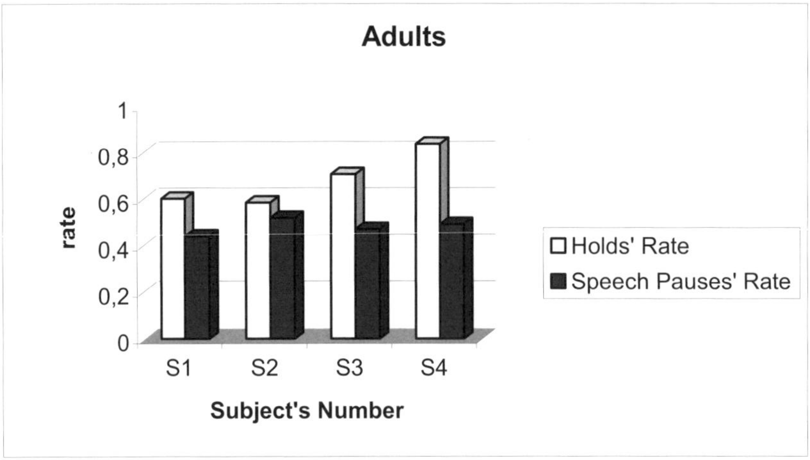

2b

Figure 2: Holds' rates against speech pauses' rates for children (2a) and adults (2b)

4. Discussion

The previous section displays two interesting results. First, a great amount of variation in speech pauses is highly correlated with holds, both in adults and children, and there is a great amount of overlaps between the two speech and gestural entities. Second, speech pauses are highly synchronized with holds and this synchronization does not depend on the speaker age.

What does this suggest about gestures and speech partnership? To answer this question it is necessary to recall the role attributed to speech pauses, in particular to cognitive speech pauses that are under examination in this work. As already pointed out in the introductive section, these speech pauses are used to "hold the floor" i.e. to prevent interruption by the listener while the speaker searches for a specific word [22] but can also serve for other functions, such as marking a successful memory search, or signalling the selection of an example, or reflecting the complexity of neural information processing. Pauses will surface in the speech stream as the end product of a "planning" process that cannot be carried out during speech articulation and the amount and length of pausing reflect the cognitive effort related to lexical choices and semantic difficulties for generating new information [7, 9, 30]. In the present paper the concepts of "new information" is interpreted as "*added information*", in contrast to "*given information*" according to the definition proposed by Chafe [11], which considers as "*added any verbal material that produces a modification in the listener's conscious knowledge*". Therefore "*given*" verbal material was intended as not producing such a modification. In summary, speech pauses seem likely to play the role of signalling mental activation processes aimed at replacing a particular attentional state with a new one.

Given the great amount of overlaps between holds and speech pauses, holds appear to be gestural entities with similar function and similar behaviour as speech pauses. Therefore, these data appear to support, between the two theories presented in the introductory section, the one that assumes non-verbal modalities and speech to have similar semantic and pragmatic functions. The data presented in this study strongly suggest that, at least in some respects, speech and gestures reflect a unified planning process, which differs in the two domains only because their implementation is realized through different modalities (the manual-visual versus the oral-auditory channel). As speech pauses seem likely to play the role of signalling mental activation processes aimed at replacing the "*given spoken content*" of a former utterance with an "*added*" one, holds may signal mental activation processes aimed at replacing "*given visible bodily actions*" (intimately involved in the semantic and/or pragmatic contents of the former "utterance" [35]) with "*added bodily actions*" reflecting the new representational and/or propositional contribution that gestures are engaged to convey in the new "utterance". Note that the meaning given here to the word "utterance" is the same used by Kendon ([35], see chapter 1, page 5) "*as an object constructed for others from components fashioned from both spoken language and gesture*".

Under the above assumption of the meaning of the word "utterance" we can speculate about how to justify the second result reported in the present work, i.e. why, as a major difference between adults and children, hold rates in the former are significantly different from speech pause rates, which makes the difference between adults and children. Here is our suggestion: gestures express the "kinetic" component of our thoughts through a signal medium which is different from speech, not in terms of differences in planning, but in terms of transmission mechanisms (visual instead of auditory perception, signals in two or three dimensions, instead of mono-dimensional signals). Due to the differences in their dynamics we would not expect that gestures and speech behave in exactly the same way. We would expect, however, that they show shared communicative features supporting each other with information that clarifies the intended meaning of the utterance being transmitted. As a result, in planning her/his utterances, the speaker may use gestures in many different ways and one of these could be punctuating the spoken discourse when lexical access does not present difficulties. This is one of the functions played by holds in adults. In fact, we observed that most of the holds performed by adults in their narrations and not synchronized with speech pauses were made at the end of clauses, as to mark out the different components of the sentence. Therefore, it seems that adults use holds also as a strategy for structuring the discourse in the absence of speech pauses, i.e. when the speech is fluent, as a mean to emphasize or underline groups of words. Children, instead, being less skilled in assembling bodily and verbal information, tend to attribute to holds the same functions as speech pauses. These considerations may explain the differences in hold and speech pause rates between the two groups. On the other hand, children may be less skilled in bodily actions than in language use since the formers start being experienced after birth, whereas language feedback is experienced during pregnancy. Furthermore, sophisticated utterances, where verbal and nonverbal entities are put together to express thoughts with the purpose to maximize the amount of information transmitted, are a prerogative of adult communication behaviour and may not be necessary in child utterances, limiting the functions and the use of gestures and consequently of holds.

As far as concerns the reported data, we have seen that also in children, as in adults, holds and speech pauses are to a certain extent synchronized and play similar functions, suggesting that they may be considered as a multi-determined phenomenon exploited by the speaker under the guidance of a unified planning process to satisfy a communicative aim. Although the present data may be relevant in assessing the partnership between speech and gestures, it should be emphasized that this is a pilot study, based on data restricted to the narration context and that further work is needed to support the above assumptions as well as to assess the functions of holds in the production of utterances.

5. Conclusions

The present paper reports perceptual data showing that both adults and children make use of speech pauses in synchronization with holds (or gesture pauses), thereby supporting the hypothesis that, at least in some respects, speech and gestures reflect a unified communicative planning process in the production of utterances. The consistency among the subjects in the distribution of holds and speech pauses suggests that, at least in the Italian language, there is an intrinsic timing behaviour, probably a general pattern of rules that speakers (in narrations) use to regulate speech flow and bodily actions for discourse organization. It would be interesting to conduct an analysis on a more extensive data set to try to model this behaviour in mathematical terms. This might help to derive a deterministic algorithm that would be of great utility for applications in the field of human-machine interaction, favouring the implementation of more natural speech synthesis and interactive dialog systems.

The analysis that has been developed in this paper sheds lights only on a subset of much richer and more subtle processes that are at the basis of the rules and procedures governing the dynamic of face-to-face communication. Among the phenomena not yet examined and worth to be investigated are the actions of the speaker and listener and their relevance to the patterning of speech and gesture pauses; the use of speech and gesture pauses in mid turn or in signalling the engagement and disengagement of the participants within the turn; their functioning and positioning at certain favourite sequential positions within conversations where they are more likely to be relaxed, such as at the end of clauses and paragraphs during a narration.

In the present study, the consequences of the listener's actions on the speaker's have not been considered. It certainly cannot be argued that speech and gesture pauses are expressions of cognitive processes internal to the speaker. However, the effects of the listener's actions in the production of some of these pauses cannot be denied. Interactions between speaker and listener are relevant and surely may lead to systematic changes in the emerging structure of the speaker's utterance and in the distribution of speech and gesture pauses. How these dynamics are implemented during interaction is a central issue for the development of interactive and user-friendly telecommunication services.

Acknowledgements

The paper has been partially supported by **COST Action 2102**: "Cross Modal Analysis of Verbal and Nonverbal Communication (CAVeNC)": http://www.cost.esf.org/index.php?id=110&action_number=2102. Acknowledgements go to **Miss Tina Marcella Nappi** for her editorial help. The authors want to express their great appreciation to Professor **Roberto Ligrone** for his useful comments and suggestions.

References

[1] K. Abrams, T.G. Bever: Syntactic Structure Modifies Attention During Speech Perception and Recognition. *Quarterly Journal of Experimental Psychology, 21,* 280-290, 1969.
[2] G.W. Beattie: Language and Non-verbal Communication: The Essential Synthesis. *Linguistics*, 19, 1165-1183, 1981
[3] R. Beaugrande: *Text Production*. Norwood NJ: Text Publishing Corporation, 1984.
[4] A. Bernstein: Linguistic Codes, Hesitation Phenomena, and Intelligence. *Language and Speech, 5*, 31-46, 1962.
[5] P. Brotherton: Speaking and not Speaking: Process for Translating Ideas into Speech. In A. Siegman, S. Feldestein (Eds.), *Of Time and Speech*, Hillsdale, Lawrence Erlbaum, N.J., 179-209, 1979.
[6] B.L. Butterworth, U. Hadar: Gesture, Speech, and Computational Stages: A Reply to McNeill. Psychological Review, 96, 168-174, 1989.
[7] B.L. Butterworth: Evidence for Pauses in Speech. In B. L. Butterworth (Ed.), *Language Production: Volume 1. Speech and Talk,* London Academic Press, 155-176, 1980.
[8] B.L. Butterworth, G.W. Beattie: Gestures and Silence as Indicator of Planning in Speech. In R.N. Campbell, P.T. Smith (Eds.), *Recent Advances in the Psychology of Language: Formal and Experimental Approaches*, Olenum Press, New York, 347-360, 1978.
[9] W.L. Chafe: Cognitive Constraint on Information Flow. In R. Tomlin (Ed.), *Coherence and Grounding in Discourse*, John Benjamins, 20-51, 1987.
[10] W. L. Chafe: The Deployment of Consciousness in the Production of a Narrative. In W.L. Chafe (Ed.), *The Pear Stories*, Norwood N.J., Ablex, 9-50, 1980.
[11] W. L. Chafe: Language and Consciousness. *Language, 50*, 111-133, 1974.
[12] D.C. O'Connell, S. Kowal: Pausology. *Computers in Language Research, 2, 19,* 221-301, 1983.
[13] J. P. De Ruiter: The Production of Gesture and Speech. In D. McNeill (Ed.), *Language and Gesture*, Cambridge University Press, UK, 284-311, 2000.
[14] A. Esposito, D. Esposito, M. Refice, M. Savino, S. Shattuck-Hufnagel: A Preliminary Investigation of the Relationships between Gestures and Prosody in Italian. In A. Esposito, M. Bratanic, E. Keller, M. Marinaro (Eds.), *Fundamentals of Verbal and Nonverbal Communication and the Biometrical Issue*, IOS press, this volume.
[15] A. Esposito: Children's Organization of Discourse Structure through Pausing Means. In M. Faundez_Zanuy, L. Janer, A. Esposito, A. Satue-Villar, J. Roure, V. Espinosa-Duro (Eds.), *Nonlinear Analyses and Algorithms for Speech Processing*, Lecture Notes in Computer Science, Springer Verlag, 3817, 108-115, 2006.
[16] A. Esposito: Pausing Strategies in Children. In Proceedings of the International Conference in Nonlinear Speech Processing, Cargraphics, Barcelona, SPAIN, 19-22 April, 42-48, 2005.

[17] A. Esposito, M. Marinaro, G. Palombo: Children Speech Pauses as Markers of Different Discourse Structures and Utterance Information Content. In Proceedings of the International Conference: *From Sound to Sense: +50 years of discoveries in Speech Communication*, MIT, Cambridge, USA, 10-13 June, C139-C144, 2004.
[18] A. Esposito, A. Natale, S. Duncan, D. McNeill, F. Quek: Speech and Gestures Pauses Relationships: A Hypothesis of Synchronization (in Italian). In *Proceedings of the V National Conference on Italian Psychology*, AIP, Grafica80-Modugno, Bari, Italy, 95-98, 2003.
[19] A. Esposito, S. Duncan, F. Quek: Holds as Gestural Correlates to Empty and Filled Pauses. In *Proceeding of the International Conference on Spoken Language Processing* (ICSLP02), Denver, Colorado, 1, 541-544, 2002.
[20] A. Esposito, K.E. McCullough, F. Quek: Disfluencies in Gestures. In Proceedings of International Workshop on *Cues in Communication*, Hawai, December 9, 2001.
[21] A. Esposito, K.N. Stevens: Notes on Italian Vowels: An Acoustical Study (Part I). *Research Laboratory of Electronic, Speech Communication Working Papers*, 10, 1-42, 1995.
[22] M.S. Erbaugh: A Uniform Pause and Error Strategy for Native and Non-native Speakers. In R. Tomlin (Ed.), *Coherence and Grounding in Discourse*, John Benjamins, 109-130, 1987.
[23] P. Feyereisen, J.D. deLannoy: *Gesture and Speech: Psychological Investigations*. Cambridge, Cambridge University Press, 1991.
[24] P. Feyereisen, M. Van de Wiele, F. Dubois: The Meaning of Gestures: What Can be Understood Without Speech? *Cahiers de Psychologie Cognitive*, 8, 3-25, 1988.
[25] P. Feyereisen: Manual Activity During Speaking in Aphasic Subjects. *International Journal of Psychology*, 18, 545-556, 1983.
[26] N. Freedman, J. Van Meel, F. Barroso, W. Bucci: On the Development of Communicative Competence. *Semiotica*, 62, 77-105, 1986.
[27] N. Freedman: The Analysis of Movement Behaviour During the Clinical Interview. In A.W Siegmann, B. Pope (Eds.), *Studies in Dyadic Communication*, Pergamon Press, New York, 177-208, 1972.
[28] J.P. Gee, F. Grosjean: Empirical Evidence for Narrative Structure. *Cognitive Science, 8*, 59-85, 1984.
[29] S. Goldin-Meadow: *Gesture: How Our Hands Help Us Think*. Harvard University Press, 2003.
[30] F. Goldmar Eisler: *Psycholinguistic: Experiments in Spontaneous Speech*. Academic press, London: New York, 1968.
[31] D.W. Green: The Immediate Processing of Sentence. *Quarterly Journal of Experimental Psychology, 29,* 135-146, 1977.
[32] M. Garman: *Psycholinguistics*. Cambridge University Press, 1990.
[33] B. Grosz, J. Hirschberg: Some Intentional Characteristics of Discourse Structure. In *Proceedings of International Conference on Spoken Language Processing,* Banff, 429-432, 1992.
[34] U. Hadar, B.L. Butterworth: Iconic Gestures, Imagery and Word Retrieval in Speech. *Semiotica*, 115, 147-72, 1997.
[35] A. Kendon: *Gesture: Visible Action as Utterance*. Cambridge University Press, 2004.
[36] A. Kendon: *Sign Languages of Aboriginal Australia: Cultural, Semiotic and Communicative*. . Cambridge University Press, 1988.
[37] A. Kendon: Current Issues in the Study of Gesture. In J.L. Nespoulous et al. (Eds.) *The Biological Foundations of Gestures: Motor and Semiotic Aspects*, LEA Publishers, Hillsdale, New Jersey, London, 23-27, 1986.
[38] A. Kendon: Knowledge of Sign Language in an Australian Aboriginal Community. *Journal of Anthropological Research*, 40, 556-576, 1984.
[39] A. Kendon: Gesticulation and Speech: Two Aspects of the Process of Utterance. In M. Ritchie Key (Ed.), *The Relationship of Verbal and Nonverbal Communication*, The Hague: Mouton and Co., 207-227, 1980.
[40] S. Kowal, D.C. O'Connell, E.J. Sabin: Development of Temporal Patterning and Vocal Hesitations in Spontaneous Narratives. *Journal of Psycholinguistic Research,* 4, 195-207, 1975.
[41] S. Kita, A. Özyürek: What Does Cross-Linguistic Variation in Semantic Coordination of Speech and Gesture Reveal? Evidence for an Interface Representation of Spatial Thinking and Speaking. *Journal of Memory and Language*, 48, 16-32, 2003.
[42] S. Kita: How Representational Gestures Help Speaking. In D. McNeill (Ed.), *Language and Gesture*, Cambridge University Press, UK, 162-185, 2000.
[43] R. Krauss, Y. Chen, R. F. Gottesman: Lexical Gestures and Lexical Access: A Process Model. In D. McNeill (Ed.), *Language and Gesture*, Cambridge University Press, UK, 261-283, 2000.
[44] R. Krauss, P. Morrel-Samuels, C. Colasante: Do Conversational Hand Gestures Communicate? *Journal of Personality and Social Psychology*, 61(5), 743-754, 1991.
[45] D. McNeill: *Gesture and Thought*. University of Chicago Press, Chicago, 2005.

[46] D. McNeill, S. Duncan: Growth Points in Thinking for Speaking. In D. McNeill (Ed.), *Language and Gesture*, Cambridge University Press, UK, 141-161, 2000.
[47] D. McNeill: *Hand and Mind: What Gesture Reveal about Thought*. University of Chicago Press, Chicago, 1992.
[48] E. Morsella, R.M. Krauss: Can Motor States Influence Semantic Processing? Evidence from an Interference Paradigm. In A. Columbus (Ed.), *Advances in Psychology Research*, New York: Nova, 36, 163-182, 2005a.
[49] E. Morsella, R.M. Krauss: Muscular Activity in the Arm During Lexical Retrieval: Implications for Gesture-Speech Theories. *Journal of Psycholinguistic Research*, 34, 415-437. 2005b.
[50] E. Morsella, R.M. Krauss: The Role of Gestures in Spatial Memory and Speech. *The American Journal of Psychology*, 177(33), 411-42, 1998.
[51] D. O'Shaughnessy: Timing Patterns in Fluent and Disfluent Spontaneous Speech. In *Proceedings of ICASSP Conference*, Detroit, 600-603, 1995.
[52] M. Oliveira: *Prosodic Features in Spontaneous Narratives*. Ph.D. Thesis, Simon Fraser University, 2000.
[53] M. Oliveira: Pausing Strategies as Means of Information Processing Narratives. In *Proceedings of the International Conference on Speech Prosody*, Ain-en-Provence, 539-542, 2002.
[54] L. L. Pedelty: *Gesture in Aphasia*. Ph.D. Thesis, University of Chicago, Department of Behavioural Sciences, Chicago, IL, 1987.
[55] F. H. Rausher, R.M. Krauss, Y. Chen: Gesture, Speech and Lexical Access: The Role of Lexical Movements in the Processing of Speech. *Psychological Science*, 7, 226-23, 1996.
[56] B. Rimé, L. Schiaratura: Gesture and Speech. In R.S. Feldman, B. Rimé (Eds.), *Fundamentals of Nonverbal Behavior,* Cambridge University Press, 239-84, 1992.
[57] B. Rimé: The Elimination of Visible Behaviour from Social Interactions: Effects of Verbal, Nonverbal, and Interpersonal Variables. *European Journal of Social Psychology*, 12, 113-129, 1982.
[58] W. T. Rogers. The Contribution of Kinesic Illustrators Towards the Comprehension of Verbal Behaviour Within Utterances. *Human Communication Research*, 5:54–62, 1978.
[59] B. Rosenfield: *Pauses in Oral and Written Narratives*. Boston University Press, 1987.
[60] J. Short, E. Williams, B. Christie: *The Social Psychology of Telecommunications*. Wiley, New York, 1976.
[61] S. Shattuck-Hufnagel, Y.Yasinnik, N. Veilleux, M. Renwick: A Method for Studying the Time Alignment of Gestures and Prosody in American English: 'Hits' and Pitch Accents in Academic-Lecture-Style Speech. In A. Esposito, M. Bratanic, E. Keller, M. Marinaro (Eds.), *Fundamentals of Verbal and Nonverbal Communication and the Biometrical Issue*, IOS press, this volume.
[62] L. A. Thompson, D. W. Massaro. Evaluation and Integration of Speech and Pointing Gestures During Referential Understanding. *Journal of Experimental Child Psychology*, 42, 144–168, 1986.
[63] E. Williams: Experimental Comparisons of Face-to-Face and Mediated Communication: A Review. *Psychological Bulletin*, 84, 963-976, 1977.
[64] Y. Yasinnik, M. Renwick, S. Shattuck-Hufnagel: The Timing of Speech-Accompanying Gestures with Respect to Prosody. Proceedings of the International Conference: *From Sound to Sense: +50 years of Discoveries in Speech Communication*, MIT, Cambridge, USA, 10-13 June, C97-C102, 2004.

A. Esposito et al. (Eds.)
IOS Press, 2007

"Unseen Gestures" and the Speaker's Mind: An Analysis of Co-Verbal Gestures in Map-Task Activities

Nicla ROSSINI[1]
Dipartimento di Linguistica Teorica e Applicata, Università di Pavia
Li.Co.T.T., Università del Piemonte Orientale

Abstract. The analysis of co-verbal gestures in map-task activities is particularly interesting for several reasons: on the one hand, the speaker is engaged in a collaborative task with an interlocutor; on the other hand, the task itself is designed in order to place a cognitive demand on both the speaker and the receiver, who are not visible to one another. The cognitive effort in question implies the activation of different capabilities, such as self-orientation in space, planning (which can also be considered a self-orientation task concerning the capability of organising successful communicative strategies for the solution of a given problem), and communication in "unnatural" conditions. The co-verbal gestures performed during such a task are quantitatively and qualitatively different from those performed in normal conditions, and can provide information about the Speaker's Mind. In particular, the recursive pattern of some metaphors can be interpreted as a reliable index of the communicative strategy adopted by the speaker: the case of the "palm-down flap" will be here analysed.

Keywords. Co-verbal gestures, map-task activities, face-to-face interaction, Speaker's Mind, "palm-down flap".

Introduction

This paper is aimed at presenting the very first results of an ongoing research on non-verbal cues in "blind" map-task activities. The analysis of non-verbal behaviour in such a quasi-experimental condition is particularly interesting for several reasons: on the one hand, in fact, the speaker is engaged in a collaborative task with an interlocutor who is sitting in front of him but is hidden by an artificial wall, while the task itself is designed in order to place a cognitive demand on the interactants. This implies the activation of different capabilities, such as self-orientation in space, planning (which can also be considered a self-orientation task concerning the capability of organising successful communicative strategies for the solution of a given problem), and communication in "unnatural" or marked conditions. The non-verbal behaviour observable in such conditions is accordingly marked, if compared to the data so far available for face-to-face interaction[2]. In particular, the role of head movements, gaze, and posture-shifting

[1]Corresponding Author: Dipartimento di Linguistica Teorica e Applicata, Università di Pavia, Li.Co.T.T., Università del Piemonte Orientale, e-mail: nicla.rossini@unipv.it .
[2] An informal comparison with the data available from other sources has already been made. In particular, the overall impression of a marked function of posture shifting, head movements and gaze is thus far

seems to be reverse if compared to that observed in face-to-face interaction, while gesticulation – which is still conspicuously present – is qualitatively different from that recorded in face-to-face quasi-spontaneous conditions. Intuitively, one would expect that the lack of a common semiotic space usually established by gaze (see for instance [1]) in normal interaction should prevent the subjects from interactive[3] behaviour: as we will see, this presumption is not completely true.

On the other hand, the co-verbal gestures recorded during "blind" map task are somehow simplified in form and more recurrent, which phenomenon is particularly interesting for the analysis of planning, or self-organization of thought in the speaker. In particular, some recurrent metaphors can be interpreted as indexing a peculiar level of what has been defined as the Speakers' Mind (see for instance [2]), namely the meta-discursive level.

The map-task experiment

This study is based on the analysis of the data collected in 2003 at the Laboratory of Computational Linguistics and Text Understanding (Li.Co.T.T.), Università del Piemonte Orientale, Vercelli, within a wider project aimed at studying the pragmatic phenomena of collaborative dialogue in route-direction. The data collection was conceived as a quasi-experimental session with two subjects at a time. One of them, (the giver of the map description. Hence, the Giver) was provided with a map showing a pre-drawn path; his or her partner (the follower of the map description. Hence, the Follower) was provided with – apparently – the same map without the pre-drawn path. The subjects were subsequently asked to sit at a special desk with an artificial wall in the middle, so that, although facing each other, they were prevented them from seeing the interlocutor. The Giver was then asked to describe the route drawn in his/her map to the interlocutor, whose task was to reproduce it on his/her map. The subjects were not informed in advance of the slight incongruence between the two maps. This element was deemed determinant for the elicitation of psycho-pragmatic mechanisms related to both orientation in space and negotiation of common ground in speech. The sessions were video-recorded. The data collected were analysed by independent transcribers, each focusing on a different aspect of the linguistic exchange. As to the non-verbal aspect, the analysis has been conducted according to a standard protocol, which is as follows: a) a first play of the video with audio in order to have an overall impression of the macro-phenomena taking place; b) a second play of the video without audio, in order to focus on finer non-verbal aspects; c) a third play of the audio-visual material in slow motion and eventually frame by frame.

Although not specifically designed for multi-modal analysis, the data collected show interesting non-verbal phenomena which are rarely observed in face-to-face interaction. Such phenomena include the use of head signs, posture-shifting, and

confirmed by means of a contrastive analysis with the data recorded in a 45-minute driven collection conducted at the Center for Speech and Gesture Research, University of Chicago, in 2001. The recordings at that time involved five Italian subjects engaged in several tasks, such as a job interview, an informal conversation, and a "guessing game". The quality of gestural performance, on the other hand, has been informally compared with the data available from subjects engaged in face-to-face route direction without a map. The data in question are extensively shown, analysed and discussed in [3]. Nonetheless, a further experimental session with the same subjects engaged, this time, in a face-to-face map-task activity would be desirable for a more reliable comparison.

[3] For the distinction between interactive, informative, and comunicative behaviour, see [4].

gesture for self-orientation in space and self-organisation of thought, although, interestingly enough, the interactive function of non-verbal cues in absence of eye-contact is dramatically reduced but does not disappear: an evident instance of this fact is the persistence of gaze towards the interlocutor, which is recorded when the subjects need for back-channeling. For the purposes of this paper, the analysis will focus on co-verbal gestures.

1. Co-verbal gestures and planning in "blind" and face-to-face conditions: an overall view

An informal comparison with the data available from other experiments on face-to-face interaction seems to confirm the impression that the number of gestures performed is strikingly comparable. In particular, the rate of gesture *phrases* recorded during the map-task tends to match that performed by subjects involved in "normal" interaction,[4] with frequent cases of communicative gestures[5]. On the other hand, the gestures recorded during the map-task tend to be simplified in form, and more recurrent, which helps to characterize them as "extra-coherent". An instance of such a simplification is visible in Figure 1, which offers a comparison between the most complex[6] gestural performance recorded during the map-task (plate a) and a complex gestural performance recorded during a face-to-face road description from [3][7] (plate b).

00:02:18-00:03:09
ff.55-77

a) LH C-shape describes a path towards left while RH goes slightly down.

b) ([3], 116): RH describes a complex shape in three strokes embedded in a single gesture phrase

Figure 1: instances of complex gestures in a) map task and b) face-to-face interaction

Such a peculiarity makes possible the isolation of recurrent gestural patterns referring to both spatial-reasoning and discourse organisation. A good instance of such a phenomenon is a recurrent gesture, which will be here named "palm-down-flap". This gesture was observed in the performance of two subjects in the role of the Giver during map-task and probably in two subjects intent in face-to-face interaction.

---

[4] A mean of 17 gesture phrases per minute was recorded for subjects in blind map-task, which is strikingly comparable to the overall average of 15 gesture phrases per minute recorded in face-to-face interaction for the Italian subjects, and with the overall average of 19 gesture phrases per minute recorded for the American subjects intent in road-description. This last datum was extrapolated by the data shown in [3], *Table 17*, page 87.

[5] Again, an interesting instance of communicative behaviour despite the fact that the subjects are not visible to each other.

[6] Complexity is here determined by means of the following parameters:

a) number of hands engaged in the performance
b) number of separable trajectories
c) number of separable hand(s) shape
d) number of strokes embedded in the same gesture phrase

[7] Thanks to Karl-Erik McCullough for his kind permission to reproduce images from his University of Chicago Ph.D. Dissertation.

2. Co-verbal gestures and planning in map-task activities: the "palm-down-flap"

The gesture in question is a downward movement performed with palm(s) flat, down and away from body (see Figure 2 [8]) recorded for two subjects (S1 and S2) in synchronisation with crucial passages during both route-direction (mainly soon after the acquisition of common landmarks) and discourse organisation (i.e., in concurrence with the confirmation of a successful communicative strategy). The downward movement depicted by the gesture shows no iconic relationship with the co-expressive speech[9]. For this reason, it is here interpreted as a metaphor, as it is defined in [5], the downward movement being interpreted as a depiction of a given state in the organisation of the speaker's communicative plan, probably identifiable as degradation of New to Given (see [7]) [10]. Interestingly, this gesture is often recorded in synchronization with adverbs signalling alignment with the interlocutor, such as "ecco!" ("Good!", when used as interjection. See Figure 4.)

00:01:18 -00:02:61
ff. 29-65

G: <eeh> { [**tu** [**pa<a>**rti e [**vai** [**dri**tto<o>]
you start and go straight

LH leaves rest position and flaps repeatedly in the air. P.A.: w S: 30°.Type: Metaphor with superimposed beat. Function: marks an easy passage. Route direction is understood.

00:02:61-00:03:73
ff.66-103
G: [# [**poi** [**gi**[**ri**]
then you turn

LH raises slightly and flaps repeatedly in the air. P.A.: w S: 45°. Type: Metaphor with superimposed beat. Function: marks second step in route-description.

Figure 2: one-handed palm-down-flap

Figure 2 shows the first occurrence of the "palm-down-flap" in S1, in the role of the Giver (in transcripts, "g"): in this case, S1 has just quickly established the basic coordinates in two brief dialogue turns in order to begin with route-direction. After having received positive feedback from the Follower (in transcripts, "f"), the Giver starts with route-direction by saying "tu parti e vai dritto" (you start and go straight) and the concurrent performance of a palm-down gesture with superimposed beat (Figure 2), which is repeated after the first pause, at 00:02:61 minutes (Figure 2, ff. 66-103) and subsequently at 08:16 seconds, after a silent pause which is undoubtedly index of perplexity (Figure 3). In this case, the gesture seems to signal the continuation of the plan already adopted before the Follower's interruption. In a sense, the Giver is here overruling the Follower's interruption in order to continue with her description of the segment of route-direction.

The same gesture is then repeated with both hands at 00:26:13 minutes, after a 2-second-hesitation in speech, and at 00:48:77 minutes, in concurrence with the

[8] The conventions for transcription follow those proposed in [5] and [6].
[9] See [5].
[10] If the interpretation attempted in these pages is correct, then the "palm-down flap" is liable to be also seen as a communicative gesture.

utterance: "ecco, allora tu…" ("ok, then you…", Figure 4), which follows a long negotiation about space landmarks and finally a positive feedback provided by the Follower.

00:08:16-00:08:86
ff.205-221

G: /[/ [**no**]

LH still holding a C-shape flaps twice. P.A.: w. Type: metaphor with superimposed beat. Function: Self-orientation in reasoning. Probably used to overrule F's interruption.

Figure 3: one-handed palm-down flap superimposed to the vestige of a C-shape stroke

00:48:77-00:50:94
ff.1195-1249

F: si **G:** [e<**e**>cco] # [**al**[**lo**ra] # [**tu**<u>]
Yes ok then you

BH spread, palms away from body, flap repeatedly down. The same movement is repeated three times. P.A.: w S.: 20° (average). Type: metaphor. The repetition suggests a superimposed beat. Function: highlights the acquisition of a common landmark in space. The repetition of the gesture indexes a landmark in the communicative strategy.

Figure 4: two-handed palm-down flap movement as index of landmark in space and reasoning

In this case, the gesture is interestingly superimposed on the vestige of a C-shape stroke-hold taking place during the Follower's dialogue turn, which is interpretable as index of an aborted idea. Another occurrence is recorded at 00:71:77 minutes, after the resolution of a wrong-footing due to the incongruence between the two maps, immediately following the performance of another recurrent metaphor used to index common knowledge.

Other occurrences of this gesture are synchronized with a request for feedback (00:76:21) and the following utterance (00:76:93; 00:77:77), probably with a superimposed beat, in a sort of gestural semantic satiation due to the fossilization of the gesture.

A palm-down flap is also observed in S2, at the very beginning of her interaction with the interlocutor, within a complex phase of negotiation about the starting-point (see Figure 5). In particular, the gesture is recorded in concurrence with the utterance "allora" – an overt index of planning an effective communicative strategy – and is soon after repeated in concurrence with the following utterance. The planning is nonetheless difficult because of several overlaying utterances by the Follower. The bad synchronization between the interactants is resolved at 00:03:40 minutes, when S2 accepts the interruption and clearly states the intention of beginning her route description from the "Start"-point. Interestingly, during this alignment, S2 performs two palm-down flaps in strict succession (00:03:40-00:04:97), with emphasis[11].

Finally, a strikingly comparable gesture was also recorded in face-to-face interaction. Figures 6 show the only occurrences of the gesture for the Italian subjects (see [8]): S3, after having received a hint, formulates a new hypothesis for the solution of a guessing game.

---

[11] Note that the gesture size is about 40° at the wrist level.

00:00:00-00:01:83
ff. 00-55
F: dai facciamo #
let's do…
G: { [allora] //
so

RH starts from rest position with palm down flat and away from body and flaps once e/w/f S: 80°e Type: metaphor Function: stresses the organization of the communicative plan to be conveyed

00:01:84-00:03:39
ff.55-101

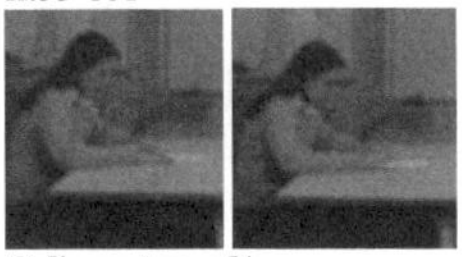

G:[la par**ten**za]/
the start
F: facciamo dalla partenza?
shall we begin from the start?

RH repeats the movement described above twice

00:03:40-00:04:97
ff. 102-149

G: [**si** ///][**dalla partenza** ok] }
yes from the start, ok

RH palms down away from body flap repeatedly down.The handshape of the right hand is probably a precision grip. P.A.: w/f S.: 40° w. Type: metaphor. Function: underlines the achievement of a common starting point with the interlocutor.

Figure 5: series of palm-down flaps in S2

00:02:35-00:02:37

S3: {[erano a chi**ca<a>**go /] [che è la win**dy** city //]
they were in Chicago, which is the windy city

BH flat flap downward twice. P.A.: e/w/f S.: 30°e Type: metaphor Function: adoption of a new plan

Figure 6: probable palm-down flap in an Italian subject intent in a face-to-face guessing game

Figure 7 shows a palm-down gesture from [3], which can perhaps be interpreted as having the same function[12].

JA: hm okay / good [so **that** helps me]

Figure 7: probable case of palm-down flap in an American subject ([3], 121).

[12] The gesture in question is semiotically more complex (McCullough, personal communication).

Conclusions

The analysis of co-verbal gestures in map-task activities has revealed interesting phenomena, which are probably due to the lack of a common semiotic space usually established by gaze. Such a condition produces – among other phenomena – a simplification in the gestural performance and allows the isolation of recurrent patterns of movement related to both spatial reasoning and discourse organisation. The "palm-down flap", which was briefly presented in this paper, is a good example of the increased gestural coherence when face-to-face interaction is not possible. The fact that this gestural pattern is recorded in more than one subject suggests some cultural-specificity of the metaphor behind it: further research is thus desirable, in order to assess its eventual cross-cultural use.

Lastly, the observed persistence of interactive and even communicative non-verbal behaviour when the interlocutor is not visible can perhaps contribute to the speculation about the complex relationship existing between behavioural patterns and language.

Acknowledgements

Thanks to Giacomo Ferrari, Adam Kendon, and Karl-Erik McCullough for discussing the data here presented with me. Any imprecision is exclusively attributable to the author.

References

[1] A. Özyürek, The influence of addressee location on spatial language and representational gestures of direction, in: D. McNeill (ed.) *Language and gesture*, Cambridge University Press, Cambridge, 2000.

[2] I. Poggi, C. Pelachaud and E. Magno Caldognetto, Gestural Mind Markers in ECAs, in: *Proceedings of the 5th International GestureWorkshop, GW 2003 Genova, Italy, April 15-17, 2003. Selected Revised Papers* (2003), 338-349.

[3] K.-E. McCullough, *Using Gestures in Speaking: Self-Generating Indexical Fields.* Ph.D. Dissertation, University of Chicago, Chicago, 2005.

[4] P. Ekman and W. Friesen, The repertoire of non-verbal behaviour: Categories, origins, usage, and coding, *Semiotica* 1(1969), 49-98.

[5] D. McNeill, *Hand and mind: What gestures reveal about thought.* The University of Chicago Press, Chicago and London, 1992.

[6] N. Rossini, The Analysis of Gesture: Establishing a Set of Parameters, in: *Proceedings of the 5th International GestureWorkshop, GW 2003 Genova, Italy, April 15-17, 2003. Selected Revised Papers* (2003), 124-131.

[7] M.A.K. Halliday, *An introduction to functional grammar.* Arnold, London, 1985.

[8] N. Rossini, *Gesture and its Cognitive Origin: Why do we Gesture?* Ph.D. dissertation. Università degli Studi di Pavia, Pavia, 2004.

Fundamentals of Verbal and Nonverbal Communication and the Biometric Issue
A. Esposito et al. (Eds.)
IOS Press, 2007

A Preliminary Investigation of the Relationship between Gestures and Prosody in Italian

Anna ESPOSITO[1a], Daniela ESPOSITO[a], Mario REFICE[b], Michelina SAVINO[c], Stefanie SHATTUCK-HUFNAGEL[d]
[a]Department of Psychology, Second University of Naples and IIASS, Italy
[b]Department of Elettrotecnica, Politecnico di Bari, Italy
[c]Department of Psychology, Università di Bari, Italy
[d]Speech Group, Research Laboratory of Electronics, MIT, USA

Abstract. This work investigates the relationship between gestures and prosodic events (such as pitch accent and boundary tones), exploiting a class of gestural movements named HITS defined for American English by [21] as: "*An abrupt stop or pause in movement, which breaks the flow of the gesture during which it occurs*". Our analysis shows that the tendency toward temporal synchronisation between these gestural units and prosodic events which is reported for American English is observable also in Italian.

Keywords. Gestures, pitch accent, hits

Introduction

Several theories have been proposed for the role of the gestures that accompany speech in face-to-face communication. For example, some investigators claim that gestures involve a separate communication channel, conveying information not directly related to the accompanying speech [1, 3-4]. Others suggest that the primary function of gestures is to support the speaker's processing effort in encoding his/her message [10, 14, 16]. Still others, however, suggest that gestures share with speech similar semantic and pragmatic functions, and reject the hypothesis that either gesture or speech alone have a primary role in the communicative act [13, 15]. All of the above theories find some support in experimental data, leaving open the question on how to integrate and evaluate these various approaches. The results we are going to present may be relevant in evaluating the relative merits of the above theories.

In previous works [7-9] we adopted McNeill's focus [15] on a hypothesis of synchronization between speech and gesture (in particular hand gestures). The theoretical framework emerging from this focus holds that, while speech clearly contributes to the semantic content of the message, gestures engage the *imagistic content*. Starting with these assumptions, we tried to answer the following questions

[1] Corresponding Author: Anna Esposito, Department of Psychology, Second University of Naples and IIASS, Via Vivaldi, 43, 81100, Caserta, Italy. e-mail: iiass.annaesp@tin.it, anna.esposito@unina2.it.
The names of the authors are in alphabetical order.

with respect to a particular set of body movements that will be defined below: 1) assuming that speech and gestures are co-involved in the production of a message, is there any gestural equivalent to pitch accent? 2) Assuming that we found such gestural equivalents, to what degree do these entities synchronize with spoken pitch accent or other prosodic units?

As an answer to our first question, we refer to some pilot studies performed by [21] which identified, for American speakers, a gestural entity, called the *hit* and showed that it seems to be temporally related to pitch accents. According to Yasinnik et al. [21] *a "hit" is an abrupt stop or pause in movement, which breaks the flow of the gesture during which it occurs. Hits appear as bouncing, jerky movements, changes in the direction of movement, or as complete stops in movement.* Moreover, hits are discrete gestures that are distinguished from continuous ones because the latter end gradually rather than abruptly. It should be emphasized that *hits* are kinematic units, identified visually by an expert human coder. Therefore, the concept of *hit* is ultimately a perceptual one and is defined in a kinematic dimension. Hits do not fall in any of the meaning-based gestural categories previously defined by other authors [12, 15] and may cross-cut these categorical features. The kinematic dimension is the most appropriate for these units since it relates them to the temporal dynamic of acoustic features (such as the fundamental frequency and speech rate) exploited in the definition of prosodic cues (such as pitch accents and boundary tones). Hits are defined and measured not only for hands but also for eyebrows, shoulders, and head movements, broadening the set of expressive gestures (i.e. gestures that contribute to the meaning of a communicative act) to wider variety of body movements.

In the present study we find further support in discourse data from two native Italian speakers for the view advanced in Yasinnik et al. [21], that gestural hits and prosodic accents are temporally aligned in spoken utterances. The two goals motivating the extension of the previous research are: 1) to determine whether the temporal relation established for English is independent of specific discourse context, as would be predicted if the relationships previously found are robust; 2) to determine whether this relationship between speech and gesture is independent of the particular language of expression, as would be predicted if (in at least some ways) speech and gestures reflect a unified planning process for producing a communicative act.

1. Materials

The video recordings on which our analysis is based are of two monologs (4 and 4.5 minutes long respectively) by two adult native speakers of Italian (1 male, 1 female). The speakers discuss freely whatever physics topics they choose to report on. The listener in each case was a silent experimenter, i.e. a female student who was a native speaker of Italian and who had no particular knowledge on the reported topics. The interlocutors knew each other in order to eliminate stranger-experimenter inhibitions from the elicitation setting. The Italian data were recorded at the Massachusetts Institute of Technology, in a silent room, using a 30 frame/second digital camera. Neither of the two participants was aware that gestures were of interest. The video was analyzed using a video analysis software (iMovie™) that allows viewing of video-shots at several velocities (real-time, slow, fast, and frame-by-frame motion), and movement forward and backward through the shots. The audio was analyzed using the Praat algorithm [http://www.fon.hum.uva.nl/praat/] which allows simultaneous viewing of the waveform, the spectrogram, and the F0 contour, as well as the insertion of tiers for prosodic and other labelling. For labelling the speech, the waveform, the spectrogram, and the F0 contour of the audio signal were considered together, in order to identify the beginning and end of each word in the utterances, the pitch accented syllables and the edge tone (phrase tone and boundary tone) positions in time.

The video and audio data were analyzed separately, the former by playing the video frame-by-frame to determine the hits, and the latter by combining perceptual and acoustic information in order to identify both pitch prominences and boundary tones. In particular, to avoid biases in *hit* identification, the video was analyzed without listening to the audio. The planned analyses required the transcription, word-by-word, of the speaker's monologs, the beginning and end time of each hit, and the absolute temporal position of pitch accents and edge tones. We labelled as a hit the extent of the hit gesture, from the onset to the end point, considering as hit onset the temporal frame where the beginning of the *bouncing, and/or jerky movement, and/or the changes in the direction of movement* is observed (the associated frame show a blurring) and as hit end point the temporal frame where the movement abruptly reached a complete stop (the corresponding frame becomes strikingly less blurry). See details in [22]. A pitch accent overlaps with a hit if its syllable overlaps with some portion of the entire extent of the hit gesture. The data reported here differ from those reported in [21] since we considered not only whether a hit overlapped with a pitch accented word (without providing information on overlaps between hits and pitch accented syllables), and considered the entire extent of the hit gesture, not just the hit end frame in the video.

2. Prosodic Labelling of the Spoken Data

Spoken Italian is presently characterised by strong regional accents [5, 20]. In the last few years, a number of studies of intonational modelling of Italian varieties within the auto-segmental metrical ToBI-like framework [2] have been produced [6, 11, 18-19] along with an attempt to develop a common ToBI-like system across different varieties [12]. The two participants in our experiment came from two different linguistic areas of Italy, namely Central (Tuscany) and Northern Italy (Veneto). Even though a complete Italian TOBI-like reference system is not currently on offer, and therefore no comprehensive prosodic modelling of the two varieties is available, we have no reason to believe that the prosodic elements described in the present work are perceptually different for different Italian dialects. However, we acknowledge that this assumption leaves unaddressed the methodological concern that the basic TOBI-like conventions used to identify prosodic events may not be the same for the two Italian varieties.

We marked the intonational prominences by aligning a star label '*' with the vowel in the accented syllable, without specifying any phonological label for a description of the pitch accent (e.g. H*, L*, !H*). These pitch accent labels were placed in the middle of the accented syllable. For the edge tones, we provided a phonological description of them when possible (typically, L-L% for falling movements, and L-H% for partially-rising movements); where this was not possible, we used the basic symbols '-' (for phrase tone, ending a lower-level Intermediate Intonational Phrase) and '%' (for a boundary tone, ending a higher-level Full

Intonational Phrase). Uncertain pitch accents and edge tone labels were flagged with a question mark. An additional set of labels was added to mark hesitation phenomena (namely empty and filled pauses) and repairs (such as *word repetitions* and *fresh starts*). Filled pauses include phrase-final lengthening of word-final vowels that occur at the ends of words in all phrasal positions - it has been statistically estimated that, in Italian, this lengthening is three times greater than the intrinsic vowel duration [17].

3. Results

The results are summarized in Tables 1, 2 and 3. Table 1 summarizes the frequency of speech events (words, empty and filled pauses, repairs, intermediate and intonation boundary tones, and pitch accented words) and gesture entities (hits) for each subject. The percentage (in parentheses) is computed over the total number of speech entities (words, empty and filled pauses, and repairs, that total 816 for the male and 603 for the female speaker respectively). Intonation and intermediate tone averages are computed over the total number of boundary tones (122 for the male and 146 for the female speaker). Note that we discarded three and seven boundary tones in the male and female speech respectively* since it was not possible to classify them as belonging to one of the two categories considered (i.e. as boundary markers for intermediate or intonation phrases).

Table 1: Summary of the data

Subject	# *Words*	# *Empty Pauses*	# *Filled Pauses*	# *Repairs*	# *Intonation Boundary Tones*	# *Intermediate Boundary Tones*	# *Boundary tones*	# *Hits*	# *Pitch Accents*
S1 (male)	554 (68)	128(16)	83(10)	51(6)	93(76)	29(24)	125*	140	258
S2 (female)	431(72)	115(19)	51(8)	6(1)	81(55)	65(45)	153*	73	277
Total	985	243	134	57	175	94	278*	213	535

Table 1 shows that there are differences between the two subjects that could be attributed to their gender difference or to individual speaker differences. The two speakers produced a substantially different number of repairs, intonation and intermediate tones, and hits (140 against 73) whereas the numbers of words, empty and filled pauses, and pitch accents they produced are comparable.

The first part of Table 2 reports the overall distribution of hits over the articulators producing them and the second and third parts provide further details, distinguishing single hits (i.e, hits that occur once in a discrete gesture) from repeated hits (i.e., hits that are repeated more than once in quick succession [21]).

From Table 2 it can be observed that the articulators involved in the production of hits differ substantially between the male and female speaker. The male produces more hits with the left or both the hands, whereas most of the hits produced by the female are made with the head. This is a clear difference between the two speakers and future analyses of utterances from more than two speakers will reveal whether it can be attributed to gender. The distribution of single and repeated hits does not seem to be different between the two speakers since the percentage of single and repeated hits is 64% (90/140) and 36% (50/140) respectively for the male and 74% (54/73) and 26% (19/73) respectively for the female speaker.

Table 2: Overall, single and repeated hit distributions

Overall hit distribution

Subject	*# Left hand hits*	*# Right hand hits*	*# Both hand hits*	*# Eyebrow hits*	*# Shoulder hits*	*# Head hits*	*Total*
S1 (male)	57	9	45	9	5	15	140
S2 (female)	1	2	0	1	0	69	73
Total	58	11	45	10	5	84	213

Single hit distribution

Subject	*# Left hand hits*	*# Right hand hits*	*# Both hand hits*	*# Eyebrow hits*	*# Shoulder hits*	*# Head hits*	*Total*
S1 (male)	30	7	28	9	3	13	90
S2 (female)	1	0	0	1	0	52	54
Total	31	7	28	10	3	65	144

Repeated hit distribution

Subjects.	*# Left hand hits*	*# Right hand hits*	*# Both hand hits*	*# Eyebrow hits*	*# Shoulder hits*	*# Head hits*	*Total*
S1 (male)	27	2	17	0	2	2	50
S2 (female)	0	2	0	0	0	17	19
Total	27	4	17	0	2	19	69

Table 3 reports for each subject, the occurrences of hits (single and repeated combined) in relation to speech and prosodic entities (over the columns) and in relation to the articulators producing them (over the rows). Notice that the number of words with hits is greater than the number of hits, because a hit (defined here as the time interval between its onset and the end point of the gestural stroke) may extend over more than one word.

The data in Table 3 show that a large proportion of the hits overlap with a pitch accent both for the male and the female speaker (109 over 140 – **78%** - for the male and 61 over 73 – **84%** for the female speaker. Notice that we included the words labelled with *? (i.e. those for which the labeller felt some doubts about whether they were accented or not).

Hits are also produced on phrase-final syllables labelled with both a boundary tone and a pitch accent, but are not found on non-pitch-accented boundary syllables. Therefore, we can assume a positive relationship between these gestural entities and prosodic units such as pitch accents and boundary tones. This relation is not confined to the articulators most frequently involved in hit production for a given speaker (i.e. hands for one speaker, head for the other (see Table 2)), suggesting that there is no favourite articulator for hits related to prosodic entities.

Table 3: Occurrences of single and repeated hits over speech and prosodic entities (columns) according to the articulators producing them (rows)

Subjects.	*# Words with hits*	*# Pitch accents with hits*	*# Empty pauses with hits*	*# Filled pauses with hits*	*# Repairs with hits*	*# Intonation tones with hits*	*# Intermediate tones with hits*	*# Pitch accented words at boundary tones with hits*	*# Only boundary tones(without pitch accent) with hits*
S1 (male)	**166**	**109**	**30**	**21**	**7**	**36**	**24**	**59**	**1**
HEAD	16	9	7	8		2	4	5	
Left Hand	66	45	11	5	3	19	11	28	1
Right Hand	13	7	1	1			1	1	
Both Hands	59	35	7	4	4	9	7	18	
SHOULDER	5	4	2	1		1		1	
EYEBROW	9	8	2	2		5	1	6	
S2 (female)	**107**	**61**	**32**	**14**	**2**	**25**	**17**	**45**	**0**
HEAD	102	58	30	14	2	23	17	43	
Left Hand	1	1	1			1		1	
Right Hand	3	1							
Both Hands									
SHOULDER									
EYEBROW	1	1	1			1		1	

Tables 4 and 5 report, for each subject, the occurrences of single and repeated hits respectively in relation to speech and prosodic entities (over the columns) and in relation to the articulators producing them (over the rows). These two tables provide more information about the results introduced in Table 3, and show that pitch accents and boundary tones with pitch accents are equally likely to occur with single and repeated hits: the percentages of pitch accents in single hits are 74% and 87% (67/90 and 47/54 – see Table 2) for the male and female speaker respectively, and the percentages of pitch accents with repeated hits are 84% and 74% (42/50 and 14/19 – see Table 2) for the male and female speaker respectively. This also suggests that there is no preferred hit-producing articulator connected to prosodic entities. Therefore, we can assume a positive relationship between these accent-like gestural entities and prosodic units such as pitch accents and boundary tones. This relation cannot be attributed to the articulators involved in the hit production, i.e. the hands and/or the head (see Table 2) since pitch accents happen with hits independently produced by hands and head (according to the speaker).

Table 4: Occurrences of single hits over speech and prosodic entities (columns) according to the articulators producing them (rows)

Subjects.	*# Words with hits*	*# Pitch accents with hits*	*# Empty pauses with hits*	*# Filled pauses with hits*	*# Repairs with hits*	*# Intonation tones with hits*	*# Intermediate tones with hits*	*# Pitch accented words at boundary tones with hits*	*#Only boundary tones(without pitch accent) with hits*
S1 (male)	**114**	**67**	**23**	**15**	**6**	**26**	**11**	**35**	**1**
HEAD	14	7	6	8		2	3	4	
Left Hand	39	22	9		2	12	4	15	1
Right Hand	11	6		1			1	1	
Both Hands	41	21	4	3	4	6	2	8	
SHOULDER	2	2	2	1		1		1	
EYEBROW	9	8	2	2		5	1	6	
S2 (female)	**78**	**47**	**27**	**11**	**1**	**20**	**11**	**33**	**0**
HEAD	76	45	25	11	1	18	11	31	
Left Hand	1	1	1			1		1	
Right Hand									
Both Hands									
SHOULDER									
EYEBROW	1	1	1			1		1	

4. Conclusions

As a result of these analyses, we can speculate that there is a tendency toward temporal synchronisation between gestural hits and prosodic events in Italian, since the majority of these gestures occurs in synchrony with prosodic entities such as pitch accents (78% of the male hits and 84% of the female hits) and pitch-accented boundary tones (see Table 2). This tendency is not confined to hits produced by certain articulators, since the synchrony is observed with hits produced with the articulator preferred by the speaker (i.e. hands or head, depending on the speaker, see Table 3). Moreover, the synchrony does not depend on whether the hits occur singly or are repeated. Differences were observed between the two subjects (e.g. different number of repairs, boundary tones, and hits (140 against 73) and different articulators preferred to produce hits). The male speaker produced more hits with the left or with both hands, whereas most of the hits produced by the female are made with the head. Further research will be necessary to determine whether these differences reflect individual speaker preferences, dialect distinctions or differences in behaviour related to gender.

Table 5: Occurrences of repeated hits over speech and prosodic entities (columns) according to the articulators producing them (rows)

Subjects.	*# Words with hits*	*# Pitch accents with hits*	*# Empty pauses with hits*	*# Filled pauses with hits*	*# Repairs with hits*	*# Intonation tones with hits*	*# Intermediate tones with hits*	*# Pitch accented words at boundary tones with hits*	*#Only boundary tones(without pitch accent) with hits*
S1 (male)	**52**	**42**	**7**	**6**	**1**	**10**	**13**	**24**	
HEAD	2	2	1				1	1	
Left Hand	27	23	2	5	1	7	7	13	
Right Hand	2	1	1						
Both Hands	18	14	3	1		3	5	10	
SHOULDER	3	2							
EYEBROW									
S2 (female)	**29**	**14**	**5**	**3**	**1**	**5**	**6**	**12**	
HEAD	26	13	5	3	1	5	6	12	
Left Hand									
Right Hand	3	1							
Both Hands									
SHOULDER									
EYEBROW									

The data reported above give further support to the hypothesis advanced in Yasinnik et al. [21] for American English, by showing a positive relationship between hits and prosodic events in Italian. This raises the possibility that such a relationship may be independent of specific languages and specific discourse contexts. Moreover, these observations also support to the idea that speech and gestures reflect a unified planning process independent of the particular language of expression. However, it should be emphasized that this is a pilot study and that more work is necessary to detail these results. For example, it would be useful to identify the acoustic landmarks in the signal [23-24] in order to align hit end points to accented syllables within the accented words (see [22]). It will also be important to distinguish which kinds of gestures (in the meaning-based typology proposed by [15], e.g. metaphoric, iconic, rhythmic and deictic) are involved in these kinematically-defined entities, and determine whether a particular subset of them occur with pitch accented syllables. Such investigations are necessary in order to assess what kinds of meaningful information these gestural entities bring to the communicative act, and the role of gestures in face-to-face communication.

Acknowledgements

The paper has been partially supported by the **COST Action 2102**: "Cross Modal Analysis of Verbal and Nonverbal Communication (CAVeNC), http://www.cost.esf.org/index.php?id=110&action_number=2102."

References

[1] G.W. Beattie: Language and Non-verbal Communication: The Essential Synthesis. *Linguistics*, 19, 1165-1183, 1981.
[2] M.E. Beckman, E.G. *Ayers: Guidelines for ToBI Labelling.* Version 3.0,Ohio State University, 1997.
[3] B.L. Butterworth, G.W. Beattie: Gestures and Silence as Indicator of Planning in Speech. In R.N. Campbell, P.T. Smith (Eds.) *Recent Advances in the Psychology of Language: Formal and Experimental Approaches*, Olenum Press, New York, 347-360, 1978.
[4] B.L. Butterworth, U. Hadar: Gesture, Speech, and Computational Stages: A Reply to McNeill. *Psychological Review, 96*, 168-174, 1989.
[5] T. De Mauro: *Storia linguistica dell'Italia unita.* Bari: Laterza, 1963.
[6] M. D'Imperio: Focus and Tonal Structure in Neapolitan Italian. *Speech Communication*, 33(4), 339-356, 2001.
[7] A. Esposito, A. Natale, S. Duncan, D. McNeill., F. Quek: Speech and Gestures Pauses Relationships: A Hypothesis of Synchronization (in Italian). In *Proceedings of the V National Conference on Italian Psychology*, AIP, Grafica80-Modugno, Bari, Italy, 95-98, 2003.
[8] A. Esposito, S. Duncan, F. Quek: Holds as Gestural Correlates to Empty and Filled Pauses. In *Proceedings of the International Conference on Spoken Language Processing* (ICSLP02), Denver, Colorado, 1, 541-544, 2002
[9] A. Esposito, K.E. McCullough, F. Quek: Disfluencies in Gestures. In Proceedings of International Workshop on *Cues in Communication*, Hawai, December 9, 2001
[10] N. Freedman: The Analysis of Movement Behaviour During the Clinical Interview. In A.W Siegmann, B. Pope (Eds), *Studies in Dyadic Communication*, Pergamon Press, New York, 177-208, 1972.
[11] M. Grice, M. Savino: Information Structure and Questions – Evidence from Task-Oriented Dialogues in a Variety of Italian. In P. Gilles, J. Peters (Eds.), *Regional Variation in Intonation*, Niemeyer: Tuebingen, 161-187, 2004.
[12] M. Grice, M. D'Imperio, M. Savino, C. Avesani: Strategies for Intonation Labeling across Varieties of Italian. In S. Jun (Ed.), *Prosodic Typology: the Phonology of Intonation and Phrasing*, New York: Oxford University Press, 362-389, 2005.
[13] A. Kendon: Current Issues in the Study of Gesture. In J.L. Nespoulous et al. (Eds.), *The Biological Foundations of Gestures: Motor and Semiotic Aspects,* LEA Publishers, Hillsdale, New Jersey, London, 23-27, 1986.
[14] R. Krauss, P. Morrel-Samuels, C. Colasante: Do Conversational Hand Gestures Communicate? *Journal of Personality and Social Psychology*, 61(5), 743-754, 1991.
[15] D. McNeill: *Hand and Mind: What Gesture Reveal about Thought.* University of Chicago Press, Chicago, 1992.
[16] B. Rimé: The Elimination of Visible Behavior from Social Interactions: Effects of Verbal, Nonverbal, and Interpersonal Variables. *European Journal of Social Psychology*, 12, 113-129, 1982.
[17] M. Savino, M. Refice: Acoustic Cues for Classifying Communicative Intentions in Dialogue Systems. In P. Sojka, I. Kopecek, K. Pala (Eds.) *Text, Speech and Dialogue*, LCNS vol. 1902, 421-426, 2000.
[18] M. Savino: Descrizione Autosegmentale-Metrica di Alcune Tipologie Intonative dell'Italiano di Bari. In E. Burr (Ed.) *Atti del VI Convegno Internazionale della SILFI* (Società Internazionale di Linguistica e Filologia Italiana), Cesati: Firenze, in press.
[19] M. Savino: Intonational Cues to Discourse Structure in a Variety of Italian. In P. Gilles, J. Peters (Eds.), *Regional Variation in Intonation*, Niemeyer: Tuebingen, 145-159, 2004.
[20] A. Tosi: *Language and society in a changing Italy*. Multilingual Matters, Clevedon, UK, 2001.
[21] Y. Yasinnik, M. Renwick, S. Shattuck-Hufnagel: The Timing of Speech-Accompanying Gestures with Respect to Prosody. Proceedings of the International Conference: *From Sound to Sense*, MIT, Cambridge, June 10-13, C97-C102, 2004.

[22] S. Shattuck-Hufnagel, Y. Yasinnik, N. Veilleux, M. Renwick: A Method for Studying the Time Alignment of Gestures and Prosody in American English: 'Hits' and Pitch Accents in Academic-Lecture-Style Speech. In A. Esposito, M. Bratanic, E. Keller, M. Marinaro (Eds.), *The Fundamentals of Verbal and Nonverbal Communication and the Biometrical Issue*, IOS press, this volume.
[23] K. N. Stevens: *Acoustic Phonetics.* MIT Press: Cambridge, MA, 1998
[24] K. N. Stevens: Toward a model for lexical access based on acoustic landmarks and distinctive features. *Journal of the Acoustical Society of America,* 111, 1872-1891, 2002.

Fundamentals of Verbal and Nonverbal Communication and the Biometric Issue
A. Esposito et al. (Eds.)
IOS Press, 2007

Multimodal Resources in Co-Constructing Topical Flow: Case of "Father's Foot"

Sari KARJALAINEN[1]
Department of Speech Sciences/Logopedics,
University of Helsinki, Finland

Abstract. The present paper examines the child's use of deictic gestures in the process of topical co-operation with his father. The method for the study is qualitative and data driven conversation analysis based on examining communicative practices in natural settings. The database is composed of videotaped naturalistic picture book conversations between the child (at the age from 1 to 2 years) and the adult. For the current paper, the sample of data is transcribed and analyzed focusing on the sequential organization of the participants' verbal and nonverbal action (gestures, gaze, vocalizations and adult's speech) focusing the analysis on referential actions in the sequence in which the topic is extended from the referent in the picture book to the referent outside the book. More specifically, the focus is on how both the verbal and nonverbal referential resources reveal the participants' orientations in the on-going interaction, creating a shared referential focus.

Keywords. Adult-child interaction, pointing, conversation analysis, topic

1. Background

The preverbal stage is acknowledged as an important stage in the child's path in socializing to language. Specifically, the child's early gestures have been acknowledged as an effective way of communicating [1,2,3] and also the interplay between gestures and later stages of language development has been acknowledged [4]. Another line of research with its sociological and ethnomethodological background has studied how the child acquires the culture involving also environmental and contextual issues [5]. Having similarities with gestures in adulthood [6], also the child's early gestures linked to the environment constitute an important class of gesture. There are studies focusing on different aspects on the interactional settings for language learning before the first spoken words [7,8,9,10], and, not until recently, a few studies on multi-modal sign complexes that encompass both gesture and phenomena in the world [11,12].

The production of the gesture can be seen as embedded within a multi-party embodied participation framework that creates for the participants a shared focus of visual and cognitive attention toward both each other and relevant phenomena in the environment [13]. Using gestures can be studied also in terms of 'joint attentional frame' [14,15] that has similarities with the participation framework, involved with the

[1] Corresponding Author: Department of Speech Sciences/Logopedics, University of Helsinki, Finland, e-mail: sari.karjalainen@helsinki.fi .

reference to the object in the surroundings. It is in these 'triadic' situations establishing shared interests between the participants in the early interactions where language development takes place. Thus, looking at the early stages of gesture use is motivated by several psychological and linguistic, and particularly, pragmatic aspects.

The present paper investigates the active and selective role of a child in bringing his experiences to the early interactions that take place in one socializing context, namely picture book environments. The picture book settings have been acknowledged as fruitful environments, especially for learning vocabulary [16]. However, this paper tries to shed light more on the actual phenomenon itself and gives micro-analytical insight of the phenomena occurring at the preverbal stage of language development, with other words, the interactional environments for language socialization at the developmental stage before the spoken words. More specifically, a child's use of gestures, particularly, deictic pointings and their interpretation by an adult are examined. One video extract of the data base for the study is selected to demonstrate the referential activity of the participants in a triadic situation created by the two participants around the picture book referents.

2. Methodological Framework

2.1. *Aim of the Study*

The major aim is to find and figure out the fundamental phenomena concerning the early developmental stage of a child using gestures as an established way of communicating with her surroundings, in other words, making referential acts to the items in the surroundings. Here, instead of examining the ties between gesture and psychological processes inside the mind of the individual the gesture is examined as part of the on-going interaction and as a part of a topical flow.

The child's pointing gesture in the sequence of topical elaboration gets a variety of meanings. The question can be put like this: How are these different context-bound meanings established in the on-going interaction? The present paper focuses mainly on demonstrating the patterning of the topical flow through certain referents and is not trying to give full insight into the organization of the participation framework. The aim is to make a detailed description of the orientation by the participants and the selected ongoing activity that the participants are accomplishing in a certain selected sequence [12]. Specifically, the aim is to demonstrate how the child makes the reference through embodied actions and how the adult displays his orientation to the child's actions mainly through certain linguistic and paralinguistic means.

2.2. *Method and Data*

The methodological approach in the present paper is qualitative and data driven conversation analysis (CA), involving the analysis of natural recordings [17,18,19]. The aim of CA is to demonstrate how the meaning making practices of participants are manifested and how successive utterances are constructed as turns in the on-going interaction and to reveal the way in which participants themselves show to be relevant to the organization of their conversation in the data collected in the real life situations. Thus, for the current purpose, the extract of the cross-sectional data is selected [20]. It concerns the boy at the stage of established use of pointing as the first means for making reference before his spoken words (at the age of 16 months). He is looking at picture book with his father. It is a sample of 50 seconds from the video database (roughly 70 hours) consisting of videotaped conversations (in pure naturalistic home settings with picture books) between a child and an adult in 6 Finnish families. The adult participant is either of the parents or the caretaker. The videotaping period is from 1 to 2 years of age of the child. Thus, the data is available for tracking the developmental line, as well, but, in the present paper, the selected extract from the certain developmental stage is examined. The practices in the focus of the extract presented in this paper are typical phenomena occurring within the picture book looking activity in the current database.

2.3. The Analysis: Case of "Father's Foot"

The task for the analysis is to make a detailed description of a selected ongoing activity and to demonstrate the orientation of the participants that they are accomplishing. Next, to explore the activity of topical collaboration by the two participants, and specifically, in order to identify the linguistic and gestural resources responsible for displaying it, we take a look at one instance, the extract labeled as "Father's Foot", whereby the child moves straight on to elaborate on the picture by making a reference outside the book. The focus of the analysis is in how the topical flow emerges from sequential organization of the participants' verbal and nonverbal actions (pointing and other gestures, gaze, vocalizations and adult's speech) establishing a certain kind of topical sequence that extend from picture referent to referent outside the picture book.

The data relevant for the research objectives are transcribed and analyzed according to the principles of conversation analysis. The transcription of the sequence is simplified for the current purpose. The items are selected so that the critical evidence, at least, bearing on the arguments being made is included. In the transcript, the participants are "F" for the father and "C" for the child. The temporal organization of overlapping activities is indicated with square brackets [. Other timing information includes (.) for pause > 0.5 seconds. Falling prosody is marked with dot. Underlining indicates stressed word. The non-vocal actions (gaze, pointing and other movements of the body) are provided with verbal descriptions. They are marked here as separate turns, indicating the beginnings of the actions. Also, they are individuated visually to make it easier to read the lines belonging to the same multimodal package of actions. Also, the line for the English gloss is included.

The sequence begins with the child pointing at the picture, which turns out to be a *foot* - as the father, puts it in line 2: *niin siin näkyy vaan jalka* ('yeah there is a foot only'). Next, in line 3, the child touches his father's shoulder. This act gets the father's attention which is displayed in turning his gaze to the child. At this point, the child starts moving away from the book, while, having a look at the direction of the child's move, the father makes a question *mitä* ('what') (lines 7-8). The child continues his moving to the father's foot, and accompanied by some vocalization, touches the foot. Then he starts getting back to the book and overlapping with this activity the father accomplishes his turn *niin just se on sun isän jalka* ('yeah, exactly it is your father's foot'). Thereafter, the child returns to the book, establishing some gesturing and sharing a laugh with his father, and making a point to the new target in the book.

"Father's Foot"

1	C	***POINTS AT THE PICTURE***
2	F	niin siin näkyy vaan jalka. se on niiden isän jalka.
		PRT ADV appear-3 PRT foot it be-3 they-GEN father-GEN foot
→		**yeah there's a foot only. it is their father's foot.**
3	C	***TURNS AWAY FROM THE BOOK***
5		***TOUCHES THE FATHER'S SHOULDER***
6		***STARTS MOVING AWAY FROM THE BOOK BEHIND THE FATHER***
7	F	[***TURNS AWAY HIS GAZE FROM THE BOOK, PARTIALLY AT C DIRECTION***
8		[mitä
		[**what**
9	C	[vocalizes
10		***GETS TO F's FOOT*** [***TOUCHES IT AND STARTS MOVING BACK TO THE BOOK***
11	F	niin just (.) se on (.) <u>sun</u> isä-n jalka.
		PRT PRT it be-3 PRN father-GEN foot
→		**that's right (.) it is (.) <u>your</u> father's foot**
12	C	***RETURNS BACK TO THE BOOK***
13		***POINTS AT ANOTHER PICTURE***
14	F	siin on Mikko. ja mis on Mulperi.
		there is Mikko. and where is Mulperi

The starting point for the analysis here is the sequence in lines 1-2 whereby the father identifies the referent as a *foot* and then, in the continuation of his turn, more accurately, topicalizes the referent as being *their father's* foot. Without going into how this is done, we only note that here the father, by referring to the layout of the current page as representing only some specific part of the referent (the foot) offers a framework for mutual examining the referent [21,22]. Beginning at this point, we will look at the forthcoming sequence of referential acts now in more detail: Thus, how the child's intentions here become transparent in on-going interaction [12].

After having touched the father's shoulder (line 5), which is the first attempt to take the turn, the child starts moving away from the book. It is during this movement when the adult makes his first attempt to follow the child's action, in line 7, by having a slight gaze shift to the direction of the child's move (which is behind the father's back). The father's further orientation establishing interest towards the child's action, is now of the one of an unknowing participant: This is manifested in line 8, where the father poses an explicit question *what.* The trouble source for this possible repair initiator [19] is the child's activity so far, including the reference by touching the father and moving away from the book. However, proposing a question *what* does not start a repair sequence here. This claim is due to the incompleteness of the on-going child's turn. To support this, it can be noted, that at the time of the adult's question, the child's attempt to construct his turn is still under way. The child's action is not projected by the father's question. Thus, the adult's question does not take the child's action forward. However, posing the question *what*, though, is an important act through which the father poses himself as a recipient of some kind of information the child is providing here. This provides us with the notion of father taking the child's view as a knowing recipient.

It can be observed that it is not until the child has touched the father's foot (accompanied with some vocalization), in line 10, when the father indicates his identification or recognition of the child's reference: *niin just se on sun isän jalka* ('yeah exactly it is your father's foot'). The father's orientation towards the child's reference indicates the father orienting to the child's point as a request to make a comment on the pointed item and establishes strong acknowledgement indicated with particle chain *niin just* ('yeah exactly') [23]. Further, by virtue of the formulation *sun isän jalka* ('your father's foot'), we notice the father's turn to do two kind of simultaneous work: First, to differentiate, or contrast, the target of the child's recent reference from the target in the previous sequence, and second, taking the child view.

The reference to the *father's foot* is first determined by the pronoun *niiden* ('their') and then, by the pronoun *sun* ('your'), indicated with the arrows in the transcript. Thus, the latter formulation in the adult turn in line 11 brings the target of the child's recent reference closely in contrast with the referent in line 2 whereby it is designated as being something belonging to someone who is referred here as *niiden isä* ('their father'). Generally, pronoun *niiden* (genetive form of the pronoun *ne*) refers to an indeterminate group of people [24], but here it denotes to the characters in the book. The father's utterance referring to the foot of the child's father as *sun isän jalka* ('your father's foot') in line 11 is marked as being lexically and prosodically contrastive with the referent identified in lines 1-2.

Using a linguistic construction that links to some previously used construction has been seen as a way of marking that the activity accomplished by the construction is part of some previous activity [22]. Moreover, an important part of the work of contrastivity is carried by intonation and certain rhythm contour in the father's turn in line 11. There is a slight preceding pause before the word *sun*. Moreover, the current word is also stressed. The appearance of these paralinguistic features supports the view that the father is underlining the work of contrastivity for reference in the current turn.

Indicated by the adult utterance (in line 14), the child's return back to the book and referring to a new target is acting as closing the current topical sequence here.

To sum up, in the current example, the referent introduced in the very beginning functions as a "framework" that continues in the following characterization of the target referent *father's foot*. The continuation of the sequence, with the active initiative contribution by the child, introduces another aspect to the target, thus resulting in explicating the existence of the father's foot.

2.4. Concluding Remarks

The child's pointing is studied here as a dynamic, situated practice [25] and, more specifically, as a part of certain topical activity. The current analysis demonstrates how the context-bound meanings of the child's preverbal actions, especially deictic pointing gestures, are established in the on-going interaction and, additionally, how this kind of context-bound gesturing by the child is understood by the adult, to be a relevant next action in the course of interaction. It is important to note that the sequence of successive utterances constructed as turns is collaboratively produced in the examined sequence. The analysis reveals the ways in which participants themselves show to be relevant to the organization of their conversation. Eventhough, the adult responds both verbally and gesturally to the child's reference and the production of the gesture is embedded within a multi-party embodied participation framework [6,13,14,25], the present paper focuses on demonstrating only the certain aspects of the participants' activity.

The analysis shows that early gestures can be used as functionable means for displaying communicative intentions. Moreover, it demonstrates how these intentions are manifested in the on-going interaction. The child is able to direct adult's attention to entities around picture book by pointing. Interestingly, the child is able to accomplish conversational moves that display his ability to make ties with the reality outside picture book. In the extract, labeled as "Father's Foot", through the identification of the target referent, the child displays his knowledge on the target referent. Particularly, here the child displays his world knowledge through explicating the existence of his father's foot in a subtle way, as it is demonstrated in the sequence in lines 5 – 10 in the extract. He specifies an earlier section of his reference (touching the father's shoulder) with the second reference that is addressed to his father's foot: It is done explicitly by tapping point (thus, touching) as the addressee is not visibly positioned to see the target.

In addition to the analysis of gestural resources as a child's part of topical collaboration, the sequence demonstrates how the adult's orientation to the child's gestures is displayed through the linguistic and prosodic means, and especially, through certain referential, or indexical, utterances [21,22], as it is indicated in line 11. These features in the adult's input are the ones that make actual ties cross the situations and, specifically, display the adult's orientation to create a coherent topical activity.

3. Discussion

The present paper gives insight to the dynamicity of 'joint attentional frame' and especially, how the topical work is collaboratively achieved through different multimodal resources, providing us with the knowledge on the phenomena concerning shared understandings as a basis for succesful communication at early stages. Future aim is not only to find these fundamental intersubjective phenomena, but, while following certain kind of topical flows, the child's ability to process cross-situational data can be traced. Hence, these findings can be elaborated in examining the child's ability to make ties with the reality outside picture book and the collaboration between the participants revealing, or not revealing, co-construction of shared knowledge.

Understanding the link between joint attention and language, and also the role and properties of the data-providing aspects in adult input and the interaction contributing to the language development at early stages, have been topics of interest [26,27]. To have more insight of these, further research is needed. The present study, with the qualitative focus only, tries to contribute also to these efforts. Hereby the data-providing aspects of the adult input are acknowledged from certain viewpoint, together with the pragmatic phenomena. It may be that it is only after finding and figuring out the actual phenomena in detail in the real interactional settings when the developmental tracking and more quantitative approaches can be applied to these multi-threaded questions about multimodality.

4. Acknowledgements

I am grateful to all Vietri people who have contributed to the ideas in this paper, and Markku Aberg and anonymous reviewers for the constructive and helpful comments on earlier versions of the draft. Finally, thanks to Travis Rose for reading the draft.

References

[1] M.-L. Laakso, *Prelinguistic Skills and Early Interactional Context as Predictors of Childrens Language Development*, Dissertation, University of Jyväskylä, 1999.
[2] S. Goldin-Meadow & M. Morford, Gesture in Early Child Language. In: V. Volterra & C. J. Erting (eds.) *From Gesture to Language in Hearing and Deaf Children*. Springer-Verlag, Berlin, 1990.
[3] M. C. Caselli & V. Volterra, From Communication to Language in Hearing and Deaf Children. In: V. Volterra & C. J. Erting (eds.) *From Gesture to Language in Hearing and Deaf Children*, Springer-Verlag, Berlin, 1990.
[4] J. M. Iverson, O. Capirci, M. C. Caselli, From Communication to Language in Two Modalities, *Cognitive Development* (1994), 23 – 43.
[5] A. Wootton, *Interaction and the Development of Mind. Studies in Interactional Sociolinguistics 14*, Cambridge University Press, Cambridge, 1997.
[6] C. Goodwin, Environmentally Coupled Gestures, To appear in: S. Duncan, E. Levy J. Cassell (eds), *Festschrift in Honor of David McNeill*, John Benjamins.
[7] C. Trevarthen, Descriptive Analysis of Infant Communicative Behaviour, In: H. R. Schaffer (ed.) *Studies in Mother-Infant Interaction*, Academic Press, London, 1977.
[8] C. L. Schmidt, Scrutinizing Reference: How Gesture and Speech Are Coordinated in Mother-Child Interaction, *Journal of Child Language*, 23 (1996), 279 - 285.
[9] F. Franco & G. Butterworth, Pointing and Social Awareness: Declaring and Requesting in Second Year, *Journal of Child Language*, 23 (1996), 307 - 336.
[10] J. M. Iverson, O. Capirci, M. C., E. Longobardi, M. C. Caselli, Gesturing in Mother-Child Interactions, *Cognitive Development*, 14 (1999), 57 – 75.
[11] A. Wootton, Pointing and Interaction Initiation: The Behaviour of Young Children with Down's Syndrome When Looking at Books, *Journal of Child Language*, 17 (1990), 565 - 589.
[12] S. E. Jones & D. H. Zimmerman, A Child's Point and the Achievement of Intentionality. *Gesture* 3:2 (2003), 155–185.
[13] A. Kendon, The Negotiation of Context in Face-to-Face Interaction. In: C. Goodwin & A. Duranti (eds.) *Rethinking Context. Language as an Interactive Phenomenon. Studies in the Social and Cultural Foundations of Language No. 11*, Cambridge University Press, Cambridge, 1992.
[14] J. Bruner, *Child's Talk*, Oxford University Press, 1983.
[15] M. Tomasello, *Constructing a Language: A Usage-Based Theory of Language Acquisition*, Harvard University Press, 2003.
[16] B. D. Debaryshe, Joint Picture-Book Reading Correlates of Early Oral Language Skill, *Journal of Child Language*, 20 (1993), 455 - 461.
[17] H. Sacks, E. A. Schegloff & G. Jefferson, A Simplest Systematics for the Organization of Turn - Taking for Conversation, *Language*, 50, No. 4 (1974), 696 - 735.
[18] J. Heritage, *Garfinkel and Ethnomethodology*, Basil Blackwell ltd, 1984.
[19] C. Goodwin & J. Heritage, Conversation Analysis, *Annual Review of Anthropology*, 19 (1990), 283 - 307.
[20] S. Karjalainen, Achieving Topic by Multimodality in Early Dyadic Conversation. In: J. Allwood, B. Dorriots, S. Nicholson (eds.) *Proceedings from the Second Nordic Conference on Multimodal Communication, Gothenburg Papers in Theoretical Linguistics*, SSKKII, University of Gothenburg, 2006.
[21] W. F. Hanks, *Referential Practice. Language and Lived Space in the Maya*, The University of Chigago, Chigago, 1990.
[22] M. Etelämäki, Context and Referent in Interaction. Referential and Indexical Dimensions of the Finnish Demonstrative Pronouns. In: R. Laury (ed.) *Minimal Reference: The Use of Pronouns in Finnish and Estonian Discourse*, SKS Suomalaisen Kirjallisuuden Seura. Finnish Literature Society, 2005.
[23] M.-L. Sorjonen, *Responding in Conversation. A Study of Response Particles in Finnish*, John Benjamins B.V. 2001.
[24] R, Laury, *Demonstratives in Interaction. The Emergence of a Definite Article in Finnish*, John Bejamins, Amsterdam, 1997.
[25] C. Goodwin, Pointing as Situated Practice. In: S. Kita (ed.) *Pointing: Where Language, Culture and Cognition Meet*, Lawrence Erlbaum Associates, 2003.
[26] D. A. Baldwin, Understanding the Link between Joint Attention and Language. In: C. Moore & P. J. Dunham (eds.) *Joint Attention: Its Origins and Role in Development*, Lawrence Erlbaum, Hillsdale, New Jersey, 1995.
[27] E. Hoff & L. Naigles, How Children Use Input to Acquire Lexicon, *Child Development, Vol.* 73, 2 (2002), 418 – 433.

A. Esposito et al. (Eds.)
IOS Press, 2007

Nonverbal Communication as a Factor in Linguistic and Cultural Miscommunication

Maja BRATANIĆ[1]
Faculty of Transport and Traffic Sciences, University of Zagreb

Abstract. Nonverbal behavior is to a great extent universal but in many ways also marked by culture-specific patterns. Being less obvious than misunderstandings in verbal communication, nonverbally induced miscom-munications are more difficult to detect. The problem is relevant for a wide range of disciplines – from lexicology and lexicography to foreign language teaching. Main categories of nonverbal behavior are briefly discussed with the focus on proxemics elaborated in examples from American cultural patterns. Further examples of culturally-conditioned miscommunication draw on an aviation-related context.

Keywords. Nonverbal communication, cultural miscommunication, lexicography, proxemics, Aviation English

Introduction

People convey meanings not only through language but also through various aspects of nonverbal communication. A great deal of human communication – verbal as well as nonverbal - is culturally molded and conditioned. Both the above assumptions have been widely recognized in modern communication studies and hardly need to be defended.

Linguistic disciplines by definition do not take into account the nonverbal component of human communication to any considerable extent. Instances of miscommunication, however, arise from both of these components. Being less obvious than misunderstandings in verbal communication, nonverbally induced miscom-munication is far more difficult to detect. Furthermore, the line between verbal and nonverbal components of communication is often almost impossible to delineate precisely. Neglecting the nonverbal component of human exchanges, or rather the extralinguistic elements pertaining to the meaning of linguistic phenomena, can thus result in less than adequate transfer of the meaning of language entities – ranging from larger units of discourse to single-sound utterances like exclamations and the like.

It is my intention to show, albeit in a limited context, that even those disciplines considered to be eminently "verbal" cannot afford to bypass nonverbal constituents of meaning, here used in the widest pragmatic sense.

---

1 Corresponding Author: Maja Bratanić, Faculty of Transport and Traffic Sciences, University of Zagreb, e-mail: marija.bratanic3@zg.t-com.hr. bratanic@fpz.hr.

1. Lexicographic Context

The fundamental hypothesis of traditional lexicography assumes that the meaning of lexical units can be explained or otherwise rendered by means of definition – in case of monolingual lexicography – or by lexical equivalents in another language – as presented in conventional bilingual or multilingual dictionaries. The concept of definition has itself been defined in a variety of ways, but a standard lexicographical definition – as opposed to an encyclopedic one – is generally assumed to be confined to strictly linguistic devices. This, in other words, implies that such a lexicographic representation of word meaning will often fall short of encompassing the full range of lexical meaning (as used by cognitive linguists) i.e. its full referential dimension.

The problem is more obvious in bilingual lexicography, since concepts called by the "same" name (translational equivalents) in two languages do not necessarily have all their features in common. Lexical meaning is very often culture-bound, and 'cultural meanings' are to a great extent embedded in linguistic meanings. As Witttgenstein implies in his *Tractatus Logico-Philosophicus*, just because something has the same name in two or more cultures it does not mean that it refers to the same thing in those cultures [1]. This has been reiterated on many occasions to support the strong interconnection of language and culture [2].

Different cultures may conceptually define a particular construct differently. These constructs may be highly abstract, such as intelligence, self-image, personality or emotion [3] but can relate to more mundane behavior or down-to-earth concepts. The consequences are, clearly, that if a concept means different things to speakers of different languages and participants in different cultures, then there is a lack of equivalence, and representations of culture-bound concepts will be somewhat misleading. The immediate consequence of a situation in which the specific culture-bound meaning is not made explicit is that a dictionary user will tacitly assume equivalence with a more familiar concept [4].

2. Form vs. Meaning

2.1. Lexical discrepancies are not always language related. Lado has proposed the following categorization of cultural phenomena overlap [5]:

1. The same form has a different meaning (or different classification) in different cultures.
2. The same meaning has a different form.
3. The same form has the same meaning, but the distribution of cultural patterns may vary considerably.

In addition to the above, the most obvious type of incongruence occurs when a particular cultural concept has no corresponding equivalent in a different culture.

On the linguistic level these cases roughly correspond to the following situations:

- Only partial or false lexical equivalence is established via lexical "equivalents" which carry different semantic and/or cultural connotations.

A lexeme like God, for example, refers to a wide variety of concepts depending on the type of culture it is used by (monotheistic, polytheistic etc.). On a more profane level, the concept of social "dating", viewed crossculturally, is associated with a whole array of different cultural behavior patterns.

- Lexical gaps result from a situation in which corresponding concepts exist in different cultures but are not lexicalized.

An example of such a case can be borrowed from Hall who claims that the Arabic language does not have a lexical equivalent for the concept of sexual violence [6].

- Referential gaps ensue from the nonexistence of corresponding cultural concepts and pose a serious barrier in translation.

An instance of such a cultural discrepancy is described by Milan Kundera in *The Book of Laughter and Forgetting* elaborating the concept of *litost*. Kundera claims that *litost* is an untranslatable Czech word expressing a uniquely Czech feeling that can be roughly defined as a state of torment created by the sudden insight into one's own misery.

2.2. Culture-specific lexis is traditionally associated with some forms of social behavior (food, clothing, tools, flora, fauna, customs, religion, social institutions ...) manifesting themselves on the denotational level as opposed to those, often implicit differences in social and cultural behavior with either no lexical expression at all or of a primarily connotative nature. In my view, however, it is the entirety of these components that make up the lexical meaning. Modern lexicography (especially the monolingual type) has recently started addressing this issue more systematically. The problem remains, nevertheless, that monolingual lexicography by definition generally tends to be less biculturally or multiculturally oriented than the bilingual type, or it at best, rather vaguely addresses an imaginary participant of a different culture.

3. The Nonverbal Component of Lexical Meaning

Segments of meaning that cannot be conveyed by purely lexical means are often of nonverbal nature. The nonverbal component of communication thus easily becomes a factor in linguistic and cultural miscommunication. The boundaries between verbal and nonverbal, linguistic and extralinguistic, are delineated by very hazy lines, and separating them into two distinct categories often proves to be virtually impossible. Following this line of reasoning, any definition of communication (as well as any definition of culture, which I do not intend to elaborate further) must include nonverbal behavior in its broadest sense.

Understanding communication events fully requires the interpretation of meanings conveyed via nonverbal channels. In the context of foreign language learning it has become common to talk about nonverbal fluency as an integral part of communicative competence. Cultural fluency can, according to Poyatos, be defined as verbal-nonverbal fluency [7].

3.1. How Much Communication is Nonverbal?

There is less accord in establishing how much of human communication is indeed nonverbal in nature. Estimates vary, and, although we know that nonverbal clues can be more important than utterances alone, some opinions tend to be farfetched and untenable.

An early study conducted by Birdwhistell [8] indicated that up to 65% of a message's meaning is communicated through non-verbal clues, while Mehrabian [9] argued that in face-to-face communication non-verbal cues convey about 93% of the meaning. His statistics (attributing 7% of meaning to the words spoken, 38% of meaning to paralinguistic features, and 55% of meaning to facial expression) have become widely popular. These findings have been quoted excessively, and rather indiscriminately and oversimplistically (mis)interpreted. Similar statistics have ever since been rather mechanically reproduced:

Westbrook Eakins and Eakins also claim that in most two-person conversations the words communicate only about 35% of the social meaning of the situation while the nonverbal elements convey more than 65% of the meaning. They also mention another estimate according to which the nonverbal message carries 4.4 times the weight of the verbal message [10].

Fromkin and Rodman [11] similarly state that up to 90% of the meaning of a message is transmitted non-verbally. In her studies on gender differences in language use, Deborah Tannen also estimates that as much as 90% of all human communication is nonverbal [12].

Other scholars argue that the percentage is much lower and tend to be more cautious in similar estimates [13]. The overemphasis probably results from the fact that verbal and nonverbal phenomena to a large extent co-occur. On the other hand, as Scollon and Wong Scollon rightly observe, no means of quantification have ever been put forward and it is hard to establish whether such quantification is possible at all [14]. These doubts, however, by no means diminish the role of the nonverbal in human communication but simply point to the fact that it is hard to define.

3.2. Defining Nonverbal Communication

Knapp and Hall approach the task rather evasively, but effectively: "To most people, the phrase nonverbal communication refers to communication effected by means other than words (assuming words are the verbal element). Like most definitions, this one is generally useful, but it does not account adequately for the complexity of this phenomenon. As long as we understand and appreciate the points listed here, this broad definition should serve us well." [15].

In my limited research in the field I have adopted a practical view according to which nonverbal behavior reflects a relation of a human being towards himself/herself, other human beings or the environment, and can accompany verbal expression or function in isolation.

The focus of my interest here are potential misunderstandings rooted in nonverbal behavior. They generally stem from our implicit and unconscious assumption that nonverbal behavior functions universally, as well as our lack of recognition of culture-specific patterns in this area. If, however, most nonverbal phenomena (at least to a point) tend to be culture-specific and therefore often interfere with the transfer of meaning, not only of a particular speech act but of lexical meaning of linguistic units as well - they should be of at least marginal concern to those involved in lexicology and lexical semantics, even beyond the scope of anthropological linguistics or pragmatics which by definition take greater interest in nonverbal signals in communication.

3.3. Constitutive Elements of Nonverbal Communication

The most important types of nonverbal behavior can be categorized according to the disciplines that study them:

Kinesics refers to body movement and posture, as well as gesture, facial expression and eye contact and is thus most closely connected with what is popularly referred to as body language.

Oculesics, or eye behavior, more specifically looks at the influence of visual contact on the perceived message that is being communicated. It analyzes eye gaze, eye contact and its avoidance etc.

Haptics, sometimes referred to as tacesics, deals with touching behavior.

Proxemics is concerned with personal space usage.

Chronemics refers to the study of how people perceive, structure, and react to time or, in other words, how we use time to communicate. Cultures are, in this respect, usually divided into two categories: *monochronic* and *polychronic*.

Olfactics is the study of the sense of smell and olfactory communication.

Paralinguistics deals with vocal communication parallel to language itself, such as non-word utterances and other non-verbal vocal clues (tone of voice, loudness, pitch etc.) rather closely related to language use.

Some of the above categories clearly overlap, proxemics and haptics in particular. A wider frame of nonverbal behavior study additionally includes some other aspects of human behavior, e.g. appearance, use of symbols and colors, and their influence on non-verbal communication and gender related issues.

All forms of behavior listed above can in certain circumstances lead to cross-cultural, as well as intra-cultural misunderstandings. The amount and length of eye contact, for example, can be confusing even to the members of the same culture, although cultures (in the broadest sense of the word) set implicit norms for all channels of communication.

Gestures, in themselves, allow a lot of room for inter-cultural confusion. Speech-related gestures (sometimes referred to as gesticulation) used mainly for emphasis, illustration and pointing are rather universal and rarely seem to impede inter-lingual and intercultural understanding. On the other hand, speech-independent gestures form a system of signs in itself with a direct "verbal translation" consisting of a word or a phrase [15].

Nonverbal behavior is learned along with verbal, and do's and don'ts rarely need to be negotiated among the participants sharing same culture patterns.

In the last decades nonverbal behavior has been recognized as an important topic, particularly in communication studies. There is a substantial body of literature exploring the influence of culture on nonverbal behavior and revealing similarities and differences in communication processes and nonverbal behavior across cultures. However, there has not been much research in the domain of applied nonverbal behavior [3], and this is precisely the type of research that might be most directly applicable to the areas I am touching upon in this presentation.

3.4. Proxemics Exemplified

The term proxemics was introduced by anthropologist Edward T. Hall to describe the study of the human use of space within the context of culture, i.e. the cultural patterning of the spatial separation individuals tend to maintain in face-to-face encounters. Cultural expectations about the use of space vary widely, so proxemics gradually came to encompass the concept of privacy, crowding, territoriality, i.e. cultural expectations about how not only personal space but also public space should be organized.

Hall pointed out that social distance is reliably correlated with physical distance so that the distances people maintain from each other depend on the nature of their relationship or involvement. According to Hall, the use of proxemic zones considered "normal" and acceptable in American culture (more precisely middle-class Americans of Northern European heritage) would approximately correspond to the following [16].

intimate distance (embracing, touching or whispering, 15-45 cm or 6-18 inches)
personal distance (interactions among good friends, 45-120 cm or 1.5-4 feet)
social distance (interactions among acquaintances, e.g. business transactions (1.2-3.5 m or 4-12 ft)
public distance used for public speaking (over 3.5 m or 12 ft)

Different cultures maintain different standards of personal space. (Latin cultures are typically contrasted with Nordic cultures.) Comfortable personal distances depend on the social situation, gender, and individual preference but still have to be understood against a given general cultural framework. This can be illustrated by a brief overview of American sociocultural norms in the domain of proxemics as opposed to something that will, for the purpose of this discussion and in a somewhat contrived manner, be qualified as a Central European norm.

3.4.1. American Cultural Patterns

Some of the essential features of mainstream American culture can be clearly viewed through the prism of the concept of space. They could be more productively interpreted against the backdrop of American history, but for the present purpose I will just mention a few that are likely to strike an outsider first.

Personal distance tends to be greater among Americans than among Central Europeans, often accompanied by broad gestures, loud communication, and lack of physical contact, while intimate and public distances tend to be similar.

Americans show a strong tendency against restricting space, especially around their homes. They seem to dislike fixed barriers; instead of walls they plant hedges of bushes or trees [17].

Freedom of movement in other people's homes is unthinkable in most European cultures; so is American readiness to allow and even invite guests into the most private parts of their homes.

Individuality and independence, right to privacy, personal freedom – all manifest themselves in the use of space. Americans like open space, surround their houses by big lawns whenever possible, and set the limits to their private space in ways much less apparent to members of most other cultures.

The role of the door in American culture is, I find, exceptionally significant. The door has much stronger symbolic meaning than Europeans are used to. An American professor working in his/her office will at times leave the door open to signal that unannounced visitors are welcome. This "open door attitude" often strikes a visitor by its literal application; it symbolizes the constant availability of professors to their students.

Knocking on a closed door, on the other hand, is far from a pro forma gesture. Many a European visitor, including myself, has experienced embarrassment upon discovering that entering someone's room after merely knocking (and thus announcing our intention to come in) is perceived as barging in, since in the American cultural pattern, knocking symbolizes asking for permission to enter and waiting for this permission to be granted. A student, encountering a professor's door closed, might even be deterred from seeking permission to enter.

Consider the scene in an old American university building that was built without central air conditioning. Only a few offices that are used year-round have air conditioning, so on hot days the office doors that usually would be open are kept closed, so that cool air does not escape into the hot corridor. On such closed doors we have seen signs: "Come in without knocking: Door closed due to air conditioning".

By the way, the type of door used in American public toilet stalls, open to about 30 cm from the floor and again above about 180 cm, allowing a high degree of outside observability, might cause a great deal of initial discomfort in newcomers.

Such apparently informal American behavior in the realm of space, as well as otherwise, can be easily misinterpreted since it is not about uncoded behavior but, simply, a differently coded one. People generally make sense of nonverbal behavior by attaching any meaning at hand to unknown or unusual forms, creating in the process a lot of room for potential miscommunication.

4. The Language Teaching Context: the Case of Aviation English

Another area where we do not expect the nonverbal element to come into play, at least in no way directly, is the area of highly standardized languages for special purposes. Drawing on my experience in the field, I'd like to provide some evidence to the contrary. Among all varieties of English for special purposes (ESP), Aviation English is very likely the most strictly prescribed one. Still, language-related air accidents are common, usually due to less than full adherence to the rules of communication.

The worst air accident in aviation history took place in Tenerife in 1977, due to a tragic chain of events resulting in the collision of two Boeing 747 airliners and 583 fatalities. The crucial error was committed when, in communication with Air Traffic Control, the KLM co-pilot used a non-standard phrase, "We are now at take-off," literally translating from Dutch. Apart from the language problem itself, there was another issue involved here – that of professional ethics and authority. In spite of being aware of the imminent danger, the third member of the crew only mildly warned the captain, who in turn recklessly dismissed the warnings. Such dismissal of possibly very important inputs by lower-ranked members of the crew, for what would in anthropological theory be described as face-saving purposes, is not an uncommon pattern of behavior [18]. Many serious accidents could probably have been avoided, had it not been considered disrespectful to question the authority of a superior.

4.1. *Crew Resource Management*

As a direct result of the experience gained from the Tenerife disaster, Crew (or Cockpit) Resource Management was introduced with the intention of modifying the existing codes of behavior.

It had been established that among the main problems leading to human error were failures of interpersonal communication, leadership and decision making in the cockpit.

CRM is intended to ensure that the pilot receives information and advice from not only the other crew members, but also from external sources. Among other things, it teaches pilots how to improve communication, prioritize tasks and delegate authority. In other words, CRM is concerned with the cognitive and interpersonal skills needed to function in the demanding aviation system.

4.2. *Cultural Conflicts*

The investigation of various accidents revealed that the factors in play extend far beyond the cockpit and the pilots and include the issue of separation of pilots and flight attendants. "The basic problem is that these two crews represent two distinct and separate cultures, and that this separation serves to inhibit satisfactory teamwork". The problem manifests itself in the cabin crew not wanting to "bother" the pilots during a critical portion of the flight, even when they feel that the matter is critical. In situations like this the flight attendant must use discretion in deciding what is critical and what is not [19].

It is interesting to note how the concept of culture, as used above, applies to a somewhat narrower context than is normally implied. This is, however, not an isolated example. As Matsumoto and Seung Hee Yoo claim, "Culture exists on multiple levels. Individuals are part of small groups, and smaller groups are part of larger and even larger groups. Each group can have its own culture and in this way culture can exist on many levels. This is true for different ethnic and community groups that live in a large country like the U.S., as well as among different departments, sections, and work units of large companies." [3]

According to Captain Steve Last, the professional culture of the pilot heavily inclines to an individualistic attitude, which CRM training generally attempts to minimize. In his view, a lot of CRM work "seems to be biased to a particular, idealized Western cultural and social model, inappropriate for many airlines". Last dubs it a "Good Buddies" concept. Its essence is related to the American social ideal based on strong respect for the individual, yet "emphasizing that all individuals are equally valuable and that teamwork needs a leader to be 'first amongst equals'. This model, claims Last, implies that individual satisfaction is critically important because it leads to the best collective results. Expert piloting (as a form of control manipulation) is, at the same time, viewed as the highest attainment [20].

This ideal is not shared by the aviation community in general. The opposite extreme to the above model can, according to this approach, be described as an "Oriental" model or the "Good Boss" concept. Such a sociocultural pattern, in turn, implies greater respect towards the most skilled and experienced team member and a more pronounced inclination towards strong leadership, as well as group satisfaction.

With the intention of bridging the cultural and other possible gaps between these two extremes, one of the main roles of CRM is to teach techniques for responsibly delegating higher authority duties to lower authority units, on one hand, and to encourage lower authority units to question and even oppose higher authority in the interest of safety.

What makes this issue pertinent to our discussion, is the fact that authority in general tends to be exerted via both verbal and nonverbal cues. In the highly culturally and linguistically mixed environment of the vast aviation community, the failure to recognize such cues can easily lead to miscommunication or lack of communication. It would, therefore, be of no little importance for pilot and air traffic controller curricula to include at least some basic skills required to develop nonverbal communication awareness. Since these skills essentially belong to a wider communication competence of an individual – part of one's language and cultural fluency - it is reasonable to assume that the responsibility for imparting them will primarily be placed upon (second) language teachers.

5. How Should Culture Be Analyzed and Taught?

Without pretensions to dealing in depth with this major question in the framework of this short paper, I would like to end my contribution with a sketchy outline of one possible approach that seems applicable in both disciplines I have touched upon. The interdisciplinary model of culturemes Poyatos has put forward [7] seems to offer a workable framework in the area of foreign language teaching and, to a certain extent, in lexicographic description as well.

The cultureme is defined as "any portion of cultural activity or nonactivity perceived through sensible and intelligible signs with symbolic value and susceptible of being broken down into smaller units or amalgamated into larger ones". This model offers a progressive transition from broader culturemes to the simplest ones. The "largest culturemes of the two basic ecological realms in advanced cultures" are URBAN and RURAL, subdivided into what characterizes EXTERIORS and INTERIORS. Through these culturemes a culture can identify itself, says Poyatos, in a rather "impressionistic fashion, and mostly sensorially", providing a true interaction with the general environment, and modeling the participation in a given culture.

In the next step, the model develops a further subdivision into ENVIRONMENTAL and BEHAVIORAL realms, and so forth. The phase of secondary culturemes and subsequent phase of tertiary culturemes, followed by derived culturemes, allow more detailed analyses of the phenomena in given cultural structures and their interaction.

Poyatos's model originally deals with nonverbal cultural systems, and what he refers to as linguistic culturemes is limited to certain pragmatic dimensions of verbal behavior. There is no reason, in my mind, why this model could not be adapted to encompass wider, verbal as well as nonverbal entities. Poyatos himself defines communication as "the emissions of signs by all the nonlexical, artifactual and environmental sensible sign systems contained in the realm of a culture, whether individually or in mutual construction, and whether or not those emissions constitute behavior or generate personal interaction" [7].

Any method broadly in accordance with the cognitive approach to communication might prove productive as long as we keep the big picture in sight: a full interpretation of the communicative meaning of lexical, as well as broader language phenomena, is virtually impossible without acknowledging and accounting for its nonverbal component or counterpart.

References

[1] L. Wittgenstein, *Tractatus Logico-Philosophicus*, ed. by D. F. Pears, Routledge, London, 1981.
[2] Y.H. Poortinga, Equivalence of cross-cultural data: An overview of basic issues, *International Journal of Psychology*, 24, 1989, 737-756.
[3] D. Matsumoto and S. H. Yoo, Culture and Applied Nonverbal Communication. In E. Riggio and R. S Feldman, eds., *Applications of Nonverbal Communication*, Lawrence Erlbaum Associates, Publishers, Mahwah, New Jersey, London, 2005, 255-277.
[4] M. Bratanić, *Rječnik i kultura*, Biblioteka SOL, knj. 1, Filozofski fakultet, Zagreb, 1991.
[5] R. Lado, *Linguistics across Cultures*, University of Michigan Press, Ann Arbor, 1957.
[6] E.T. Hall, *The Hidden Dimension*, New York, Doubleday, 1966.
[7] F. Poyatos, *Nonverbal Communication across Disciplines*, Vol. I, John Benjamins Publishing Company, Amsterdam/Philadelphia, 2002.
[8] R.L. Birdwhistell, *Kinesics and context, Essays on body motion communication,* University of Philadelphia, Philadelphia Press, Philadelphia, 1970.
[9] A. Mehrabian, *Nonverbal communication,* Aldine-Atherton, Chicago, 1972.
[10] B. W. Eakins, R. G. Eakins. *Sex Differences in Human Communication,* Houghton-Mifflin, Boston, 1978.
[11] [V. Fromkin, J. Rodman, *An Introduction to Language*, CBS College Publishing, New York, 1983.
[12] D. Tannen, *That's not what I meant: How communication style makes or breaks relationship*, Ballantine, 1986.
[13] J. Neulip, *Intercultural Communication: a contextual approach,* Houghton-Miflin, Boston, 2000.
[14] R. Scollon, S. Wong Scollon, *Intercultural Communication, A Discourse Approach,* Blackwell Publishing, Oxford, 2003.
[15] M. L. Knapp, J. A. Hall, *Nonverbal Communication in Human Interaction*, Thomson Wadsworth, Belmont, CA., 2002
[16] E.T. Hall, *The Silent Language*, Doubleday & Company, Inc., New York, 1959.
[17] R. Carroll, *Cultural Misunderstandings, The French American Experience,* The University of Chicago Press, Chicago and London, 1987.
[18] C. Linde, The quantitative study of communicative success: Politeness and accidents in aviation discourse, *Language in Society* 17, 1988, 375-399.
[19] R. D Chute, E. L. Wiener, Cockpit/cabin communication: I. A tale of two cultures, *The International Journal of Aviation Psychology*, 5 (3), 1995, 257-276.
[20] S. Last, Eliminating "cockpit-caused" accidents, Error-tolerant Crew Procedures for the Year 2000, at http://www.crm-devel.org/resources/paper/last/last.htm.

A. Esposito et al. (Eds.)
IOS Press, 2007

The Integrative and Structuring Function of Speech in Face-to-Face Communication from the Perspective of Human-Centered Linguistics

Krzysztof KORŻYK
Jagiellonian University, Kraków, Poland
korkling@lingua.filg.uj.edu.pl

Abstract. This paper illustrates the need for study of the interdependencies between verbal and nonverbal behavior treated as a unified form of activity, manifesting itself in face-to-face communication. Invoking the principles of Human-Centered Linguistics the author treats communication not as something passed on via language, but rather as something to which language merely contributes. One of the consequences of such an approach to this issue is a reassignment of focus. Rather than attention being drawn to linguistic phenomena, the spotlight is on the communicative properties of the interlocutors, creatively utilizing various elements of the interactional "symbolic spaces." With reference to the above, the communicative action is perceived as a function of choices correlating verbal and nonverbal signs and signals. Light is also shed on the advantages stemming from an integrated modeling of communicative phenomena.

Keywords. Conduit Metaphor, verbal and nonverbal communication, interaction, Human-Centered Linguistics

By Way of Introduction

A strong trend in contemporary research into verbal and nonverbal communication[1] is a gradual departure from a modular treatment in favor of an integrated comprehension of communicative phenomena. Therefore, instead of investigating interpersonal communication as has been the case heretofore – as if it took place via speech accompanied by other, less essential, non-linguistic semiotic means – it treats communicative behavior homogenously.

---

[1] Nonverbal communication (or nonverbal behavior) is here understood, after Kendon [11], as "all of the ways in which communication is effected between persons when in each other's presence, by means other than words. It refers to the communicational functioning of bodily activity, gesture, facial expression and orientation, posture and spacing, touch and smell, and of those aspects of utterance that can be considered apart from the referential content of what is said. [...] It will be seen from this that there are three main limits governing the use of the term, its literal meaning notwithstanding. First, it is used mainly in reference to communication that occurs when people are able to respond directly to each other's actions and are able to directly affect one another through such responses. [...] Second, 'nonverbal communication' is generally considered to refer to communication as it is effected through behavior whose communicative significance cannot be achieved in any other way. [...] A third characteristic of 'nonverbal communication' is that messages that are at the center of interest (whether in fact conveyed by words or not), are typically those messages that are not given explicit formulation. They are the messages that may be inferred from or are implied by a person's actions [pp. 3-4]."

If one were to elucidate the behavior of partners in the interactive "symbolic space" by referring to one sort of "grammar" or another, one would have to resort not to some specific semiotic system of non-linguistic communications, but rather to a multidimensional structure correlating verbal and nonverbal signs and signals. This would permit the modeling of linguistic, cognitive, communicative, and interactive conduct occurring between interlocutors as a function of choices which they make. Accordingly, in order to facilitate a full-fledged and realistic account of communicative phenomena a thoroughly different perspective and conceptualization must be adopted [8, 9, 17, 18, 24, 25].

Invoking the standards of empirically-oriented linguistics (and related disciplines) the text at hand points to the typical threads which appear in the accepted, yet flawed wisdom about human communication. Common errors lie in an overly literal treatment of the semiotic or linguistic models or a metaphorical projection of these models on phenomena with reference to which one should use other instruments [16]. Although the insights presented herein are, on the whole, of a theoretical nature and not the results of unique research into a specific thesis, they constitute a deeper survey of what should be common knowledge, yet is often overlooked. Ultimately, the unbiased reader will find directions – words to the wise, as it were – towards the right path.

1. The *Conduit Metaphor* – Accepted Wisdom on Communication

The man on the street, unconcerned with analyses of his own and others' behavior in social interactions, is generally convinced that the most important, if not the sole instrument of understanding and communicating is language. Such a common sense conviction is, to a great extent, linked to the presence (in many world languages) of expressions embodying the metaphorical image schema which Reddy named the *conduit metaphor*.[2] Nevertheless, despite their seeming obviousness, the images and concepts sanctioned by this linguistic-cognitive structure exemplify the correlation between specific elements of the communicative-interactional situation in such a way as to distort its very essence.

According to Reddy [19], this metaphorical image schema takes one of two forms: primary and secondary. Expressions of the primary type treat communication in various ways. Firstly, it treats communication as if language functioned as a sort of conduit (similar to electrical wiring, fiber-optic cables, or a system of pneumatic postal tubes) facilitating the translocation of some "substance" between speakers. Or, secondly, writing and speaking are seen as dependent on the placement of images, thoughts, and feelings into "containers" of linguistic expressions or their graphic equivalents. Thirdly, communication is seen as if the storage of this "substance" was untouched until someone decided to look into the "containers." And finally, fourthly, listening or reading is seen as if it hinged on the addressee's opening of a consignment delivered via a "communicative conduit" and removing the "package" contents.

This imaginary procedure is naturally connected with yet another condition. In order for interlocutors to exchange thoughts, emotions, and images via a system such as

[2] A few examples of the *conduit metaphor* from Reddy's work [19] include: "Try to *pack* more *thoughts into* fewer *words*"; "Your *words are hollow* – you don't mean them"; "You still haven't *given me* any *idea* of what you mean"; "Try to *get* your *thoughts across* better"; "That *concept* has been *floating around* for decades"; or "How many different *concepts* can you *get into* your *head* in one evening?" See also [14].

this, each must have the possibility of "exteriorizing" the contents of their consciousness. Elements are "removed" from their "heads," then "apportioned" and placed in appropriately chosen "containers" of linguistic expressions. The logic of this metaphor also requires recognition of the autonomy – as much from their home environs as from the conditions of their "storage" and "transport" (in audio, visual, tactile, etc. "containers") – of these cognitive constructions.

The secondary form deals with this aspect in more detail. First of all, communication is here understood as if speaking or writing entailed "releasing," "discarding," or "shooting" bullets of thoughts, feelings and images out of one's "head" into some external "realm of ideas." This is, of course, connected with a reification of these constructs which are treated as if their existence did not depend upon whether someone at a given moment thinks these thoughts, has these or other feelings, or imagines something in one way or another. Finally, these entities can naturally "make their way" into someone's mind, but can also exist independent of anyone's consciousness. Both these patterns of imagination can overlap; in fact, this often occurs because their structures can be harmonized.

It is not difficult to note that both variants sanction and favor a manner of thinking about linguistic communication which turns attention away from the cognitive, communicational, and interactional real feats of the interlocutors, directing them instead towards some imaginary "sender-conduit-receiver" activity [14, 19]. The consequences are crucial. In the communication model suggested by the conduit metaphor, the speaker or writer is responsible for the effectiveness of understanding. If, however, this means "sending" untainted thoughts and images packed into "containers" protecting them from outside influences (and keeping the signifying "substance" from getting out), then the addressee at the other end of the conduit can "remove" from the container no more than was "inserted," in its original condition. Responsibility for any misunderstanding thus falls on the sender who clearly must have forgotten something, leading to the loss or change in the properties of the "meaningful substance" in the communicative process. In accordance with the strange logic of this metaphorical structure, the linguistic effectiveness of communication is not the result of anyone's interpretative efforts: success is automatic, guaranteed by the very "technology" of communication. Nonetheless, it is obvious that symbolic communication is impossible without the creative, interpretative and interactive cooperation of the partners [3, 5, 8, 12, 20].

2. The Human-Centered Linguistics Perspective[3]

In reality, in the physical space between interlocutors or between writers and readers, there are no "meanings," "images," or "cognitive contents" – neither existing

[3] Human-centered linguistics is, according to Yngve [25], hard-science linguistics which "studies objects given in advance, namely people both individually in terms of their linguistic-communicative properties and collectively as united in social linkages. It also studies the sound waves of speech along with the physical means of communicative energy flow and other relevant parts of the domain of concrete reality accessible through experience. To study these objects in a hard-science manner it is necessary to study them directly, not indirectly through concepts of language and grammar. It thus becomes necessary for the practitioners of linguistic sciences to leave the soft-science domain of language understood in terms of definitional models or theoretical constructs depicting abstract interrelationships between speech sounds and their meanings from the logical-philosophical standpoint [pp. 3]."

independently, nor "connected" with any kind of perceivable object-carrier. There are only *signals*, maintaining a structural identity of forms of energy, relatively stable spatial forms, events, etc. They are not "accompanied by," nor are they "connected with" any "content," "information," "message," or "communiqué" in some time-space. They are neither "transferred" nor are they "conveyed." This could only be said of *signals* which, however, as their very name indicates, do not "convey," "communicate," nor "contain" and cannot "inform" anybody about anything. They are simply specific physical phenomena, interpreted by interacting partners as meaningful thanks to their reference to binding cultural and social standards or conventions [7, 8, 22, 23, 24].

Nevertheless, it is worth bearing in mind that the codes which assure understanding are not – as they are often presented – sets of signs or signals; they are *transformations* which project one set of elements on another. The functioning of a code can never mean the "exchange" of one sign for another or change into another. Likewise, strictly speaking, thoughts cannot be changed into linguistic expressions, linguistic expressions into graphic equivalents, or the graphic equivalents of sounds into electronic impulses. No – each of these objects remains no more, no less than what it is. A code itself – as a relationship between two different sets or systems of elements – only guarantees the preservation of the order of the one system projected onto another. This, in turn, permits the addressee to interpret the sequence of signals he or she has received by referring it to a predetermined pattern. Interpretation is, nonetheless, the result of cognitive acts by the addressee who must assess its probability, taking into consideration not only the aforementioned communicative invariants but also other factors influencing the course of this process [3,5].

Undoubtedly, the metaphorical conceptualizations presented earlier bear little resemblance to the realities of linguistic communications [7,14,16,19]. Yet there is more: these conceptual structures distort the image of language itself, treating it as if it could be "mastered," "wielded," or "born." Regardless of evident incongruities in the expressions used (how can someone "wield" that which they "bear"?), it is plain to see that these are natural and conventionalized means of not only everyday but also scholarly thinking and speaking of language as an autonomous system of signs passing for a model or prototypical instance of a semiotic system. Still, keeping facts and not the theoretical constructs of one discipline or another in mind, let us note the following:

1. In concrete communicative interactions we are dealing not with language, but with the verbal and nonverbal behavior of people who have learned these not as the result of some sort of acquisition of some grammar existing "out there," but rather thanks to the participation in the life of a community whose members use precisely such behaviors when entering into social interactions.
2. Doing what is normally called "speaking such and such language," partners in an interaction strive to attain their desired goals – most often changes of mind, feelings, intentions, or ultimately behaviors of their interlocutors. These, in turn, lead then to changes in the social reality. It would be difficult to call conduct of this type merely "using language" or "uttering words or sentences"; speaking is always, if not above all, a social activity.
3. No meanings are conveyed between communicating subjects via language or any other semiotic system. Naturally, changes – key from the perspective of

> communication sciences – do take place in the interlocutors. These shifts, however, appear under the *influence* of the partners, not as their direct *effect* because at play are always the interpretations and sovereign decisions of each of the participants in the interaction.

Of course, all of the above constitutes only the linguistic part of the truth about communications. People do not correspond solely through some singular, unique system of signs – even if it were as distinct and unique as natural language. Individuals communicate with everything at their disposal. More than that, inasmuch as they wish to influence the thoughts, feelings, and actions of their partners, speakers utilize any and all elements of the physical, cultural, and social space which could serve as "bearers" of information or a means by which some social relationship would be created between them. At play here are more than just the verbal acts with their linguistic components (phonetic, morphological, syntactic, semantic, and pragmatic); there are also the nonlinguistic vocal phenomena associated with voice modulation, varying execution of specific sounds, latent responses, hesitations, and other supra-segmental traits of live speech.

This "signalling" activity is, not surprisingly, accompanied by various body movements which form a sequence encompassing postural-gesticulation, facial expressions, gazes, as well as ways space is used – physical distance, touch, or even scent. In the well-known classification referring to the usage, origin, and coding of this body language, Ekman and Friesen [4] categorize its elements as *emblems*, *illustrators*, *regulators*, *affect displays*, and *adaptors*. But this is not all because concrete and real verbal and nonverbal behavior appears in specific *situational frames*.

From the point of view of Hall [6], who sees culture as a form of communication, these frames are both the material of which communicative actions are made as well as the context which endows them with meaning. In particular, these situational frames constitute modules of a sort, forming the pragmatic milieu in which diverse codes, meanings, and contexts can only be seen as aspects of the same communicative event thanks to the assignment of a place in the interactive-interpretative coordinate system.

A particularly clear cut example of such a framework is *speech genres* – i.e., language-behavioral structures, or the codified forms of interactive behavior (typical of a specific culture or society) to which speech contributes [1,20]. Speech genres organize everyday occurrences such as questions, requests, commands, threats, warnings, appreciation, congratulations, condolences, apologies, complements, praise, complaints, conversations, discussions, conflicts, arguments, explanations, information, announcements, flirting, teasing, jokes, toasts, speeches, lectures, chats, and many, many others. Their elements are:

1. mutually correlated communication roles "played," as it were, by the participants;
2. default thinking, feeling and intending paired with utterances characteristic of a given activity;
3. a goal which could be achieved via such behavior;
4. scenarios for such an interaction, etc.

Although the whole is described as speech genres, what is emblematic of these phenomena is that they are not solely linguistic. They comprise, as already mentioned, patterns of interactive behavior in whose structure utterances fulfill foremost a function signaled in the title of this article: the structuring and integration of other, no less crucial components of these communicative-interactive modules.

In concrete and real interactions realized through means embodying various behavioral and contextual modes, the behavior of partners takes on a structure both *holistic* and *gestaltic*. Human exchanges cannot be otherwise: each and every human communicative-interactive action is inherently multimodal and speech activity is no exception [8,10,12,13,17,18]. Although its functional universality and exceptional "informativeness" makes this the most distinct form of human behavior, a moment of reflection suffices to lead one to the realization that communication does not take place *through* linguistic means, but rather *with* their *contribution*. It would be impossible to speak without modulating one's voice, or without a whole range of other interactive behaviors, starting with the facial, motor, postural, and gestural and ending with the proxemic and chronemic. In fact, even if one were to exert every effort to limit all such nonverbal behavior, the efforts themselves – in the face of standards known by each member of a given community – would be treated as significant and interpreted in accordance with circumstances.

3. Ockham's Razor – By Way of Conclusion

A consequence, on the one hand, is that we can, in specific cases, analytically derive distinct verbal and nonverbal aspects of some behavioral, communicative-interactional configuration. On the other hand, there is no possibility of drawing the global sense of this whole from a simple sum of its presumed parts.

Again, human beings do not communicate via the means of one system or another, as these means only participate in the communication. Depending on the character of the interaction, some of them can (at any stage) demand more attention of the interlocutors, yet even the most palpable communicative actions are usually accompanied by other behavior which modifies them, or delineates their meanings. As a result, the sense of a configuration of communicative-interactive behavior emerges as a *global structure* whose elements – though of different provenience – are characterized by a fundamental *functional unity* [10,12,18].

Therefore, if one wished to remain in accord with these facts and not multiply entities unnecessarily, then one should seek principles key to the course of the communicative interaction, not taking hypothetical structures of some autonomous semiotic system (e.g., natural language) as a base, but rather the similarities and differences observed in the concrete, verbal, and nonverbal behavior of human subjects. Road signs pointing towards a serious and empirical treatment of communication [5,8,9,17,18,24,25] might include the following:

1. The study of communicative phenomena should be conducted from an appropriate, integrative, non-reductionist, and non-deterministic perspective. Maintenance of such a view is as significant and crucial as the generation of various empirical data. (It is especially important to remember that a

conventional demarcation between verbal and nonverbal means of communication comprises less a reflection of some key difference between them and more an expression of a unique methodological consciousness and the research practices sanctioned by it.)

2. Realistic explanation of communicative phenomena demands substitution of abstract, linguistic-semiotic theoretical constructs (such as language and its grammar, or a system of signs and the communiqués created via this means) with models which take into consideration the real relationships and dependencies between partners in concrete symbolic interactions.
3. Accompanying this should be an integrative analysis of the actual communicative behavior between partners while its "symbolic components" should be treated as aspects of functionally homogenous communicative activity despite their "belonging" to conventionally distinguished "semiotic systems."
4. Taken altogether, the crux of the matter would be investigation of human behavior from the communicative and interactive point of view and not researching them via the means of some language or code (regardless of whether it was verbal or nonverbal), and certainly not through exegesis of subsequent theories into the "functioning" of such semiotic systems.

Next, instead of subsequent views on the theoretical constructs of a semiotic, logical, or linguistic type, one would arrive at empirically-oriented theories encompassing interlocutors and their attributes viewed from a communicative and interactive perspective. These, in turn, could be tested with reference to data extracted from ensuing observations and experiments [22,24,25].

Such theories might not provide "explanations" for the above-presented phenomena in categories "justifying" or "grounding" them thanks to hypothetical "encoded rules" or "deep structures of meaning" of some "internal grammar" or other object "rooted" in one reality or another. In fact, all that one could reasonably expect would be a series of descriptions demonstrating what the interlocutors were really doing when they communicate – a sorting and arranging of the verbal and nonverbal behavior at hand in a manner clear enough as to facilitate the perception of connections which had previously escaped the attention of the researcher. The outcome of such a systematization would be the disclosure of a vital aspect of the phenomena under study.

In this way, setting hypostatized "language" and "grammar" aside and heading for that which can be directly observed, linguistics – and along with it other disciplines of similar status – would migrate to the realm of the hard sciences, towards more realistic and empirically justified models of communication.

References

[1] M. Baxtin: *Speech Genres and Other Late Essays*. Austin, TX: University of Texas Press, 1986.

[2] R.M. Brend, Ed.: *The Eighteenth LACUS Forum*. Lake Bluff, IL: LACUS [Linguistics Association of Canada and the United States], 1992.

[3] C. Cherry: *On Human Communication: A Review, A Survey, and A Criticism*. Cambridge, MA: MIT Press, 1968.

[4] P. Ekman, W.V. Friesen: The Repertoire of Nonverbal Behavior: Categories, Origins, Usage, and Coding,, *Semiotica*, vol. 31, pp. 49-97, 1969.

[5] J.G. Greene, Ed.: *Message Production: Advances in Communication Theory*. Mahwah, NJ: Lawrence Erlbaum, 1997.
[6] E.T. Hall: *Beyond Culture*. Garden City., NY: Doubleday, 1976.
[7] R. Harris: *The Language Myth*. London: Duckworth, 1981.
[8] R. Harris: *Signs, Language and Communication*. London: Routledge, 1996.
[9] R. Harris: *Introduction to Integrational Linguistics*. Oxford: Pergamon, 1998.
[10] A. Kendon Ed.: *Nonverbal Communication, Interaction, and Gesture: Selections from Semiotica*. The Hague: Mouton, 1981.
[11] A. Kendon: Introduction: Current Issues in the Study of 'Nonverbal Communication. In A. Kendon Ed.: *Nonverbal Communication, Interaction, and Gesture: Selections from Semiotica*. The Hague: Mouton, pp. 1-53, 1981
[12] M.R. Key, Ed.: *The Relationship of Verbal and Nonverbal Communication*. The Hague: Mouton, 1981.
[13] M.R. Key, Ed.: *Nonverbal Communication Today: Current Research*. Berlin: Mouton, 1982.
[14] G. Lakoff, M. Johnson: *Metaphors We Live By*. Chicago: University of Chicago Press, 1980.
[15] V.B. Makkai, Ed: *The Twentieth LACUS Forum 1993*. Chapel Hill, NC: LACUS, 1994.
[16] A. Ortony, Ed: *Metaphor and Thought*, Cambridge: Cambridge UP, 1995.
[17] F. Poyatos: *Paralanguage: A Linguistic and Interdisciplinary Approach to Interactive Speech and Sounds*. Amsterdam: John Benjamins, 1993.
[18] F. Poyatos: *Nonverbal Communication across Disciplines*. Amsterdam: John Benjamins, 2002.
[19] M. Reddy: The conduit metaphor: A case of frame conflict in our language about language. In A. Ortony, Ed: *Metaphor and Thought*, Cambridge: Cambridge UP, 164-201, 1995.
[20] A. Wierzbicka: *Cross-Cultural Pragmatics: The Semantics of Human Interaction*. Berlin: Mouton de Gruyter, 1991.
[21] G. Wolf and N. Love, Eds.: *Linguistics Inside Out*. Amsterdam: John Benjamins, 1997.
[22] V.H. Yngve, Linguistics as a Science, Bloomington: Indiana UP, 1986.
[23] V.H. Yngve: What Can We Believe This Days? – Can Linguistics Actually Be Scientific. In V.B. Makkai, Ed: *The Twentieth LACUS Forum 1993*. Chapel Hill, NC: LACUS, 33-50, 1994
[24] V. H. Yngve: *From Grammar to Science: New Foundations for General Linguistics*. Amsterdam: John Benjamins, 1996.
[25] V.H. Yngve, Z. Wąsik, Eds.: *Exploring the Domain of Human-Centered Linguistics from a Hard-Science Perspective* (Workshop). Poznań: Motivex, 2000.

Section 2

Nonverbal Speech

Fundamentals of Verbal and Nonverbal Communication and the Biometric Issue
A. Esposito et al. (Eds.)
IOS Press, 2007

How Speech Encodes Affect and Discourse Information

Conversational Gestures

Nick Campbell
[1] *National Institute of Information and Communications Technology*
[2] *ATR Spoken Language Communication Research Laboratory,*
Keihanna Science City, Kyoto 619-0288, Japan
`nick@nict.go.jp & nick@atr.jp`

Abstract. This paper presents an analysis of several recorded conversations and shows that dialogue utterances can be categorised into two main types: (a) those whose primary function is to impart novel information, or propositional content, and (b) those whose primary function is to relay discourse-related and interpersonal or affect-related information. Whereas the former have characteristics that are closer to read speech, the latter are more varying in their prosody and present a considerable challenge to current speech synthesis systems. The paper shows that these apparently simple utterances are both very frequent and very variable, and illustrates with examples why they present such a difficult challenge to current speech processing methods and synthesis techniques.

Keywords. Speech Technology, Discourse, Affect, Non-verbal Speech Communication

Introduction

It should be common knowledge that "The act of sending and receiving messages is a process of negotiation of meaning wherein both the sender and the receiver are mutually responsible for the creation of this meaning" [1].

In the domain of speech technology, there are already well-developed devices for the recognition of verbal speech and for the processing of its propositional content, but very little is yet known about methods for processing the non-verbal content in a dialogue speech signal. We can process broadcast news easily, but we still perform poorly when faced with ordinary dialogue speech.

This paper describes some current work towards the processing of non-verbal speech utterances in a dialogue context. The work is based upon an analysis of a very large corpus of everyday spoken interactions captured under extremely natural situations [2]. The paper presents a view of speech interaction as not only facilitating the exchange of linguistic or propositional information, but also facilitating the display of affect, interpersonal stances, and social relationships. By incorporating such non-verbal content in a model of speech communication it may become easier to model the subtle two-way interactions between both speaker and listener that are necessary for facilitating the transfer of meaningful discourse content.

Concurrent work [3] using the same model is being carried out towards the production of a 'conversational' speech synthesis system for use in interactive dialogues, such as might take place between a person and an information system, a robot, or a speech translation device. There are several types of response and feedback utterances that are currently very difficult to implement using traditional speech synthesis methods, yet these non-verbal speech sounds or 'conversational gestures' function to provide status-updates in an interactive discourse. Such (often phatic) utterances include laughter and grunts as well as many common phrases and idioms, and their choice and variety can reveal much information about the speaker's (i.e., the current listener's) states in an interactive discourse.

This model of information exchange incorporating non-verbal backchannel speech utterances shows how feedback from the listener is used to help the speaker both to deliver content more efficiently, and at the same time to be reassured of the degrees of success in the flow of information transmission. It assumes that both the sender and the receiver are equally responsible for the mutual creation of meaning for each segment in a discourse and that they do this through the mediation of non-verbal cues. This paper will concentrate on those cues that are signalled by speech, which we refer to here as the 'audio landscape'.

1. The Audio Landscape

The 'audio-landscape' of a discourse enables a participant or observer to estimate the types of interaction and to make guesses about the relative status of participants without the need for a content-based analysis of any given utterance or sequence of utterances. In other words, even a foreigner who has no understanding about the specifics of what is being said can often make an intelligent guess about the functional states, i.e., about what is happening in a dialogue at the interpersonal level.

By simply watching what is happening in a conversation, without even any sound information at all, we can often see who is doing what; not just who is speaking (which can be determined relatively easily from the amount of bodily movement, for example), but also who is listening (which can be determined from the synchrony of movements related to events in the speech) as has been shown by e.g., the early work of Kendon & Condon [4,5] and the more recent 'meetings' findings [6,7,8].

Furthermore, if in addition to the visual information we also have access to the sound, then we can make an intelligent guess about how the participant listeners are reacting to the content of each utterance, even if (like the foreigner) we do not understand the content of the speech itself. Laughs, nods, grunts, and other such speech gestures serve to indicate the degrees to which the listener is attentive, synchronised with the content of the discourse, and in relative states of agreement with it. This much can be determined from the non-verbal content [9,10].

We are currently performing research into technology to process this audio landscape in order to detect the main speaker in a given discourse situation, both in a meeting environment [11] and in general two-person conversations, to categorise the competing forms of speech in a given situation. Several speech gestures such as laughter, agreement, and feedback-responses can be recognised, isolated, and used to determine the progress of the meeting and the degrees and types of participation status among the members present.

2. Data Collection

As part of the JST/CREST Expressive Speech Processing (ESP) project [12], a series of conversations were recorded between ten people who were not initially familiar with each other and who had little or no face-to-face contact but who were paid to meet once a week to talk to each other over the telephone for thirty-minutes each over a period of ten weeks. The content of the conversations was completely unconstrained. These recordings constitute the ESP_C subset of the ESP corpus.

The volunteer speakers were paired so that each conversed with a different combination of partners to maximise the different types of expressiveness in the dialogues without placing the speakers under any requirement to self-monitor their speech or to produce different speaking styles "on-demand". The ten speakers were all recorded in Osaka, Japan, and all conversations were in Japanese. Since the speakers were not familiar with each other initially, little use was made of the local dialect and conversations were largely carried out in the so-called 'standard' Japanese. Again, no constraints on types of language use were imposed, since the goal of this data collection was to observe the types of speech and the variety of speaking styles that 'normal' people used in different everyday conversational situations.

Four of the ten speakers were non-native; their inclusion was not so that we should have foreign-accented speech data, but rather that we should be able to observe changes in the speech habits of the Japanese native speakers when confronted with linguistically-impaired partners. Two were male, two female, two Chinese, and two English-language mother-tongue speakers. These and the two Japanese who spoke with them formed Group A in our study. Group B is the 'baseline' group, consisting of a male and a female Japanese native speaker who conversed in turn with the each other and with the Japanese native speakers of both sexes from Groups A and C. Group C similarly consisted of a

```
female          male

( cfa efa    cma ema )      (foreign)
   /              \          Group A
 jfa       -      jma
  |                |
 jfb              jmb        Group B (baseline)
  |                |
 jfc       -      jmc
  |                |         Group C
(fam)            (fam)      (intimate)
```

Figure 1. Showing the form of interactions between participants in the ESP_C corpus. The first letter of each participant identifier indicates the mother-tongue (Japanese/Chinese/English) of the speaker, the second letter indicates the speaker's sex (female or male), and the third letter is the group identifier. (fam) is short for family; indicating intimate conversations with relatives.

```
CFA JFA C01 200.369 0.491 #
CFA JFA C01 200.860 0.808 laugh
CFA JFA C01 201.668 0.869 あとは
CFA JFA C01 202.537 1.099 変わりました
CFA JFA C01 203.636 1.868 laugh
CFA JFA C01 205.504 0.670 うん
CFA JFA C01 206.174 0.744 #
CFA JFA C01 206.918 0.917 はい
CFA JFA C01 207.835 2.691 #
CFA JFA C01 210.526 0.602 はい
CFA JFA C01 211.128 2.791 #
CFA JFA C01 213.919 0.749 @S
CFA JFA C01 214.668 2.685 そうです結構もう四年間なりました
CFA JFA C01 217.353 0.785 はい
CFA JFA C01 218.138 0.561 #
CFA JFA C01 218.699 0.731 はい
CFA JFA C01 219.430 1.384 #
CFA JFA C01 220.814 1.088 行ってます
CFA JFA C01 221.902 0.738 #
CFA JFA C01 222.640 0.784 はい
CFA JFA C01 223.424 1.107 #
CFA JFA C01 224.531 1.356 あの一歳です
CFA JFA C01 225.887 0.525 #
CFA JFA C01 226.412 0.600 はい
CFA JFA C01 227.012 2.795 #
CFA JFA C01 229.807 0.443 はい
CFA JFA C01 230.250 0.941 #
```

Figure 2. Transcription was performed by hand, using the Transcriber software package. The first 3 columns in the figure identify the speaker, partner, and conversation number. The numbers represent the start time of each utterance in the conversation (in seconds) and its duration. Laughs, non-speech noises, and silences are also transcribed along with the text.

male and a female Japanese native speaker who conversed with each other and with the members of Group B, but who also telephoned their own family members each week and spoke with them for a similar amount of time. Figure 1 illustrates these pairings graphically.

The corpus thus obtained allows us to examine the prosodic characteristics and speaking habits of Japanese native speakers when confronted with a range of different partners on the spectrum of familiarity, and to observe changes in their speech as this familiarity changes over time. Our principal targets for this series of recordings were the six Japanese native speakers (three male and three female) who came to separate rooms in an office building in Osaka once a week to speak with each partner over the telephone for a fixed period of thirty-minutes each time. All wore close-talking, head-mounted, Sennheiser microphones and recordings were taken directly to DAT with a sampling rate of 48kHz. The offices were air-conditioned, but the rooms were large and quiet, and no

```
JFA CFA C01 203.276 1.362 4456 ==> <[ laugh ]>
JFA CFA C01 204.638 0.902 0 ==> <[ @S ]>
JFA CFA C01 205.540 1.927 0 +->  <[あーそうなんですか]>
JFA CFA C01 207.467 0.322 0 ==> <[ @S ]>
JFA CFA C01 207.789 0.401 0 ==> <[ はい ]>
JFA CFA C01 208.190 0.227 0 ==> <[ @S ]>
JFA CFA C01 208.417 1.744 0 ==> <[ あのー ]>
JFA CFA C01 210.976 0.393 814 ==> <[ え ]>
JFA CFA C01 211.369 0.260 0 ==> <[ え ]>
JFA CFA C01 211.629 1.139 0 --> お,ご結婚をきょ
JFA CFA C01 212.768 0.264 0 ==> <[ え ]>
JFA CFA C01 213.032 1.566 0 --> 何時なさったとおっしゃいまし
JFA CFA C01 216.356 0.687 1757 --> 四年目
JFA CFA C01 217.043 0.301 0 ==> <[ @S ]>
JFA CFA C01 217.344 1.498 0 +-> あ,<<そうです>>か
JFA CFA C01 218.842 0.422 0 ==> <[ @S ]>
JFA CFA C01 219.264 0.241 0 ==> <[ え ]>
JFA CFA C01 219.505 1.193 0 --> お子さんは
JFA CFA C01 221.686 0.283 987 --> X
JFA CFA C01 221.969 0.819 0 --> あ,いらっしゃる
JFA CFA C01 223.180 0.360 392 ==> <[ あ ]>
JFA CFA C01 223.540 1.248 0 --> お幾つです.か
JFA CFA C01 225.571 0.749 783 --> 一才
JFA CFA C01 226.320 0.347 0 ==> <[ @S ]>
JFA CFA C01 226.667 1.235 0 --> あっそう,じゃ
JFA CFA C01 227.902 1.891 0 +-> こういう<<ときは>>どういう風に
JFA CFA C01 229.793 1.494 0 +-> お子さんはされてる<<んですか>>
JFA CFA C01 231.287 0.746 0 ==> <[ @S ]>
JFA CFA C01 232.033 0.798 0 --> お家
JFA CFA C01 234.539 0.424 1707 ==> <[ あ ]>
```

Figure 3. Part of a dialogue, showing frequent utterances (n>=100) in <[square]> brackets, and frequent segments (N>=100) as part of longer utterances in <<angle>> brackets, which may be embedded. Speaker, listener, conversation number, start time and duration (in seconds) and delay (milliseconds) from end of previous utterance are also shown. "@S" indicates a sharp sucking intake of breath, a common speech gesture in Japanese. The paper argues that these very frequent interjections carry a separate stream of information through their prosody

unwanted noises (or acoustic reflections) were present in the recordings.

The speakers were all mature adults who were employed part-time by the recording agency and were paid for their participation in the recordings. They were initially unfamiliar with each other, but the degree of familiarity naturally increased throughout the period of the ten conversations. All have signed consent forms allowing the contents of the recordings to be used for scientific research. The ultimate purpose of the data collection was not made specific to the participants who were only told that their speech would be recorded for use in telecommunications research.

10073	うん	467	ズーー	228	ううん	134	へー.ーー
9692	@S	455	スー	227	えっ	134	はい.はい.はい.はい
8607	はい	450	んーー	226	へ.ーー	134	そう.です
4216	laugh	446	うーーーん	226	ハハハ	133	@E
3487	うーん	396	ねー	225	う.んー	133	あ.そう.な.ん.です.か
2906	ええ	395	あ.あー	200	そうですね	130	そう.な.ん.です.か
1702	はーい	393	はい.はい.はい	199	ほ.ーー	129	は.ー
1573	うーーん	387	あー.はい	193	ハー	129	い
1348	ズー	372	ねえ	192	その	127	ほ.ー
1139	ふん	369	ふーーん	190	え.えー	125	ハハハハハ
1098	あのー	369	だから	188	あ.あーー	119	はい.はい
1084	あっ	368	あー.ん	187	ね	119	は.ーーー
981	はあい	366	ああ	180	ん.はい	114	ハハ
942	あの	345	あの.ーー	180	あの.ーーー	113	は
941	ふーん	337	なんか	173	ん.ん	113	で.ー
910	そう	335	え	172	アハハハ	113	て
749	えー	311	でも	168	はい.ー	112	は.あー
714	あーー	305	スーー	164	う.うーん	110	フフフ
701	あ	274	うん.うん.うん	161	は.ーー	110	そのー
630	あーーー	266	ハハハハ	160	@K	110	もう
613	あ.はい	266	て.ー	159	そう.です.ねー	109	ふーーーん
592	うん.うん	266	え.ーー	151	あーーーー	108	はあ.ーー
555	あー	258	で	143	だから.ー	106	そうですね.え
500	んー	248	う	139	アハハハハ	105	んー.ん
469	ん	242	へー	137	そう.そう.そう	104	いや

Table 1. The hundred most frequent single utterances in the ESP_C corpus. The numbers indicate the count of each word or phrase when it occurs as a single utterance in the transcriptions. Since duration is usually considered as distinctive in Japanese, the lengthening (an extra mora beat is indicated by a dash) may be significant. Note the highly repetitive nature of many of these utterances, very few of which can be found in any standard dictionary of Japanese. Note that these few samples alone account for more than a third (n=72,685) of the 200,000 utterances in the corpus. Less then half (n=92,541) of the utterances were unique.

3. Data Characteristics

Figure 2 shows part of a typical dialogue segment from Chinese speaker CFA, talking with her Japanese partner JFA during their first conversation. We can see even from this very short sample that there is considerable repetition; in this case of the word 'yes' (or its Japanese equivalent), interspersed with occasional longer content utterances. Table 1 lists the 100 most-frequent expressions from a corpus of 200,000 such dialogue utterances transcribed from recordings of the six people's telephone conversations. We (even those of us who cannot yet read Japanese) can see from this table that repetition is a common identifying characteristic of these frequently-repeated utterances. The same syllable (Japanese character or character sequence) repeats in more than half of the cases. If we expand this list to include the less frequent common utterances, then we will find that they differ primarily in the number and type of repeats.

Among these repeats, we can discern several different patterns or types: one uses progressive lengthening (hah: はー, はーー, はーーー) (hmm: ふん, ふーん, ふーーん, ふーーーん), another simple repetition (ha!: ハ, ハハ, ハハハ, ハハハハ, ハハハハハ),

or complex repetition (umm: うん, うんうん, うんうんうん), and yet another increasing complexity(so: そうです, そうですね, そうですねー, そうなんですか, あそうなんですか). The hundred utterance types listed in Table 1 account for more than a third of the total number of utterances in the corpus. If we include their less frequent (typically longer) variants, then we find that more than half of the utterances in the corpus are of this non-verbal type (not usually found in a standard dictionary).

If we exclude these feedback utterances (i.e., just listen to those utterances marked with "–> " in Fig.3), then we can still understand the propositional part of the discourse, almost without change, but we lose the 'landscaping' information. Alternatively, if we just listen to those primarily non-verbal utterances ("==> " in Fig.3), then we can follow much of the social interaction (in 'foreigner mode') without knowing anything about the content of the discourse. i.e., we can interpret the prosody to make an inference about the function of each utterance without knowing its specific lexical meaning.

3.1. Features of Non-Verbal Speech

Unlike regular lexical items which have a fixed form and a variable prosody depending on contextual information, these non-verbal 'speech gestures' rather seem to have a fixed prosodic identity (or underlying prosodic dynamic) and a variable form, extending to meet the requirements of the prosodic dynamics that they function to substantiate. Like bodily gestures, which have a few basic finite forms but considerable freedom of gestural expression, or dynamics [13,14,15], these sounds perhaps function primarily to express the feelings, states, and attitudes of the speaker [9,16] and then secondarily to support the text, or at least to function in parallel with it.

Being very frequent, and effectively 'transparent' with respect to the propositional content of the discourse, the prosodic features of these speech gestures can be easily detected and compared. In addition to obvious variation in duration and intonation they are also marked for 'tone-of-voice' or phonatory voice-quality characteristics. Being so frequent, they can be compared 'like with like' as the speaker's and listener's affective and discoursal states and relationships change and progress throughout the discourse. As we have shown previously [10], the prosodic aspects of these non-verbal speech sounds share much in common across different cultures and languages, and they may represent a basic form of pre-linguistic human communication.

Figure 3 shows the corresponding part of the dialogue segment presented in Figure 2 which has been bracketed to highlight the frequently-repeated speech segments. Here we see the Japanese speaker's utterances and can combine them with those of her Chinese partner to reproduce the conversation segment. Some potentially ambiguous utterances can thereby be disambiguated by use of the textual content of the surrounding utterances, but a large number remain functionally indeterminate from the transcription alone. They are not at all ambiguous when listening to the speech, and carry a considerable amount of discourse information.

The text in Figure 3 has been annotated by a computer program to indicate which utterances are unique (and therefore presumably convey more propositional content) and to bracket those which are subject to frequent repetition and hence act as potential carriers of affect or discourse-control information. Two types of repetition have been bracketed: (a) whole phrases that occur more than a threshold number of times in the corpus, and (b) phrasal chunks that form part of a larger, possibly unique, utterance but which are frequently repeated anyway. The current setting of the repeated-pattern recognition program, arbitrarily takes more than 99 repeats throughout the corpus as the minimum threshold for bracketing, and yields 74,324 untouched utterances, 72,942 marked as repeated phrases, and 49,136 utterances including repeated phrasal segments. These thresholds were determined by trial and error and are not intended to be more than examples.

Taking some of the frequent repetitions from one of the corpus speakers as an example, we notice different strategies of usage according to differences in partner. This speaker (JFA) makes considerable use of "ah", "ano", "hai", and "un", but not equally with all partners (see Table 2). For example, when speaking with foreigners, she uses "hai" frequently (はい= yes (with all of the ambiguity that it can have in an English conversation)), but significantly less so when speaking with Japanese partners. She uses "demo" (でも=but) much more frequently with Japanese partners, and "ah" much less when conversing (in Japanese) with the English-native-speaker partners.

Such differences may reflect interpersonal relationships, personal characteristics, or cultural peculiarities, but perhaps more interesting is the considerable variety of pronunciations within each utterance type, reflecting the speaker's interest, state-of-mind, and types of participation in the discourse.

3.2. Physical Characteristics of Repeated Segments

It is a central tenet of this paper that these repeated segments function to carry affect-related and interpersonal information in parallel to the linguistic content of the message. They do this by means of small but consistent variations in such acoustic characteristics as tone-of-voice, spectral tilt, pitch range and excursion, speaking rate, laryngeal and phonatory setting, etc. In this section we will examine some of these physical characteristics. By being so frequent and repetitive, the transparent speech gestures allow a listener (even one not yet familiar with the particular speaker's traits or habits) to make comparative judgements about the speaker's emotional and affective states and stances and to interpret subtle nuances in the speech by means of the prosodic cues hereby revealed.

Table 3 and Figure 4 illustrate some differences in pitch range (i.e, the amount of variation in the f0 or fundamental frequency of the voice throughout the utterance) and voice energy (signal power in decibels) for three representative but randomly-selected sample speech gestures taken from speaker JFA's conversations with the six different partners. We see significantly lower pitch for "ano" (a hesitation) when used with Chinese partners, and equivalently higher pitch for "umm" (agreement) with the English partners. These differences appear partner-related rather than linguistic.

Table 2. Counts for some frequently-repeated simple utterances from one speaker to six partners. The table illustrates differences in usage strategies for these utterances. The first letter of each speaker's ID indicates their native language, followed by sex and group

JFA:	CFA	CMA	EFA	EMA	JFB	JMA
a,a–	143	145	88	89	138	170
ano	224	277	221	176	209	266
demo	41	24	31	17	89	134
e–	48	51	37	25	74	94
hai	2932	2234	2181	3239	72	33
un,un	1029	546	585	1190	909	1037

Table 3. F0 range (f0r) and mean (f0m) values in Hz and Power range (pwr) and mean (pwm) values in dB for three sample utterances (ah, umm, and ano) from speaker JFA according to differences in conversational partner

"a,a–"	CFA	CMA	EFA	EMA	JFB	JMA
f0r	125	181	266	232	234	241
f0m	201	214	220	192	206	198
pwr	28	29	29	28	31	31
pwm	38	39	36	35	42	41
"un,un"	CFA	CMA	EFA	EMA	JFB	JMA
f0r	154	152	182	181	161	141
f0m	172	175	162	145	198	174
pwr	28	29	27	26	29	27
pwm	37	40	36	35	42	39
"ano"	CFA	CMA	EFA	EMA	JFB	JMA
f0r	106	113	161	154	169	155
f0m	131	136	142	133	156	149
pwr	27	28	28	27	31	29
pwm	38	40	37	36	42	39

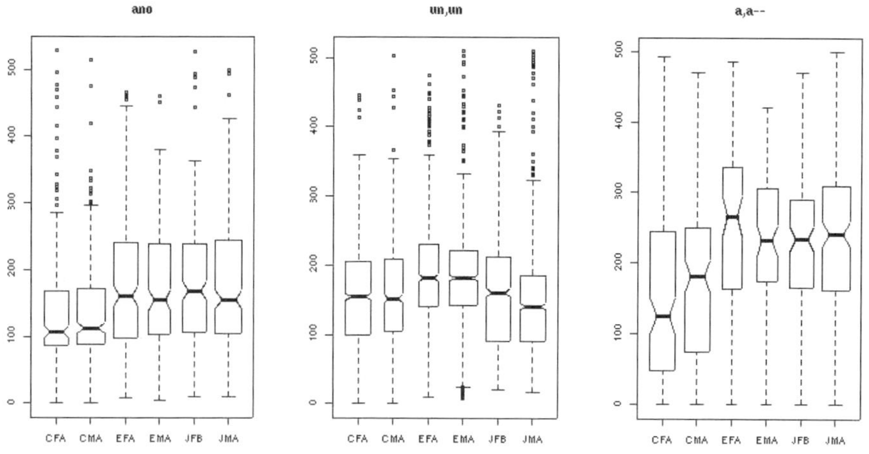

Figure 4. Plots of Pitch Range (amount of variation in the fundamental frequency of the voice) for three utterances from speaker JFA when conversing with six different partners. The width of the boxes is proportional to the number of tokens. Differences are significant at the 5% level if the notches do not overlap. The vertical axis shows pitch range in Hz.

Figure 5 shows a subset of this data (fundamental frequency contours for the utterance "un,un") and plots a representation of the 'shape' of each utterance by showing averaged f0 values for each progressive third of the utterance. Again we see that there is considerable variation, but that the variation between contours for different types of conversation partner is greater than that between utterances within a given set of conversations.

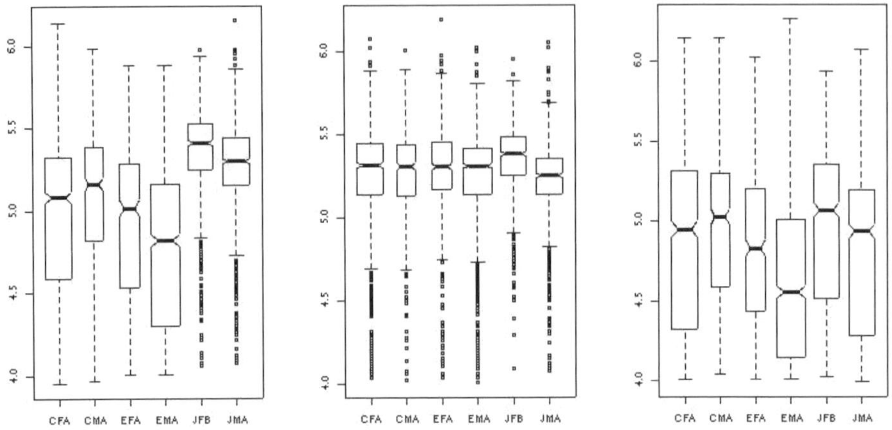

Figure 5. Fundamental frequency contours differ according to the listener. The left-hand plot shows average f0 values for the initial third of the utterance, the middle plot for the middle third, and the right-hand plot shows average f0 values for the final third of each utterance. Plots show'averaged contours' for all samples of the utterance "un,un" factored by partner. Japanese partners evoke a high initial contour, and English-native-speakers a lower fall at the end, though all countours appear to pass through the same high range of values mid-utterance.

These data indicate that the speaker's basic acoustic settings and the amount of physical energy used in each utterance vary not just per utterance, as could be expected, but also according to the listener (and presumably according to the content and smoothness of flow of the conversations). We see that not just the overall pitch ranges, but also the dynamics or shape of the utterance contours is also varied.

It is apparent that Japanese partners evoke a higher utterance-initial contour, and English-native-speakers a lower fall at the end, though all contours appear to pass through the same high range of values mid-utterance. The fact that these differences appear more related to partner than to local contextual differences implies that a higher-level of socially-inspired prosodic processing may be taking place, i.e., that a level of social interaction is influencing the basic overall prosodic contours in much the same way as the linguistic, semantic, and prosodic content influence them at a finer granularity.

4. Discussion

"In human communication a great deal of failure comes about not because information has been lost in transmission but because the sender is unable to express what he has to say, or because the receiver is unable to interpret the message in the way intended." [18]

In written communication, great care is usually taken so that the structure and wording of a text will clearly and unambiguously portray the meaning intended by the author. In spoken communication, on the other hand, the interaction is in real-time, two-way, and often put together on the spur of the moment. Little time is available for a careful planning of the structure of a spoken utterance, and the resulting 'text' is often broken up and spread out among several sequential utterance segments that are interspersed with discourse-control and interpersonal stance messages expressed non-verbally.

There is no guarantee that the speaker is optimally expressing her intended meaning, not that the listener is optimally comprehending the speech stream. Instead, a constant stream of feedback and feedback-elicitation is necessary so that the information transfer may be optimised. I speculate that the greatest failures of communication come about when this secondary stream is ineffective.

Allwood's theory of *Communication as Action and Cooperation* [19] prescribes the communicative activities of a sender and a receiver and provides such a framework for their interconnection. However, in current speech processing technologies, only the primary stream (i.e., the linguistic or propositional content) is currently in focus. The notion of interactive speech being a series of mutually-dependent communicative acts comes secondary (if at all) to that of speech as a sequence of 'packets of textual content' being passed from speaker to listener.

Such a text-based form of information processing may well be adequate for the analysis of broadcast news, where the speaker is transmitting to a plurality of listeners as a remote audience which has no interactive potential in real-time. However, future speech technologies must incorporate *both channels* of information (verbal as well as non-verbal) if they are to process real-time interactive human speech communication efficiently.

5. Conclusion

The auditory landscape of a dialogue contains fluctuating surfaces of sound whose characteristics provide cues to the interpersonal relationships and discourse participation of the conversing partners. The speech data from the ESP_C corpus of conversational dialogues show that there is considerable prosodic variation on what are seemingly very simple but also very frequent utterances. This variation may therefore serve primarily to indicate the speaker's relationships with the listener, since it seems to depend more on differences between conversational partners than between different utterance types.

This background provides an element of the discourse in which *how* something is said is more important than *what* was said, and where the prosody of the non-verbal speech components provides a dynamic expression to the simple 'umms' and 'ahhs' that are more normally considered as noise. By the interplay of such feedback comments and their elicitation, conversational speech takes on its characteristic forms of expression and the interactive transfer of knowledge is achieved.

This paper has described how the lowest level of discourse information can be processed in a speech signal for the automatic annotation of discourse progress and for producing an estimate of speaker participation status. In a semi-formal round-table meeting situation there is typically only one main speaker at any given moment, but several participants may be speaking simultaneously, expressing agreement (or otherwise), chatting, translating, etc., in addition to the main speaker. We are currently performing research into technology to process this audio landscape in order to detect the main speaker and to categorise the contributions of all forms of speech in any given situation.

Acknowledgement

This work is partly supported by the Ministry of Public Management, Home Affairs, Posts and Telecommunications, Japan under the SCOPE funding initiative. The ESP corpus was collected over a period of five years with support from the Japan Science & Technology Corporation (JST/CREST) Core Research for Evolutional Science & Technology funding initiative. The analysis was carried out while the author was employed by the National Institute of Information and Communications Technology. The author also wishes to thank the management of the Spoken Language Communication Research Laboratory and the Advanced Telecommunications Research Institute International for their continuing support and encouragement of this work.

References

[1] Harrigan, J.A. & Rosenthal, R., "Nonverbal aspects of empathy and rapport in physician-patient interactions", pp. 36-73. In P.D. Blanck, R. Buck, & R. Rosenthal (Eds.). Nonverbal communication in the clinical context. University Park, PA: The Pennsylvania Univ. Press. 1986.

[2] The SCOPE 'robot's ears' project homepage: http://feast.atr.jp/non-verbal

[3] Campbell, N., "Conversational Speech Synthesis and the Need for Some Laughter", in IEEE Transactions on Audio, Speech, and Language Processing, Vol 14, No.4, July 2006.

[4] Kendon, A.,"Movement coordination in social interaction: Some examples described". Acta Psychologica, Amsterdam, 32(2): 101 125. 1970.

[5] Condon, W. S., "Communication: Rhythm and Structure. Rhythm in Psychological, Linguistic and Musical Processes", J. R. Evans and M. Clynes. Springfield, Illinois, Charles C Thomas Publisher: 55-78. 1986.

[6] I. McCowan, D. Gatica-Perez, S. Bengio, G. Lathoud, M. Barnard, and D. Zhang, "Automatic analysis of multimodal group actions in meetings", IEEE Trans. on Pattern Analysis and Machine Intelligence, vol. 27, no. 3, pp. 305 317, Mar. 2005.

[7] Zhang, D.,et al., "Multimodal group action clustering in meetings"., VSSN'04, 54-62, 2004.

[8] Campbell, N., "A Multi-media Database for Meetings Research", pp 77-82 in Proc Oriental COCOSDA, 2006, Jakarta, Indonesia.

[9] Campbell, N., "Expressive Speech - Simultaneous indication of information and affect", pp.49-58 in *From Traditional Phonology to Modern Speech Processing"(Festschrift for Professor Wu Zongji's 95th birthday)*, eds G.Fant, H.Fujisaki, J.Cao & Y.Xu, 2004.

[10] Campbell, N., & Erickson, D., "What do people hear? A study of the perception of non-verbal affective information in conversational speech", pp. 9-28 in Journal of the Phonetic Society of Japan, V7,N4, 2004.

[11] Campbell, N., "Non-Verbal Speech Processing for a Communicative Agent", Proc Eurospeech, pp. 769–772, Lisbon, 2005.

[12] The JST/CREST Expressive Speech Processing project, introductory web pages at: http://feast.his.atr.jp

[13] D. McNeill, "Gesture, Gaze, and Ground", in Proc 2nd Joint Workshop on Multimodal Interaction and Related Machine Learning Algorithms, Royal College of Physicians, Edinburgh, UK. July 2005.

[14] Condon, W. S., "Synchrony Demonstrated between Movements of the Neonate and Adult Speech", Child Development 45: 456-462. 1974.

[15] Tanenhaus, M. K., Spivey-Knowlton, M. J., Eberhard, K. M. & Sedivy, J. E., "Integration of visual and linguistic information in spoken language comprehension". Science, 268, 1632-1634. l995.

[16] Campbell, N., "Getting to the heart of the matter; speech as expression of affect rather than just text or language", pp 109-118, Language Resources & Evaluation Vol 39, No 1, Springer, 2005.

[17] Campbell., N, & Suzuki, N., "Working with Very Sparse Data to Detect Speaker and Listener Participation in a Meetings Corpus", in Proc Language Resources & Evaluation Conference, Genova, Italy, 2006.

[18] Warner, T., "Communication Research", Vol. 19 No.1, p. 52-90 Communication Skills for Information Systems. London. Pitman Publishing, 1996.

[19] Allwood, J., "Linguistic Communication as Action and Cooperation", Goteburg Monographs in Linguistics, Goteborg University, Department of Linguistics, 1976.

Beats for Individual Timing Variation

Eric KELLER[1]
IMM, University of Lausanne, Switzerland

Abstract. We subjectively experience humans to speak with a certain regularity – which creates perceived rhythm within speech – at the same time as we expect them to display variation, mostly for emphasis and to satisfy personal preferences. Synthesized speech that does not exhibit these perceptual qualities is often classified as "robotic" and "unnatural". The search for the objective bases of the perceived regularity in speech is old and has produced less than satisfactory results. In 1977, Ilse Lehiste, in an extensive review of the issue of isochrony (acoustic evidence for rhythmicity in speech) came to the conclusion that there were no direct acoustic correlates of rhythmicity [1]. This view, supported by a number of further studies, has formed the consensus for spontaneously produced speech since then. However, Robert Port and his colleagues have in recent years suggested that some parts of perceived regularity may actually be directly dependent on the suddenness and the relative strength of voice onsets (so-called "beats"). This hypothesis was examined here with respect to continuous speech by a series of analyses performed in two languages, and it was found that indeed, beats do provide a minor temporal organizational effect within the speech phrase, but that the effect is so minor that it is of no or only circumscribed value to such applications such as speech synthesis or speech recognition.

Keywords. Speech timing, rhythmicity, voice onsets, beats

Introduction

Human speech has both shared and individual components. The *shared* components of speech make it possible to transmit ideas between speakers, while the *individual* parts identify and distinguish those who are speaking. Scientific analysis of speech has in the past explained primarily the shared parts of language, those that encode systematic linguistic information and are thus held jointly by the various members of a language community. Till a few years ago, speech research had nearly "forgotten" about the individuality of language and speech.

But detailed analyses of speech have recently called much more attention to the individual aspects of speech. Individuals show considerable variation with respect to the verbal material that they use to meet similar communicative challenges (Campbell, this volume). Also, when using a particular expression, individuals employ a certain liberty in how they structure their material in their own individual manner. It may be useful to characterize individual variation in the following simplistic, but initially helpful manner: When information is new and when the communicative situation demands good understanding at first presentation, speakers tend to use hyper-articulated and prototypical forms of speech. On the other hand, when basic agreement

[1] Corresponding Author: IMM, University of Lausanne, Switzerland, e-mail: eric.keller@unil.ch .

between speakers is assumed, or if the communicative situation is more relaxed, or if the speaking party shows low concern about being well understood, hypo-articulated and greatly reduced and individually varied forms of speech tend to predominate.

If one were to record all human speech in all possible combinations of face-to-face, telephoned or televised speech, one would probably find that by far most oral communications are variations of the latter type. This poses considerable challenges to various types of speech analysis. For example one central issue in recent speech recognition work has been the following: "How can relevant information be extracted from strongly divergent forms of speech?" And a well-posed question in current speech synthesis work concerns the types and limits of individual variation that could or should be implemented to obtain realistic and individually coloured forms of speech. Also of interest is the question of whether there might be portions inside stretches of speech with less individual variation, and some others where there is more. If that is so, speech recognition work could initially be directed at the more stable sections of speech, and could return to the more varied portions in subsequent analysis steps.

In this article, we wish to address a subcomponent of these issues, the individual time structure of phrases, and explore the question of more or less stable points in the speech chain. It has been known for some time that phrasal time structure has both shared and individual components. This emerges from prediction experiments where the systematic (information-bearing) timing components of phrases are predicted by common statistical methods. In these experiments, the average timing of the various linguistic units (segments, syllables, minor/major phrases, etc.) is predicted from the components and structures of various grammatical and phonetic contexts (see summary view [2]); this process leads to a fairly adequate modelling of such phenomena such as utterance-final lengthening, word-initial lengthening, polysyllabic shortening, etc. [3]. However, depending on the speaker, the speech material and the predictive algorithm, model adequacy for this type of variation generally shows rates of between 0.7 and 0.9 out of a possible 1.0, leaving a good proportion of the overall variation still unaccounted for. A good part of this remaining variation is likely to be individual variation, while some other parts may be due to "noise" in measuring technique and data analysis.

The present study examines such non-systematic individual variation in the context of the effects of *beat*. "Beat" is the term given by Robert Port, Fred Cummings, Keichi Tajima and various other colleagues to rapid onsets of sonorous information in speech. In their studies, beats were shown to be an important temporal organizing concept for short, repeated sentences [4, 5, 6]. In an experimental setup known as "speech cycling", subjects had to repeatedly produce short sentences (like "dig for a duck" or "take a pack of cards") in a time frame set up by two external sounds. After listening to the synthesized sentences once without speaking, they had to repeat the sentences in time with subsequent stimulus sentences. Subjects had to temporally align the first main stress in the sentence ("dig", "take") with a first sound, which reoccurred every 1.5 seconds in the study reported in [4]. The second main stress (as in "duck" or "cards") had to coincide with the second sound. If the second sound was randomly chosen in the range of about one half of the repetition cycle, the second main stress peaked fairly exactly around the mid-point (50%) of the repetition cycle, and did not show the expected random distributions around the middle area of the repetition cycle. Randomly chosen shorter durations between the acoustic signals tended to congregate fairly precisely at about one third of the repetition cycle, while random longer durations peaked at about two thirds of the repetition cycle (Figure 4).

In an extensive theoretical article of these effects, Port motivates this and similar results in chaos terms [6]. He relates these findings to "coordinative gestures", a theoretical formulation about motor actions that has been applied to speech for a number of years by scientists from Haskins Laboratory (e.g., Kelso & Tuller [7]). It has been shown in a variety of motor tasks that coordinative cyclic actions tend to subdivide at half-way points, and often as well at other low harmonic frequencies, such as at one third or one quarter of the entire movement. Port notes that with respect to a large number of activities, humans in the presence of other humans show spontaneous emergent coordination, and he argues that this is the effect that manifested itself in his speech cycling data.

One of the curious aspects of this type of coordination is that no physical link is necessary for the production of harmonic subdivisions. For example, Schmidt, Carrillo & Turkey (1990) showed that when two people sit on the edge of a table and swing one of their legs, they found it easiest to swing in phase, somewhat less easy to swing in alternating phase, and most difficult to swing out of phase [8]. This breaks up a perceptually coordinated phase into a harmonic at one half of the phase. In similar fashion, Port argues, speakers will find it easier to make vowel onsets occur at low harmonics of the repetition cycle, such as at halves or thirds of the cycle. Stated in chaos terms, this says that the natural harmonics in the phase constitute "attractors" for the speech behaviour. These are sites where a certain set of behaviour is more likely than in-between "attractors".

Beats can be calculated quite easily (see below), so it was inviting to examine their effect on timing in speech phrases that were more typical of continuous speech than Port's simple, memorized and repeated phrases. We had some carefully-recorded and transcribed corpora of read French and English texts in our laboratory, so we posed the following question: "Given the documented time structuring effect of beats within the sentence, do beats have an effect on individual temporal variation in speech read aloud from text?" By examining the effects of beat on various portions of speech produced in similar phonetic manner by the various speakers, we could examine the potential organizing effects of beats on speech in the two languages.

We did not have any firm expectations about the results of this experiment. On the one hand we estimated that beats might indeed reduce the variation between the speakers, at least to some degree, since there should be a natural tendency to form coordinate structures at common beat locations. Given the results from Port's laboratory, we expected that the locations of strongly beat-marked sound transitions would show somewhat less inter- and intra-subject variation than un-marked or weakly marked sound transitions.

On the other hand, it was not certain that our experimental conditions would permit beats to fully develop their coordinating effects on speech. Continuous speech, spoken freely or read aloud as in these experiments, is quite different from the memorized and repeated utterance of a just a few words or syllables. Phrases – however we wish to define them – are of different duration and syllabic structure, and there may not be enough similarity between phrases for a phase-locked pattern to evolve. The two corpora were examined successively during the summers of 2005 (French) and 2006 (English) to clarify these issues and to probe for evidence of the presence of beats. We shall present the method and the results separately for the two corpora.

1. Study 1: French Inter-Subject Experiment

1.1. Method

Task and Segmentation. Nine student and faculty members (2 F, 7 M) of the University of Lausanne recorded the same reading-aloud task of a paragraph of 218 words (11 sentences) in quiet surroundings. The recordings were performed at 96 kHz mono, were normalized against the peak volume in the file, and downsampled to 16 kHz using anti-aliasing software (Pristine Sounds 2000 Pro). Sentences varied considerably in duration and in underlying phrase and syllable structure (Appendix 1). The same text was also read by our French speech synthesis system ("subject 10") and its output was subjected to the same types of analysis as were the human sentences.

As a reference object comparable to Port's and colleagues' repetition cycle we elected the phrase, defined as an acoustically identified, prosodically unified group of temporally and grammatically and semantically coherent linguistic materials (see Appendix 2). All temporal positions for beats and segment transitions were calculated in terms of proportions of the containing phrase. In the material in this experiment, it was appropriate to consider that an on- or offset of a phrase had occurred if any of these three conditions was encountered: (a) punctuation, such as a period, interrogation mark, comma or suspension mark in the input text, (b) the coincidence of a pause of 50-150 ms and an intonational reset, or (c) any pause in excess of 150 ms. When unvoiced plosives initiated the phrase, the phrase was considered to start with the portion of the sound considered audible to both speaker or listener, which was generally the onset of the burst (and not the silent period preceding the burst). Likewise, the end of phrases with slowly waning amplitude was judged to have occurred at a point where the signal-to-noise ratio no longer permitted the audible distinction of the speech signal.

Recordings were manually segmented by experienced segmenters for both segmental and phrase duration using the Praat software[2], and were spot-checked by two other segmenters. After training [9], segmenters agreed on segmentation placement within an average of about 5-10 ms.

Beat detection. An adapted version of the method used by Port and colleagues for identifying beats was applied. It consists of the following steps:

Low-frequency information in speech signal: (a) Filter low-frequency information: Obtain the sound file and bandpass it with cutoff frequencies of 800 and 200 Hz[3]. This filtering step can be performed very satisfactorily with Praat's pass Hann band filter. (b) Obtain intensity: Take the intensity curve of the filtered signal (with Praat: 100 ms limit).

Smoothing and peak location: (a) Take a spline of the intensity curve. The spline's tension is initially adjusted in such a fashion that peaks can be measured on the basis of *continuous* vowel onset intensity rises (no double-peaks measured from double-hump rises). In our version of the spline algorithm, the tension setting was 0.001 for French (0.0001 for emphatic English speech [second experiment]). Once set for a given speech

---

[2] *Praat: Doing Phonetics by Computer*, phonetics analysis software, www.praat.org.

[3] We implemented the limits of 200-800 Hz reported on by Port in 2003 [6], rather than the 300-2000 Hz limits used by Cummings and Port in 1998 [4].

condition, the spline tension was not changed any more (1 setting for French, 1 for English). (b) Locate peaks in the derivative of the splined intensity curve (in ms) and identify potential segment transitions.

Data cleaning: Only beats situated within 25 ms of a segment transitions were scored. The 25 ms limit was arrived at by examining the error fall-off of beats situated close to segment transitions. Also several beats often "competed" for a segment transition. If two or more beats were identified for the same segment transition, the stronger beat was taken.

Strong and weak beats: Beats vary considerably, and it was argued that the "stronger" beats (those with greater derivative amplitude, i.e., those with more rapid and more definite onsets of low-frequency information) might have more temporal structuring effect than the "weaker" beats. For this reason, beats were initially divided in such a manner that about half the beats lay above average strength ("strong beats") and half below average strength ("weak beats"). Because of the non-linear distribution of beat strengths, average strength lies considerably below the halfway point of the distance between minimum and maximum amplitude.

The Hypothesis: A time structuring hypothesis was formulated on the basis of Port's hypothesis for beats. If a number of speakers read aloud the same paragraph, we expected under a generalized hypothesis for beats that variation in timing should be lower at strongly beat-marked transitions than at other places in the speech chain.

The Test: To test this hypothesis, beat-marked segment transitions situated between >0.0 and <1.0 of the containing phrase were compared for the 9 subjects. If two subjects used the *same phrase* in their reading, and if they used the same *beat-marked segment clusters* (the same *triphones*), the segment transition positions were included in the comparison. It is to be noted that points 0.0 and 1.0 were excluded from the test. Many phrases begin with some form of vocalic onset, and a coincidence on point 0.0 is therefore not meaningful for the test. Also, all phrases end in point 1.0, and this point is therefore excluded as well. Only points between 0.0 and 1.0 for which different speakers could indeed show different locations were compared in the test.

1.2. Results

Some initial observations. As Port and colleagues indicate, beat peaks cluster somewhat to the left of transitions towards what are generally voiced segments. Most peaks (about ¾ of sound transitions) marked vowel or semi-vowel onsets. The remaining ¼ generally marked the onsets of voiced consonants (/n, m, l, r, d/). Beats tend to be anticipatory. Also, subjects show more coherent behaviour with respect to stronger than to weaker beats (figures 1 and 2). In addition, peak clusters in some subjects show an incline to the left (more anticipation for the stronger peaks). Some relationship with fluidity of speech was observed in the 13 speakers seen in the two corpora: the more fluid a speaker appeared to the ear, the more incline was shown. On the basis of these initial observations, it was considered important to systematically distinguish results obtained from stronger and weaker beats in this study. Furthermore, inter-subject variation for beat-marked sound transitions was examined for all inter-subject comparisons (Figure 3). Absolute inter-subject differences between beat-marked phrase points averaged between about 2% and 4% of phrase, but notable variation was observed.

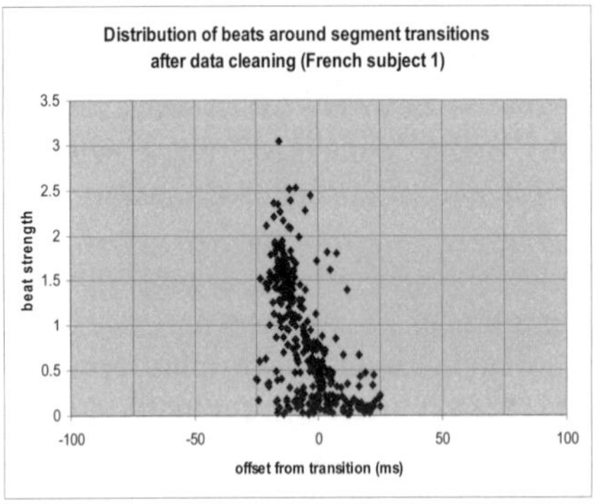

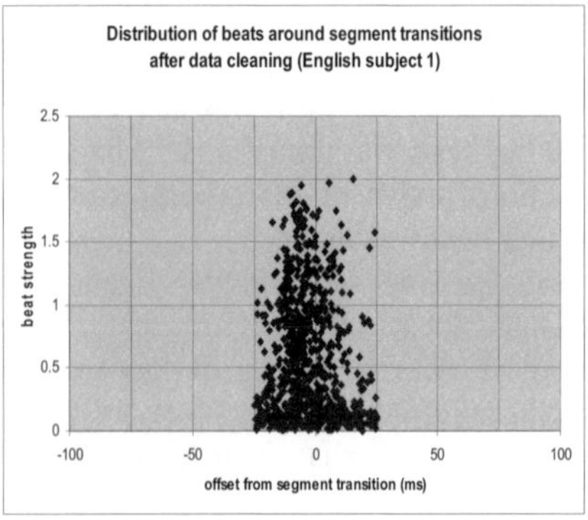

Figures 1 and 2. Beats measured for individual speakers in the two corpora. The zero-point represents the transition point between two speech segments (typically the onset of a vowel or a voiced consonant).

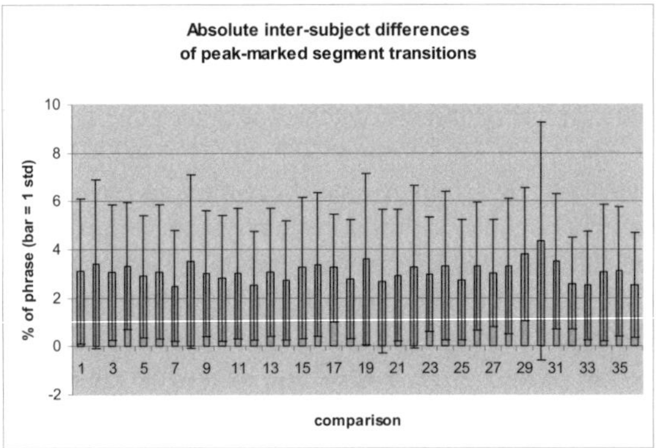

Figure 3. 36 absolute inter-subject differences obtained by comparing transitions for all 9 voices [9*8/2=36 pairs]) at positions > 0% and <100% in phrase as marked by beat peaks. In these positions within the phrase, average agreement between subjects ranged between about 2% and 4%, but the one-standard deviations were generally well above 1%, suggesting considerable variation within the data.

Onset distribution. To examine the relationship between beats and temporal position, the distribution of beats within the phrase was examined[4]. Beats were divided into approximately equal numbers of strong and weak amplitude exemplars ("strong" and "weak" peaks). In this analysis, only vowel-onset information was examined (no voiced consonant onsets). Combined data from the nine human speakers were placed into the same positional bins that had been created for the first figure by Cummings and Port, 1998 (as reported by Port in 2003) (Figures 4 and 5). The number of tokens was similar between the two studies (Our study: $N_{strong} = 703$, $N_{weak} = 671$, $N_{total} = 1374$, Cummings & Port $N = 484$).

Minor peaking was found at 50% of the phrase for both strong and weak vowel onsets (Figure 5). However, other peaks of similar sizes were found in other parts of the phrase which were not evident harmonic positions. We thus conclude that for our

[4] These results were first reported in [10].

French-language readers, "harmonic positional attraction" was barely a factor in the determination of sites where beats were recorded in continuous (read) speech.

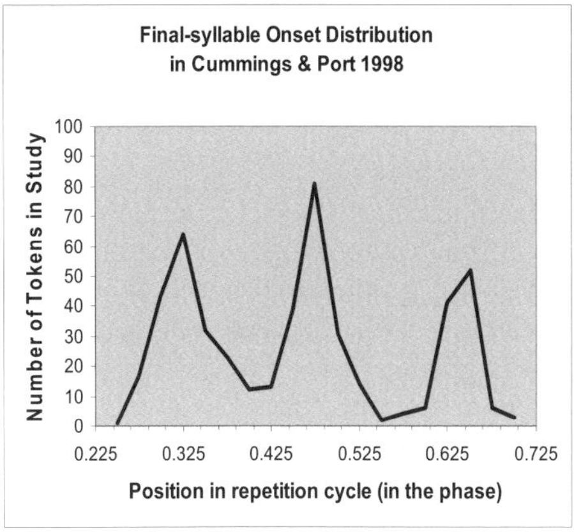

Figure 4. Data recreated from Figure 1 in Port (2003) [6]. Final-syllable onset distribution in terms of a phase angle in the repetition cycle created by repeating a 4-syllable phrase like "Dig for a duck" in time with two metronome tones. The first tone marked the onset of the phrase and the second tone varied randomly from 0.20 to 0.80 of the phrase's duration.

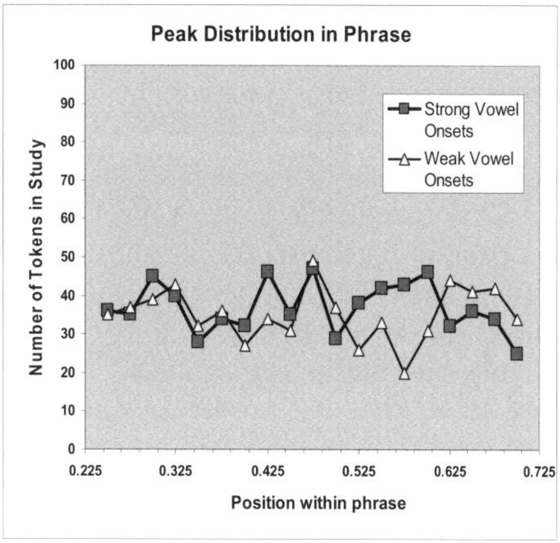

Figure 5. Data from our experiment 1. Although there is a minor peak close to the 50% mark in the phase for both strong and weak vowel onsets, the hypothesis that beats have a strong "attracting force" for temporal structures in continuous speech (reading aloud) did not receive strong support.

Degree of variation according to strength of beat. As indicated above (Figures 1 and 2), some differences were observed with respect to the temporal position of strong and weak beats. In particular, strong beats showed a more coherent temporal position than weak beats with respect to the onset of the succeeding sound segment. If beats exert an organizational "attractor effect" within the phrase, it was possible that it could be more reliably demonstrated for stronger than for weaker beats.

In this analysis, similarity in the 36 inter-subject comparisons was examined for different degrees of beat strength. In Figure 6, the upper line shows the effects of beats below average beat strength ("weak beats), and the lower line shows those for beats above average beat strength ("strong beats"). Average beat strength was found to be 3.266. The degree of cut-off was varied along the x-axis in tenths of the average strength. At the leftmost position, data were included from all beats. In this position, only minor differences in inter-subject agreement on temporal position in the phrase were noted. But as we excluded more data and compared stronger and stronger with weaker and weaker beats, substantially more positional stability emerged for temporal positions marked by strong beats. Differences reached the statistical significance level of $p < .05$ at condition 7, and the greatest difference (condition 8) was different at $p = 0.0175$ (Paired t-Test, Mean of Paired Differences = 0.462, t = 2.49 w/36 df). At the extreme position at the right, it was found that average temporal differences between subjects were at 2.9% for strong beats, and at 3.4% for weak beats.

This average difference of one half a percent must be put in perspective. As seen in Figure 3, there was considerable variation in the agreements on beat-marked positions of segment transitions (average s.d. well over 1%). Even though the half percent average difference is statistically significant and is of great theoretical importance when evaluating the work by Port and colleagues, the practical value of using beats to identify high-agreement areas in speech remains circumscribed, because the effect is generally lost in the high signal-to-noise ratio established by the beat effect vs. average difference between the individuals. We conclude from this examination that although segment transitions with strong transitions between unvoiced and voiced sounds tended to be in somewhat more similar position in the phrase than those with weaker transitions, the effect was minor and regrettably not of notable practical value for applied purposes like speech recognition or speech synthesis.

This conclusion is also supported by an examination of the relationship of beats and position in the speech synthesis rendition of the passage (Figure 7). Strong beats did not show the same temporal organizing effect for synthesized speech as it had in human speakers. There was no decrease in variability with increased beat (points 12, 13, 14), and the human-synthesis variability was greater (3.4 - 4.6%). This suggests once more that the "beat attractor" phenomenon described here is a real human phenomenon, even though its practical effect in continuous speech is probably minor.

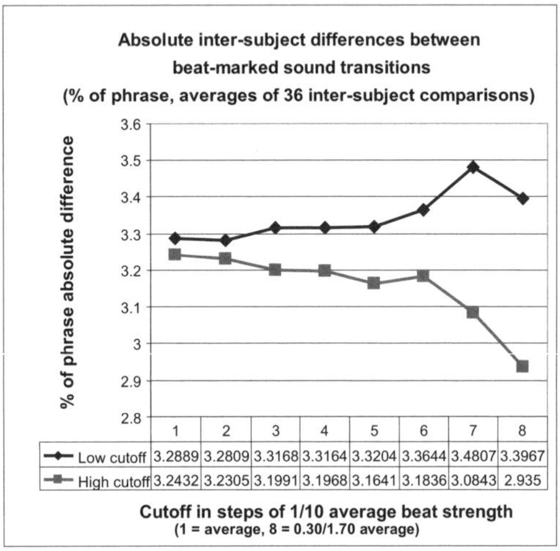

Figure 6. Comparisons between human speakers, study 1, French. On the average for these human speakers, strongly beat-marked positions show 0.5% more agreement than weakly beat-marked positions. The pattern is not evident for the speech synthesis system shown in Figure 7. In the two graphs, the data cutoffs move from average cutoff (left) to high/low cutoffs (right). Specifically, at the *left extreme*, inter-subject variation is calculated for all parts of data above and below the beat average, and at the *right extreme,* variation is calculated only for very high and very low beats, and the intermediate beat data is eliminated. At the extreme right position, average positional differences between subjects were 2.9% for strong beats, and at 3.4% for weak beats.

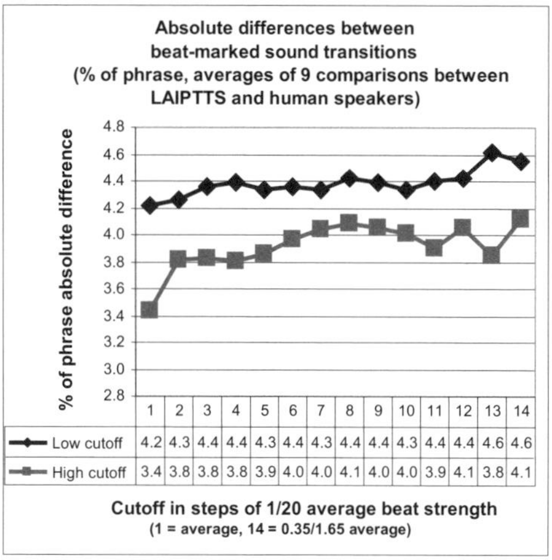

Figure 7. Comparisons between human speakers and the speech synthesis system, study 1, French. In contrast to human speakers (Figure 6), the output of the speech synthesis system did not show any reduction in variance at stronger beat locations.

1.3. Conclusion for Study 1

In this experiment with French speech read aloud, we were able to demonstrate that strong transitions from unvoiced to voiced segments imposed some weak temporal organization effects. At points of unvoiced stop-vowel transitions for example, subjects tended to agree more on the exact placement for initiating the post-stop vowel than in other positions in the phrase. Although theoretically interesting, plausible, and demonstrated to be statistically significant in a fine-combed analysis, the practical effects were not excessive, since the reduction in subject-to-subject variation was not very great. It is interesting to note that the increase in temporal similarity with greater beat strength was not found in the speech-synthesized data.

This first study left some questions open, notably (a) whether the temporal organization effect would be greater if we examined material from another language, particularly a language with strong stress patterns, and (b) whether we would obtain compatible effects in a second and theoretically independent verification of the coordinative gestures hypothesis. We thus decided to examine another database which permits these verifications, a database for *English* where in addition to *inter*-subject verifications we also had *intra*-subject comparisons. This study will be presented next.

2. Study 2: English Intra- and Inter-Subject Experiment

2.1. Method

Corpus and segmentation. A speech database was examined for English with a well-known speaker, Winston Churchill, recorded in (a) a prepared war speech presented under emotionally charged circumstances to free country representatives and Dominion high commissioners in London on June 12th, 1941 (the "St-James Palace speech"), (b) a post-war BBC re-recording of the same speech performed under less straining circumstances (probably in 1947). Furthermore (c), we made a re-recording in 2003 of the same speech with another UK-English speaker (initials PW), who is currently of the same general age group as Winston Churchill was in 1941, and who speaks the same dialect variant (British Received Speech). The analyses of the historical recordings were performed with the permission of the recording firm.

The historical recordings presented some minor difficulties in the analysis, due to elevated noise levels (e.g., judgment of initial /h/), echoes in the original environments (e.g., some minor difficulties of establishing phrase duration; where does a terminal /n/ end and where does the echo begin?), and due to microphone reverberation. Also, some slight slurring of speech was noted in some of the BBC re-recordings, suggesting a few local effects of inebriation. Churchill's liberal use of alcoholic beverages is well-attested, particularly for the post-war period [12]. Manual segmentation was performed by an experienced phonetician trained in the LAIP segmentation procedure [9] to mark for segments and phrases in the three recordings. A complete segmental verification was performed by the author.

Calculation of voice onset peaks. Peaks more than 25 ms from the peak-marked transition were removed prior to automatic beat analysis. Comparable phrases were identified by automatic comparisons of phonetic transcriptions in the three re-recordings, and beats were identified for each segment-to-segment transition by the procedure indicated in Study 1. If there were several candidate beats relating to the same transition, the stronger beat was chosen. The speech contained 980 comparable tokens of beat-marked segment transitions.

2.2. Results

Initial observations. As in study 1, peaks clustered around segment transitions, but no incline was observed in the data.

Inter-subject and intra-subject variation. For both analyses, similar patterns of positional precision were identified as in Study 1. For both inter- and intra-subject variation, agreement on position in phrase increased as stronger beat exemplars were chosen. It was found that (a) the *inter-subject* similarity between Churchill and PW was greater than the average French inter-subject similarity (1.5% average difference at point 18 for Churchill-PW vs. 2.9% for the comparable French analysis) and (b) that curiously, Churchill shows **less** *intra-subject* (3.5% difference) than *inter-subject* (1.5% difference) similarity (Figures 7 and 8). In other words, the difference in *style of speech* between the two Winston Churchill recordings introduced greater positional differences in the phrase than the differences between the *two speakers*. Similar overall variability in the data was shown as in Study 1.

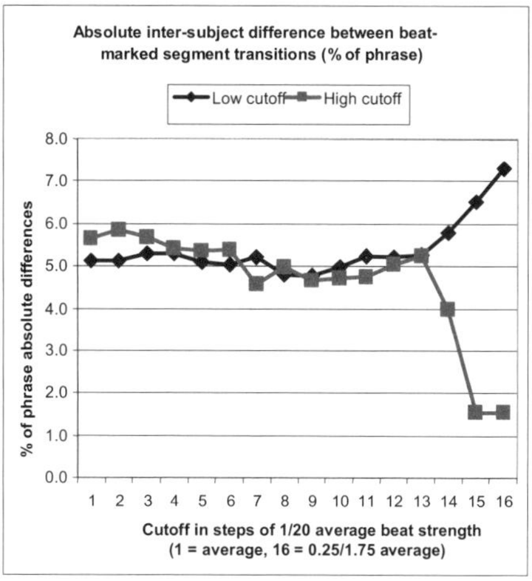

Figure 8. Study 2, English, ***inter***-subject differences between positions of beat-marked segment transitions in the phrase. The tendencies are similar to those found in the French study, but are more pronounced. As in Figure 6, the data cutoffs move from average (left) to high/low (right).

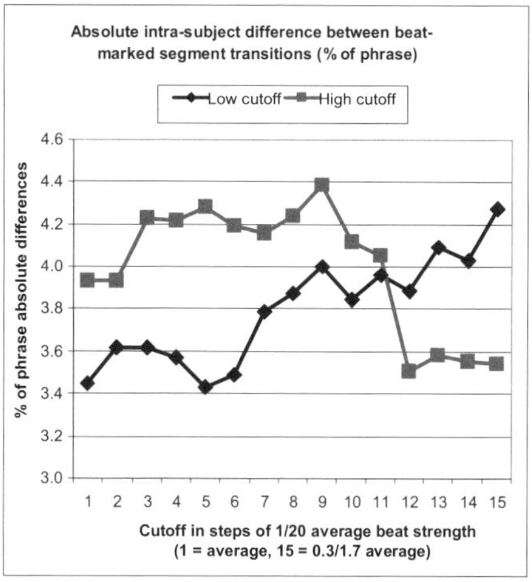

Figure 9. Study 2, English, ***intra***-subject differences. Again, stronger beats found with high cutoffs are related to greater positional agreements within the phrase.

Relationships with non-phonetic parameters. A variety of further examinations were undertaken to examine possible relationships with non-phonetic parameters (data reported in [11]). For example, earlier speech productions in the phrase were expected in the psychologically stressed speech condition, while delayed productions in conditions under less psychological stress. The relative timing of peak-marked transitions was thus compared between the original (stressed) and the re-reading (less stressed) of the same speech. However, no significant pattern was observed. Also, no relationship was found between specific phonetic transitions and anterior and posterior beat-marked position. Finally, beats were found to relate to nothing else than the type of phonetic transition between unvoiced and voiced status; for example, they showed no relationship with grammatical class.

3. Conclusion

Support was found in corpora from two languages with different accentual patterns for the temporal organizing effects of beats, in that strong beats were correlated with reduced temporal inter- and intra-subject variation in the phrase. The variation effects noted for French, where accents are difficult to define and to measure, were also found in the English corpus, where there is widespread agreement on the placement of accentual ("stress") patterns. Furthermore, this effect was absent in synthesized speech, which suggests in turn that the absence of beat patterns may to some degree be responsible for the "robotic" impressions that speech synthesis often evokes in human listeners.

A number of questions remain open. First of all, it is possible that the elusive "rhythmic effects" frequently perceived in continuous language may indeed be related to the beat effects that were found by Port and colleagues and that were again demonstrated in this study. However, the experimental demonstration that this is indeed so has yet to be made. For example, a listener experiment could be imagined in which perceived rhythmicity is related to degree of inter- / intra-subject variation and the presence of beat phenomena.

Also, this study raises some interesting new questions concerning the relationship between beat and accent. Although accentual patterns are quite different for French and English, the similar patterns of beat dependency shown here raise the question of the exact relationship between beats and accent marking in these and other languages. A question of further interest is the issue of how beat effects may be related to known systematic timing effects such as utterance-final lengthening, word-initial lengthening, polysyllabic shortening, etc.

Finally, the differences between natural and synthetic speech could be more systematically investigated. In speech synthesis, durations are obtained with a multidimensional average calculation of elements determing segment and syllable duration, word-central reduction, phrase-final lengthening, etc [2]. Since increased temporal organizing effects were documented here for beat positions, individual variation could be reduced in beat positions (i.e., one would apply a strict multidimensional average calculation there), while in non-beat positions, individual variation would be given more latitude (i.e., average computations for those places could be varied more freely). One would have to investigate with detailed data sets whether that approach improves the temporal calculations for speech synthesis applications.

4. Appendix 1: The text for Study 1

« Pour fêter ses vingt-cinq ans, le Festival d'automne invitait, en 1996, tous ceux qui, au cours des années, façonnèrent sa légende, dans la musique, le théâtre, la danse et les arts, à l'initiative de Michel Guy. L'édition 1997 s'inscrit dans une autre logique, en forme de questions : où trouver la nouveauté ? Où puiser, à l'approche de cette fin du siècle, de quoi nourrir la réflexion et l'imaginaire ? Comment rendre compte des mouvements qui traversent le monde improbable d'aujourd'hui ? En allant là où le soleil se lève : au Japon. Sous la direction d'Alain Crombecque, le Festival d'automne consacre une part majeure de son programme à ce pays à double face, où la splendeur du kabuki, du nô et du bunraku, polie par des siècles de tradition, côtoie les quêtes formelles menées au théâtre. Autre invitée de marque : l'Egypte, la folie de ses nuits, ses chants millénaires, les transes soufies et les ballades amoureuses du delta du Nil. Bien sûr, il y a des retrouvailles dans le programme du Festival. Il y a aussi des "curiosités" - indispensables rendez-vous insolites proposés cette année par Jérôme Nicolin ou Christian Boltanski. Il y a enfin, qui donne son sens à l'ensemble, le désir de trouver en l'art une "voûte de lumière" sous laquelle poser son regard. »

5. Appendix 2: Definition of a phrase

Phrases are difficult to define, and yet the precision, validity and reproducibility of the current experiments depend on the use of precise definitions. Here are the guidelines used in the current set of experiments.

Definition. A phrase is an acoustically identified, prosodically unified group of temporally and grammatically and semantically coherent linguistic materials.

Explanation 1. "Acoustically identified, prosodically unified, temporally coherent": the phrase is defined as an uninterrupted acoustic waveform incorporating a unified and uninterrupted temporal and intonational structure.

Explanation 2. "Grammatically and semantically coherent": a single and coherent grammatical and semantic structure is associated with the acoustic structure. Speech interruptions ("uh", etc.) and speech corrections are considered to break up the phrase, resulting in two or more "phrases" for this experiment. Also, acoustic material without evident grammatical and/or semantic structure was eliminated from the experiment.

Questions on the definition of phrase duration. (a) Do ingressive respiration, or clearing the throat prior to the start of phrase form part of the phrase? Decision: no, this is considered part of the planning phase for the phrase. (b) Does the pre-burst part of unvoiced plosives form part of the phrase? Decision: no, we do not generally know how long this pre-burst lasts. Only audible events are part of a commonly defined phrase for both listener and speaker. The phrase is thus considered to start with the burst. (c) Do pauses end a phrase? We distinguished three types of pauses: (i) Short and not interrupting intonational flow; this type of pause does not end the phrase. (ii) Longer and interrupting intonational flow; this type of pause ends the phrase. (iv) Longer pauses. This type of pause ends the phrase. What are the cutoffs for these pauses? That depends on the speaker, but we used 50 ms for (i/ii) and 150 ms for (ii/iii) in the French project.

6. Acknowledgements

We wish to thank Dr. G. Peter Winnington (UNIL, English Department) for re-recording the Churchill passages, Prof. Jürg Schwyter (UNIL, English Department) for help with the Churchill project, and Dr. Brigitte Zellner Keller (UNIL, Psychology) for help with the design, the segmentation and the initial analysis of the French corpus. The transcriptions for the French and Churchill corpora were performed by Lukas Wiget, and were verified by Brigitte Zellner Keller for the French corpus and by the author for the English corpus. We also wish to thank the students and staff at the University of Lausanne who participated in the recordings for the French Corpus. The Winston Churchill recordings were used for the present scientific purposes with the written permission of ARGO, the producers of the Winston Churchill CDs. Work reported here was in part financed by the Swiss Staatssekretariat für Bildung und Forschung SBF under its programme in support of COST projects (COST 277).

7. References

[1] Lehiste, I. (1977). Isochrony reconsidered. *Journal of Phonetics, 5,* 253-263.

[2] Keller, E. & Zellner Keller, B. (2003). How much prosody can you learn from twenty utterances? Linguistik online. 17, 5/03, 57-78. http://www.linguistik-online.de/17_03/kellerZellner.html, ISSN 1615-3014.

[3] White, L. (2002). *English Speech Timing: A Domain and Locus Approach.* Ph.D. Thesis, The University of Edinburgh.

[4] Cummings, F., & Port, R. (1998). Rhythmic constraints on speech timing. *Journal of Phonetics, 26,* 145-171. http://www.asel.udel.edu/icslp/cdrom/vol4/437/a437.pdf

[5] Tajima, K., & Port, R. (2003). Speech rhythm in English and Japanese. In J. Local, R. Ogden, & R. Temple (Eds.), *Phonetic interpretation: Papers in laboratory phonology VI* (pp. 317-334). Cambridge, UK: Cambridge University Press.

[6] Port, R.F. (2003). Meter and speech. *Journal of Phonetics, 31,* 599-611.

[7] Tuller, B., & Kelso, S. (1990). Phase transitions in speech production and their perceptual consequences. In M. Jeannerod (Ed.), *Attention and performance VIII* (pp.429–451). London: Academic Press.

[8] Schmidt, R., Carello, C., & Turvey, M.T. (1990). Phase transitions and critical fluctuations in the visual coordination of rhythmic movements between people. *Journal of Experimental Psychology: Human Perception and Performance, 16,* 227-247.

[9] Schwab, S., Keller, E., Zellner, B., Connan, P.-Y., & Siebenhaar, B. (1998). *Conventions de segmentation pour la construction de diphones.* Rapport interne, laboratoire LAIP, Université de Lausanne.

[10] Keller, E. (2005). A phonetician's view of signal generation for speech synthesis. *Studientexte zur Sprachkommunikation, 36.*

[11] Keller, E., *Individual speech rhythm variation within the plosive structure of speech*, Part 1: The beat structure for speech timing, Part 2: Verification of Hypotheses. Nato Advanced Study Institute Advanced Study Institute (Asi) - Summer School " E. R. Caianiello" on The Fundamentals of Verbal and Non-verbal Communication and the Biometrical Issue, September 2 -12, 2006 - Vietri sul Mare, Italy. Available at http://homepage.sunrise.ch/mysunrise/ekeller00/Kellerdoc.html.

[12] Owen, D. (1996), Diseased, demented, depressed: Serious illness in heads of State, *Quarterly Journal of Medicine, 96.* 325-336. http://qjmed.oxfordjournals.org/cgi/content/full/96/5/325.

Fundamentals of Verbal and Nonverbal Communication and the Biometric Issue
A. Esposito et al. (Eds.)
IOS Press, 2007

Age as a Disguise in a Voice Identification Task

Ruth Huntley BAHR[1]
University of South Florida, Tampa, FL USA

Abstract. Listeners are quite reliable in their judgments of speaker age; however little is known about the use of vocal age as a disguise. Actors provided voice samples of varying ages. Listeners were asked to estimate speaker age and to estimate the age relationship of voice pairs within and across speakers. Results indicated that actors were able to produce voices of different ages; however none of the vocal age simulations were judged to be as old as the elder control group. Auditors exhibited considerable difficulty estimating the age relationships in voice pairs from the same talker. These findings will be discussed in terms of aging stereotypes and cues to speaker identity.

Keywords. Vocal age, aging stereotypes, voice identification

Introduction

Voice production is not a static phenomenon. It is highly variable, both within and across speakers. Obvious vocal differences are evident among people that can be attributed to gender, age and dialect, to name a few. Within speaker variations, such as differences related to situation (i.e., vocal register), fatigue or illness also influence voice production, but sometimes are overlooked in the forensic speaker identification process (1, 2). Of particular interest for this project are the identification of vocal age, activation of stereotypes associated with age, and the production of vocal disguise. Previous research has clearly demonstrated that auditors are capable of making direct age estimations from voice with considerable accuracy (3, 4; among others). It would appear that age is one factor that listeners can use to identify speakers. But is it possible for a speaker to modify his/her voice in such a way as to sound older and deceive the listener?

1. Vocal Aging and Aging Stereotypes

The characteristics of vocal aging have been well-described (see 5 for a complete review). Much attention has been directed towards the role of fundamental frequency (f0) in the aging process. The voice is known to change in systematic ways with age

[1]Corresponding Author: Ruth Huntley Bahr, 4202 E. Fowler Ave. PCD 1017, University of South Florida, Tampa, FL 33620; E-mail: rbahr@cas.ufs.edu.

and gender (6, 7, 8, 9; among others). The f0 of males lowers until middle age and then begins to climb slightly in old age, while the female f0 remains fairly constant until middle age and then gradually declines. As a result, there can be considerable overlap in f0 across gender in old age (i.e., the male-female coalescence model) (10).

Other researchers have described a slower rate, lower intensity, a potential for increased hoarseness, and less precise articulation as other acoustic factors associated with the aging process (11, 12, 13, 14). It would appear that these characteristics also are used by listeners to identify speaker age, but auditors can be deceived by these cues. For instance, Ramig and Ringel (15) have demonstrated that older speakers in good physiological condition are judged to sound younger than older speakers in poor physiological condition. In addition, it is possible that older speakers adopt patterns of speech that they believe are expected of a person of their age. Ramig, Titze, and Scherer (16) demonstrated that the chronological age of the speaker was judged to be younger during isolated vowel productions than in running speech. They explained that the isolated vowel condition revealed physiological age, while running speech was more susceptible to societal influences or what they called a sociolect of age. In other words, these speakers conformed to the expected old age stereotype in running speech.

So, just what is this stereotype? Research has suggested that older talkers are judged to be less competent, frail, and in need of special accommodations (like a slower rate and louder vocal presentation on the part of their communication partner) in order to facilitate comprehension (17, 18, 19). Their speech production is presumed to be slower, more hoarse, tremorous, and softer (20, 21, 22). These studies suggest that the prevailing stereotype is basically negative. However, more knowledge is needed as to how listeners and speakers respond to vocal age and the potential influence of aging stereotypes.

Even though age is presumed to be an important cue, few investigators have considered its potential as a voice disguise (23, 24). These investigators have indicated that an age disguise can interfere with a listener's judgment of speaker identity. The problem with these studies is that the effectiveness of the age disguise was not considered. In other words, when someone produces a voice that is supposed to be representative of another age, is that voice really identified as being older or just disguised? Therefore, it is the purpose of this investigation to assess the ability of individuals to produce a voice of a different age and the influence of this type of disguise on age identification. Two experiments were developed to analyze these abilities. The following questions were asked:

1) Is an individual capable of producing a voice that simulates a person of a different age?
2) Are listeners able to establish vocal age relationships even when the talker is using an age disguise?

2. Method

2.1. Participants

Four young actors (2 males and 2 females) volunteered to participate in this project. Their mean age was 22.5 years and each participant had experience portraying older individuals on stage. They were currently working on a degree in Fine Arts, but had also been compensated for their acting skills. To serve as exemplars of vocal age, the voices of four healthy elders were selected from an existing database. They were matched to the younger talkers in terms of gender. The group's mean age was 65.25 years.

Twenty-nine untrained listeners participated in the listening experiment. They all passed a brief hearing screening [25], which assessed their ability to hear and understand word productions by asking them to select a target word from a group of minimal pairs. All auditors passed this test with 90% or greater accuracy. This pool of listeners was then quasi-randomly assigned to one of two listening tasks.

2.2. Voice Stimuli

All talkers were asked to read the *Rainbow Passage* [26] in their normal speaking voice. The younger adults then were asked to provide their rendition of the same paragraph read as a middle-aged and as an older person. These recordings were made in a sound-treated booth with laboratory quality equipment. The talkers were provided an opportunity to practice reading the passage with their age simulated voices prior to recording. These voice recordings were utilized in two listening tasks:

1) A direct age estimation task, where the listener was asked to judge the age of the person talking. They were instructed to estimate the age that they thought the person was portraying.
2) A paired comparison task, where the auditor had to judge if the first voice in the voice pair was younger, older, or the same age as the second voice in the pair.

3. Experiment 1

3.1. Experimental Tape for Direct Age Estimation Task

Two text-independent voice samples were selected from the reading passages of the younger and older speakers, as well as from the age-disguised voices of the younger speakers (N = 32 sentences). Approximately half of the sentences were repeated for reliability (mean rating consistency = 85%). These sentences were then quasi-randomized to form a 48-item listening task, which included both real and simulated age voice productions. Four practice samples, without experimenter feedback, were provided to assist the listener in developing a response set for this task. The entire listening session lasted about 35 minutes.

3.2. Procedures

Auditors were seated in a semi-circle in quiet room so that each person was equi-distant from the speaker system. After completing the brief hearing screening and practice trials, they were asked to make direct age estimations of the voices presented to them. Specifically, they were asked to provide their best estimate of the age that the speaker portrayed.

3.3. Results

Data from the first experiment were analyzed to determine if the talker was able to alter the age of their voice. A multivariate analysis of variance (MANOVA) revealed a significant main effect for age, $F(3,78) = 118.907$, $p < 0.001$. Post hoc testing with t-tests and a Bonferroni correction indicated that five out of six possible pairwise comparisons were significant ($p < 0.001$). The only voice pairs that did not significantly differ in age from one another were the simulated middle-age and simulated older voices. As Figure 1 indicates, these young individuals were able to make their voices sound older; however, none of the talkers were able to produce a voice that sounded as old as an actual older person.

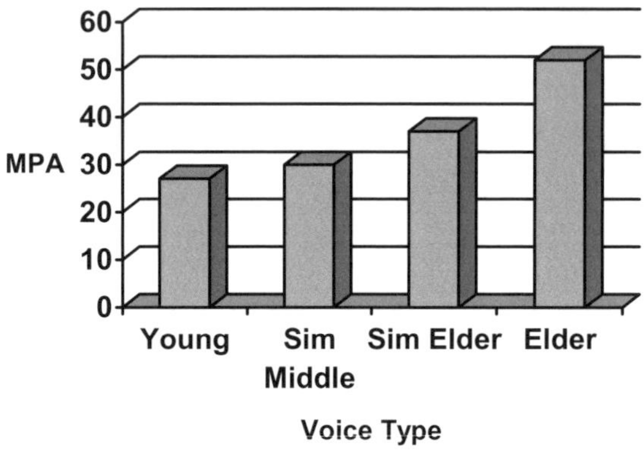

Figure 1. Mean Perceived Ages (MPAs) **by Voice Type.**

These data also revealed that the mean perceived age (MPA) was somewhat higher than the chronological age for young voices and lower than the chronological age for the older talkers. This is not surprising since this type of relationship at the ends of the age continuum is consistent with previous research (3, 4; among others). However, the results for the elder voices support the idea that these individuals were in good health and would be expected to sound younger than their less fit peers (15). Therefore, it might be difficult to separate the role of listener expectancies in age estimation from the physiological effects of aging.

These data also suggest that the age disguises are perceived as older than the actual age of the talker, but not as old as an elder's voice. An acoustic analysis was attempted to determine why the MPAs of the simulated voice productions did not approach the actual vocal ages of the older talkers. In other words, was the talker altering the features of the aging voice in the same directions as previous research would indicate or was the talker responding in a stereotypical manner?

All voice samples presented in the listening test were analyzed for speaking fundamental frequency (SFF), average intensity, duration and speaking rate, and resonance (which considered the vowel space between formant 1 and 2). The preliminary results indicated that the young speakers relied on temporal slowing to achieve the older sounding voice. Changes in SFF were more subtle, but appeared to be in the right direction. These findings are consistent with the research literature (11, 12, 27, 28). Hence, the young talkers were able to rely on their knowledge of what an older voice sounds like to mimic an older voice. Their attempts resulted in a voice that sounded significantly older than their real voice, but not what one would consider to be an old voice (i.e, mean MPA for the middle-age voice was 30 years and for the older voice, mean MPA was 37 years). The talker's reliance on rate cues to convey age may have indicated to the listeners that the voice portrayal was actually a disguise. In other words, the auditor knew the voice sounded older, but the acoustic information provided in the voice sample did not adequately trigger their vocal stereotype of an older individual. More research is needed to begin to tease out the effects of acoustics, physiology, and stereotypes on MPA.

4. Experiment 2

4.1. Procedures

In the second phase of the experiment, the listeners were asked to determine age relationships within voice pairs and to identify age relationships in the presence of an age disguise. The participants, stimuli and procedures followed those outlined for Experiment 1; however, this time a paired comparison task was utilized. The sentences used in Experiment 1 were randomized into all pairwise comparisons. Half of the pairs were replicated to assess listener reliability (mean rating consistency = 83%). The result was a 107-item test, which lasted about 35 minutes. Listeners were asked to determine if the first voice in the pair sounded older, younger, or the same age as the second voice.

4.2. Results

In determining the listener's ability to recognize age relationships, the voice samples were divided into three separate tasks. The first task assessed the listener's ability to recognize voice pairs of the same age as being produced by a particular speaker and if listeners rated the elder's voices as sounding older than the young voices. In other words, could the listeners judge vocal age with any consistency?

As indicated in Figure 2, the listeners were quite accurate in their identification of age relationships within a talker when the ages were the same and across the two age groups (young vs. old). These data demonstrate that the listeners could perform the task. Across speaker comparisons were considered next. To do this, the rest of the listener responses were divided into two categories: across-age pairs involving the different persons and across-age pairs involving voices taken from the same individual. As Figure 2 would suggest, listeners were able to estimate the age relationships across speakers. In fact, auditors were rarely confused when the voice pairs involved different people.

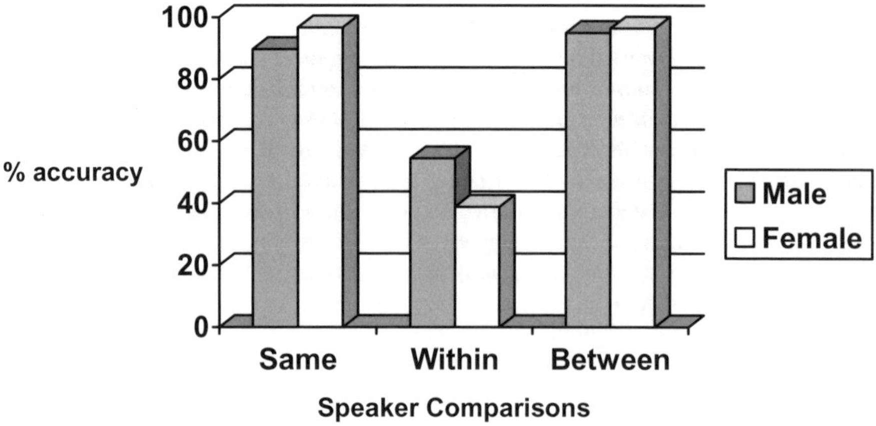

Figure 2. Listener Judgments of Vocal Age Within and Across Speakers.

On the other hand, listeners did have a difficult time estimating age relationships in the presence of disguise (i.e., simulated middle age--SM or simulated old age—SO). When responses to the specific voice pairs involving age simulations were considered, listeners performed poorly on all pairings (see Figure 3), with the effect being more pronounced with the female voice. Hence, altering vocal age might be effective in disguising a person's voice. At least these auditors found estimating the age of a simulated voice production to be quite difficult. In fact, the presumably older voices were considered to be the same age or younger than the other voice in the pair.

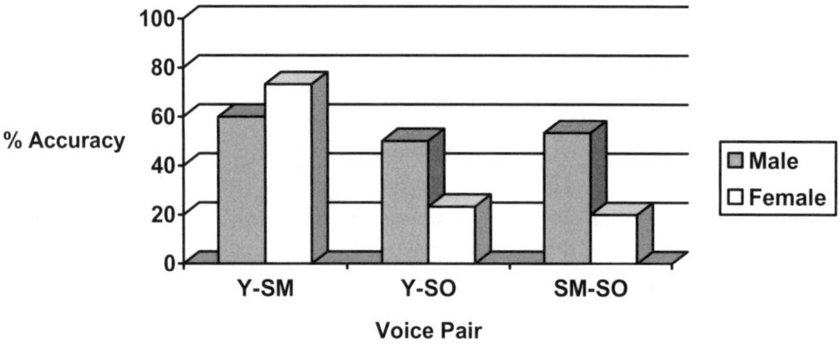

Figure 3. Mean Listener Response Accuracy to Within Speaker Comparisons of Vocal Age.

To further illustrate listener ability to identify age relationships, listener responses were analyzed using the d' measure of signal detection theory (see Macmillan & Creel (29), for a thorough introduction) to determine their level of skill in identifying the age relationships in voice pairs using "real" voices versus voice pairs involving a simulated voice production (i.e., "real" to simulated voice and simulated voice to simulated voice age judgments). In this case, the d' measure assigns a numeric value to the listener's ability to identify the same voice in different stimuli, taking into account the listener's response bias. Listener bias is taken into account by considering both the listener's hit rate and false alarm rate. The hit rate was the percentage of responses where the participant judged the age relationship portrayed in the voice pair correctly. The false alarm rate was the percent of responses where the volunteer responded "same" when the two voices represented different age categories. To compute d', each listener's hit and false alarm rates were determined. These percentages were then converted to standard normal distribution z-scores. The d' measure is the difference between Z (Hit rate) and Z (False Alarm rate). An individual that showed no difference in performance based on whether the age relationships were judged correctly would have a d' of zero. Individuals that are more likely to correctly judge the age relationship would have a positive d'. For example, a participant with a 90% hit rate and a 33 % false alarm rate would have d' = Z(0.90) – Z(0.33) = 1.29- (-0.43) = 1.72.

To analyze the effect of speaker comparisons, differences in mean d' between conditions were analyzed using a MANOVA. This analysis revealed a significant effect for speaker comparisons, $F(2,28) = 44.13$, $p < 0.001$, $\eta^2_p = .759$. Post hoc testing with the Tukey procedure indicated that performance in the condition that compared "real" voices was far superior to the conditions that involved a simulated age production (see Figure 4). While listeners are able to identify a speaker despite attempts to alter vocal age (i.e., d' values greater than 1.0), they are far superior at judging age when the voice samples are not disguised (d' = 8.66).

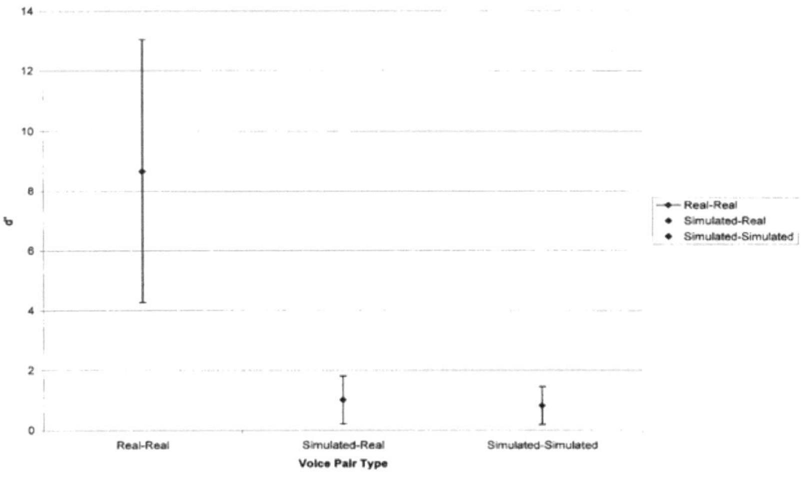

Figure 4. Illustration of the d' values for the three types of voice pairs.

5. Discussion

The results of these experiments have demonstrated that listeners are responsive to acoustic cues that are associated with vocal aging. Auditors were able to distinguish between voices of differing ages. The findings for the young and old voices were consistent with previous research—the age of the younger voices was overestimated (MPA = 27 years and mean chronological age = 22 years) and the age of the older voices was underestimated (MPA = 52 years and mean chronological age = 65 years). The age discrepancy for the young adults was within expected limits (3, 4) and the larger age difference noted for the older adults could be attributed to the fact that these older individuals were in good health (15). The more interesting finding was that pertaining to the simulated age productions. These voices were judged to be significantly older than the younger voices and significantly younger than the older voices. Interestingly, the simulated voice productions for the middle-age and older voices were not judged to be different from one another. Hence, the young actors were able to alter their voice sufficiently enough to sound older, but they were not effective enough in their simulations to actually sound old.

But what does this say about the presence of an aging stereotype? It would appear that the young talkers relied on selected parameters of that stereotype in their voice productions. The acoustic analyses suggested that the talkers relied heavily on rate reduction and less so on changes in f0 to achieve the target vocal age. These changes are consistent with the literature (11, 12, 13, 14); however, manipulation of these factors appears to only approximate vocal aging. There are other aspects of the vocal age stereotype that appeared to be ignored, like vocal hoarseness, tremor, and lower intensity to name a few (5, 20, 21, 22). It would appear that the young adults either were not able to mimic these vocal qualities or chose not to include them. Perhaps this result is related to the physiological aspects of aging. In other words, there are unique contributions to voice production that are attributable to physiological age and are difficult to replicate in younger speakers.

It is difficult to determine if the listeners were using an aging stereotype to make their judgments of vocal age. They did respond appropriately to the noted changes in speech rate across the samples—they judged the simulated productions to be older than the talker, but younger than the older comparison voices. However, the simulated voice productions were not judged to be significantly different from one other, suggesting that the talkers were not able to produce voices that clearly portrayed differences in age. There are two possible reasons for this finding. First, the talkers appeared to be fairly unidimensional in their use of speech rate as indicative of vocal aging. In this case, the manipulation of rate was effective for altering vocal age, but not in producing voices of different ages. It is possible that it is the interaction of several acoustic cues that results in the perception of true vocal age. This hypothesis needs further testing.

The second factor that appeared to influence the perception of vocal age was the physiological age of the talker. It would appear that the alteration of acoustic parameters performed by the young talkers was not enough to simulate a true elder's voice. These findings suggest that the acoustic parameters associated with the aging voice interact with aging vocal physiology in ways not yet understood. In the case of this project, it would appear that physiological age played a significant role in the judgments of vocal age, supporting the notion that physiological age is a critical factor in influencing social age [30].

Finally, the results of this project do support previous research. Lass and his colleagues [31] had college-age individuals without acting experience produce voices that sounded much younger and much older than their "normal" voice. Like the current experiment, their listeners judged the age relationships appropriately, but the actual differences in age among the groups were relatively small (i.e., between 3-8 years). In the current experiment the range was between 3-10 years, so the use of trained actors did not appear to result more effective presentations of vocal age.

5.1. Applications to the Forensic Situation

While the first experiment was designed to assess the role of stereotypes in determining vocal age, the second experiment was directed toward a more practical application of vocal stereotypes. Previous research has suggested that people can adjust their vocal register as they speak and that these alterations can influence the perception of speaker identity (1, 2). These studies have considered non-intentional disguises or changes in the voice that occur naturally. However, what would happen if someone chose to disguise their voice using vocal age stereotypes? Instead of altering the voice to just sound different, the talker could choose to make changes that sound more natural, like those associated with aging. Would these vocal alterations deter the listener from identifying the speaker?

The first analysis determined whether the listeners were able to make age judgments when the voice pair representing a particular vocal age was taken from the same talker. In other words, how accurate were the listeners in making "same" judgments? The results indicated that auditors were quite adept at this task, with no real difference attributed to speaker gender.

Once the listeners' ability to perform the task was established, the abilities of listeners to identify age relationships across age categories and within and across speakers were considered. The results here would suggest that the listener can be confused by age disguises, with the effect being somewhat more evident with female voices (though statistically nonsignificant). It was interesting to note that the listeners accurately estimated the age relationships when the age comparison involved different talkers, but when the age judgments involved samples from the same speaker, the listener's ability to identify speaker age was adversely affected. The old age disguise yielded the most incorrect responses for the females, while the age simulations were similarly difficult for the males (see Figure 3). In all cases, the percent accuracy was below 75% and listener accuracy was the worst (< 55%) when the simulated older voice was involved. It is obvious that vocal disguises involving old age can significantly impact a listener's perception of speaker age, especially when voice samples from the same person are used.

To further analyze this finding, these results were subjected to a signal detection analysis. While listeners were able to judge vocal age when comparing real and simulated productions and across simulated voice productions, this ability was significantly poorer than their ability to identify speaker age in non-simulated conditions. This finding can be interpreted as indicating that voice simulations (or disguise) can interfere with listener perceptions. So, while the speaker did not appear to successfully alter their voice to be reflective of a truly older person, he/she was able to change their voice enough to interfere with age estimation and potentially speaker identification.

It should be noted that the present results also support the idea that listeners were aware of discrepancies in the voice productions that did not match with their aging stereotypes. When listeners were asked to describe their reactions to the age estimation tasks, they reported that they knew the voice was altered, making age estimation difficult. This uncertainty associated with estimating vocal age when disguise is suspected could negatively impact listener skill in a voice identification task.

6. Conclusions

In summary, talkers are able to produce voices that can be perceived as being different from their chronological age. However, these vocal disguises are not perceived as being representative of actual older voices. So, the effectiveness of an age disguise is called into question. What does this mean in the forensic setting? While a listener may be deceived by an age disguise, their perceived age responses will probably not cause the talker to jump over broad age categories (i.e., young, middle age, older). If they are young, their voice will still sound relatively young. It is important to keep in mind here that some individuals may not be good at an age estimation task, so caution is suggested [32].

A second, more significant, finding is the effectiveness of the age-disguised voices in resulting in false eliminations. It is interesting to note that listeners did not have trouble judging age relationships across speakers, but these same auditors had significant difficulty estimating age when presented with another sample from the same person. This is a significant concern in voice identification. The following conclusions can be drawn:

1) Individuals appear to be able to utilize vocal age as a disguise. In fact, some people may be more effective in its use than others.
2) The listeners were able to tell different speakers apart with relative ease.
3) Auditors had a difficult time determining the age relationships of the voice pairs produced by the same individual; hence age disguises could interfere with speaker identification.

References

[1] Bahr, R. H., & Frisch, S. (2002). "The Problem of Codeswitching in Voice Identification," In Braun, A. & Masthoff, H. R. (Eds.) *Phonetics and its Applications: Festschrift for Jens-Peter Koester on the Occasion of his 60th Birthday*. Stuttgart: Steiner, p. 86-96.

[2] Bahr, R. H., & Pass, K. J. (1996). The Influence of Style-Shifting on Voice Identification, *Forensic Linguistics*. 3:24-38.

[3] Shipp, T., & Hollien, H. (1969). Perception of the aging male voice. *Journal of Speech and Hearing Research, 12*, 704-710.

[4] Huntley, R., Hollien, H., & Shipp, T. (1987). Influences of listener characteristics on perceived age estimations. *Journal of Voice, 1*, 49-52.

[5] Linville, S. E. (2001). *Vocal aging*. San Diego, CA: Singular.

[6] Hollien, H., & Shipp, T. (1972). Speaking fundamental frequency and chronologic age in males. *Journal of Speech and Hearing Research, 15*, 155-159.

[7] McGlone, R., & Hollien, H. (1963). Vocal pitch characteristics of aged women. *Journal of Speech and Hearing Research, 6,* 164-170.

[8] Russell, A., Penny, L., & Pemberton, C. (1995). Speaking fundamental frequency changes over time in women: A longitudinal study. *Journal of Speech and Hearing Research, 38*, 101-109.

[9] Stoicheff, M. (1981). Speaking fundamental frequency characteristics of nonsmoking female adults. *Journal of Speech and Hearing Research, 24*, 437-441.

[10] Hollien, H. (1991). The older voice. *Proceedings of the 12th International Congress of Phonetic Sciences, 5*, 90-93.

[11] Amerman, J., & Parnell, M. (1992). Speech timing strategies in elderly adults. *Journal of Phonetics*, *20*, 65-76.
[12] Hartmann, D., & Danhauer, J. (1976). Perceptual features of speech for males in four perceived age decades. *Journal of the Acoustical Society of America*, *59*, 713-715.
[13] Linville S. E., & Rens, J. (2001). Vocal tract resonance analysis of aging voice using long-term average spectra. *Journal of Voice*, *15*, 323-330.
[14] Shipp, T., Qi, Y., Hollien, H., & Huntley, R. (1992). Acoustic and temporal correlates of perceived age. *Journal of Voice*, *6*, 211-216.
[15] Ramig, L., & Ringel, R. (1983). Effects of physiological aging on selected acoustic characteristics of voice. *Journal of Speech and Hearing Research*, *26*, 22-30.
[16] Ramig, L. Scherer, R., & Titze, I. (1985). The aging voice. In V. Lawrence (ed.) *Transcripts of the Fourteenth Symposium on Care of the Professional Voice*. New York: The Voice Foundation.
[17] Coupland, J., Coupland, N., Giles, H., & Henwood, K. (1988). Accommodating the elderly: Invoking and extending a theory. *Language in Society*, *17*, 1-41.
[18] Feezel, J. & Hawkins, R. (1988). Myths and stereotypes: Communication breakdowns. In: C. Carmichael, C. Botan, & R. Hawkins (Eds.) *Human communication and the aging process*. Prospect Heights, IL: Waveland Press.
[19] Montepare, J. M., & Zebrowitz, L. A. (1998). Person perception comes of age: The salience and significance of age in social judgments. In: Zanna, M. P. (Ed.). *Advances in Experimental Social Psychology*, *Vol. 30*. San Diego, CA: Academic Press.
[20] Ryan, E. B., & Cole, R. L. (1990). Evaluative perceptions of interpersonal communication with elders. In: H. Giles, N. Coupland, & J. M. Wiemann (Eds.). Communication health and the elderly. Manchester, England: Manchester University Press.
[21] Ryan, E. B., & Laurie, S. (1990). Evaluations of older and younger speakers: Effects of varying speech rates. *Journal of Language and Social Psychology*, *1*, 514-519.
[22] Stewart, M. A., & Ryan, E. B. (1982). Attitudes toward younger and older speakers: Effects of varying speech rates. *Journal of Language and Social Psychology*, *1*, 91-109.
[23] Endres, W., Bambach, W., & Flosser, G. (1971). Voice spectrograms as a function of age, voice disguise, and voice imitation. *Journal of the Acoustical Society of America*, *49*, 1842-1848.
[24] Reich, A. R., & Duke, J. E. (1979). Effects of selected vocal disguises upon speaker identification by listening. *Journal of the Acoustical Society of America*, *66*, 1023-1028.
[25] Griffiths, J. (1967). Rhyming minimal contrasts: A simplified diagnostic articulation test. *Journal of the Acoustical Society of America*, *42*, 236-241.
[26] Fairbanks, G. (1960). *Voice and articulation drillbook*. New York: Harper & Row.
[27] Harnsberger, J., Shrivastav, R., Brown, W. S., Rothman, H., & Hollien, H. (in press). Speaking rate and fundamental frequency as speech cues to perceived age. *Journal of Voice*.
[28] Manning, R., Fucci, D., & Dean, R., (2002). College-age males' ability to produce the acoustic properties of an aging voice. *Perceptual and Motor Skills*, *94*, 767-771.
[29] Macmillan, N. A., & Creel, C. D. (1991). *Detection theory: A user's guide*. Cambridge: Cambridge University Press.
[30] Mulac, A., & Giles, H. (1996). "You are only as old as you sound": Perceived vocal age and social meanings. *Health Communication*, *8*, 199-215.
[31] Lass, N., Justice, L., George, B., Baldwin, L., Scherbick, K., & Wright, D. (1982). Effect of vocal disguise on estimations of speakers' ages. *Perceptual and Motor Skills*, *54*, 1311-1315.
[32] Bahr, R. H., & Pass, K. J. (1995). Task influences on earwitness reliability. In: Braun, A. & Koester, J-P., *Studies in Forensic Phonetics*. Trier: WVT Wissenschaftlicher Verlag Trier, p. 121-131.

Acknowledgements

Gratitude is expressed to Kimberley Pass who assisted in data collection and analysis for this project. Portions of this paper were presented at the 1994 American Association of Forensic Sciences meeting.

A cross-language study of acoustic and prosodic characteristics of vocalic hesitations

Ioana VASILESCU [a] and Martine ADDA-DECKER [a]
[a] *Spoken Language Processing Group, LIMSI-CNRS, BP 133*
91403 Orsay Cedex, FRANCE

Abstract. This contribution provides a cross-language study on the acoustic and prosodic characteristics of vocalic hesitations.One aim of the presented work is to use large corpora to investigate whether some language universals can be found. A complementary point of view is to determine if vocalic hesitations can be considered as bearing language-specific information. An additional point of interest concerns the link between vocalic hesitations and the vowels in the phonemic inventory of each language. Finally, the gained insights are of interest to research in acoustic modeling in for automatic speech, speaker and language recognition.

Hesitations have been automatically extracted from large corpora of journalistic broadcast speech and parliamentary debates in three languages (French, American English and European Spanish). Duration, fundamental frequency and formant values were measured and compared. Results confirm that vocalic hesitations share (potentially universal) properties across languages, characterized by longer durations and lower fundamental frequency than are observed for intra-lexical vowels in the three languages investigated here. The results on vocalic timbre show that while the measures on hesitations are close to existing vowels of the language, they do not necessarily coincide with them. The measured average timbre of vocalic hesitations in French is slightly more open than its closest neighbor (/œ/). For American English, the average F1 and F2 formant values position the vocalic hesitation as a mid-open vowel somewhere between /ʌ/ and /æ/. The Spanish vocalic hesitation almost completely overlaps with the mid-closed front vowel /e/.

Keywords. Vocalic hesitation, Filler words, Automatic language identification, Formant extraction, Speech alignment.

Introduction

Substantial progress in speech processing research over the past decade has led to a variety of successful demonstrators involving spontaneous, or at least unprepared speech. To increase the usability of speech systems the challenges of multilinguality and spontaneous speech must be addressed efficiently. In this context there has been growing intrest in studies on hesitation phenomena. Hesitation phenomena have been initially viewed as "speech disfluencies", globally noisy and irregular events, in opposition with "fluent" or "well formed" speech [13]. However, recent corpus-based studies have shown that hesitation phenomena exhibit regularities, carry multiple functions in speech and contribute to the elaboration of the verbal message [6,14].

Hesitations can result in different vocalic realizations, varying from simple events such as silent and filled pauses, and word lengthening, to more complex phenomena involving repetitions and speech repairs. In this study we focus on the very common phenomenon of filled pauses, which correspond to autonomous vocalic hesitations. Such vocalic hesitations occur without lexical support and thus are to be distinguished from vocal lengthening of segments belonging to lexical items (generally function words). According to Clark and Fox Tree [6] the main role of vocalic hesitations is *to announce the initiation of a delay in speaking*. As for other types of hesitations such as silent pauses, repairs, restarts or repetitions, the vocalic hesitation may mark a speaker's effort to build a verbal message and can carry different additional functions in speech interactions: keeping or passing the floor, indicating the affective state of the speaker, etc. Vocalic hesitations are widely encountered in the world's languages and consist of the insertion of a relatively long, stable vocalic segment, considered as a type of 'filler'. The orthographic transcription of vocalic hesitations varies across languages suggesting differences in their perception by native speakers, for example *uh/um* in American English, *er* in British English, *euh* in French, *eh* in Spanish. Fillers can have other possible realizations, as for instance lengthened nasal consonants (*mm* in Mandarin Chinese [18]) or demonstratives (*ano*, *eto* in Japanese [17]). The density in speech of vocalic hesitations varies with the speaking style: in semi-prepared speech corpora (broadcast news-type) they are relatively infrequent, rarely exceeding 1% of the speech data whereas in corpora collected in more variable conditions (i.e. speakers under stress) they can easily reach more than 5% of the data [16].

If the role of vocalic hesitations in speech is now commonly accepted, comparative studies on their language-specific characteristics are still lacking. While most of the recent studies conducted on large speech corpora have focused on English or French [2,7,8,14,15], descriptions can now be found for other languages [17,18]. The acoustic and prosodic parameters generally studied are vocalic quality, pitch and duration. Pitch and duration appear to exhibit reliable patterns across the analyzed languages, i.e. flat and stable pitch and long duration compared to intra-lexical vowels [4,7,14]. Observations on the vocalic quality are less consistent: according to [14] the timbre of vocalic hesitations varies predictably with the vowel inventory of the language. For instance, in American English the hesitation vowel has a neutral (close to schwa) timbre and appears to be almost homophonic with the determiner *a*. The neutral quality of the hesitation vowel has been correlated with a possible "rest" position and articulatory settings of a language. The universal or language-specific properties of the articulatory settings are still unclear, even though recent experimental manipulations with X-ray data seem to support the second option [11]. If the language-specific hypothesis of the articulatory settings is confirmed, the vocalic hesitations might have articulatory targets of their own which differ across languages.

In some of our earlier studies comparing hesitations in 8 languages (American English, Middle Oriental Arabic, Mandarin Chinese, French, German, Italian, South-American Spanish and European Portuguese [5]) we have shown that vocalic hesitations can be observed in all these languages and that they can be characterized by duration, pitch and timbre. The present study aims to contribute both to speech processing research as well as to linguistics. From a linguistics perspective, this work addresses the question of whether some language universals can be found for hesitation vowels across languages. Seen from a complementary perspective, this study also investigates whether hesitation vowels bear language-specific information. An additional point of interest is then the link between hesitation vowels and intra-lexical vowels for each language. Does the timbre of the vocalic hesitations correspond to a vowel of the phonemic inventory of the languages or does the vocalic hesitation tend to a given, potentially universal schwa-like "rest" position? The gained insights may also be of interest to research in speech technology fields such as acoustic modeling for automatic speech, speaker, language and accent recognition.

1. Corpus and methodology

Some of the most readily available sources of transcribed speech corpora in different languages are journalistic shows and parliamentary debates. These data are available at LIMSI in the framework of different projects. For French, 20 hours of speech from several national radio and TV broadcast channels (*France Inter, France Info, France2...*) have been selected. The American English data is comprised of about 10h of recordings from broadcast channels (CNN, VOA, ABC, etc.) distributed by LDC. For Spanish, 10h of European Parliamentary debates are used. For all three languages, several dozens of different speakers contribute the majority of the audio data. Male speakers are roughly twice as frequent as female speakers in these types of corpora [11]. Although the Spanish corpus is of a different type than the French and English ones, the global acoustic properties investigated throughout this paper should not be significantly affected by this change.

In journalistic shows and parliamentary debates, the articulation is usually quite careful, so that speech can be understood by a broad audience. It is comprised mostly of prepared speech, so the speaking style cannot be qualified as fully spontaneous. There are substantially fewer hesitations here than as are observed in more informal, spontaneous speaking style corpora. However, the observed hesitations can be considered as relatively standard and prototypical, which is an interesting property for first cross-language comparisons.

The LIMSI speech transcription system [9] has been used for speech alignment. Using orthographic transcriptions and pronunciation dictionaries, the alignment system locates phones, hesitations, silences, breath and other noisy segments. For the class of extra-lexical events, such as silence, breath and filler words, the alignment system permitted an insertion at any word boundary. Since all aligned filler words are potential vocalic hesitations, before considering them as autonomous hesitation vowels, they were manually verified in order to avoid selection errors.

Parameter analysis (pitch, duration, timbre with first, second and third formants F1, F2, F3) was conducted using Praat [3]. For pitch and formant values, measures have been carried out on a frame by frame basis every 5ms. For each segment, a voicing ratio was computed as the ratio between the number of voiced frames and the total number of frames. Only segments with a voicing ratio > 0.4 are used in the analysis. For each eligible segment, mean values have been computed for all parameters using different strategies: (1) over all voiced frames of the segment; (2) over the three central voiced frames; (3) over three voiced frames in the first half (respectively the last half) of the segment. If not specified otherwise, the described measures correspond to the mean of the segment center (strategy 2). Differences between segment initial and final measures allow

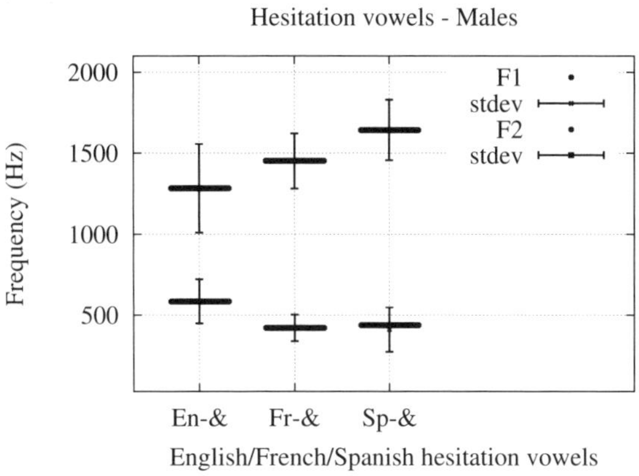

Figure 1. Average F1/F2 measures (and standard deviations) of hesitation vowels ("&") in English, French and Spanish (male speakers).

the slope of pitch and formants of hesitation and intra-lexical vowels to be estimated. In particular they have been used to measure the slope of F0 in hesitations.

2. Timbre

In this section we first investigate the timbre of hesitation vowels across languages in terms of F1 and F2 measures. Next, for each language, the hesitation vowel is examined with respect to its intra-lexical vowel system. Vocalic hesitation selection using automatic alignment resulted in approximatively 1300 items in French (respectively 2000 in English and 1700 in Spanish). Results here are presented for male speakers since they are better represented in the data, however the observations remain valid for female speakers.

2.1. Timbre of hesitation vowels

As already mentioned in the introduction, the transcription of vocalic hesitations varies across languages, for example *uh/um* in American English, *euh* in French or *eh* in Spanish. In order to measure acoustic differences between vocalic hesitations, formant values have been extracted using PRAAT. These differences can be seen in Figure 1, which presents the mean F1/F2 values obtained for American English, French and Spanish, suggesting distinct timbres for each language. The standard deviations of vocalic hesitations are very similar to those measured on intra-lexical vowels. Vowel timbre moves from mid-open in English to mid-closed in Spanish with a /œ/-like realization in French. The measured F1/F2 values are consistent with the expected formant values given the orthographic transcription of hesitation vowels in each language.

2.2. Link between hesitation vowels and vocalic systems

The formant measures given here show important differences between hesitations across languages. In this section we aim to clarify how the hesitation vowels are positioned in

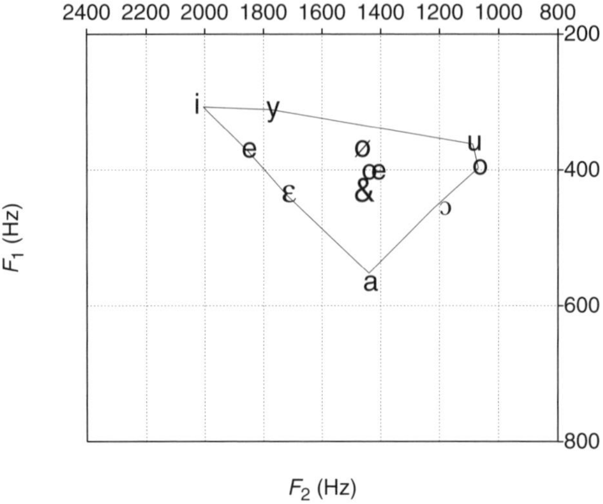

Figure 2. French F1/F2 mean values for intra-lexical vowels and hesitation vowel ("&"). The hesitation corresponds to a central mid-open vowel close to the French /œ/.

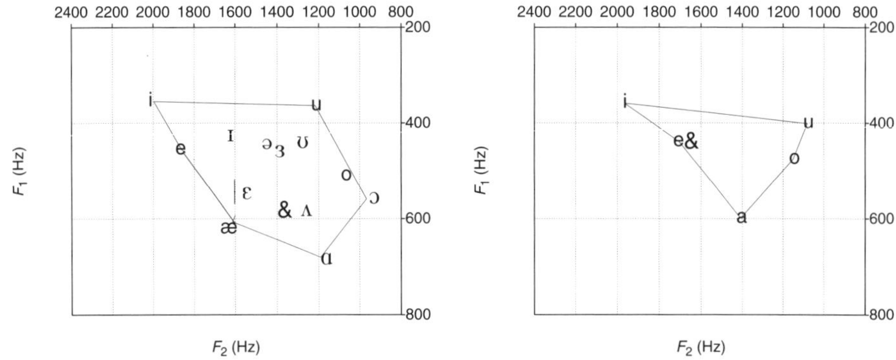

Figure 3. English and Spanish (from left to right) F1/F2 mean values for intra-lexical vowels and the hesitation vowels ("&"). For American English the mean position of the hesitation vowel corresponds to an open vowel, close to /ʌ/, whereas the Spanish hesitation is closest to the mid-closed front vowel /e/.

their vocalic system. To do so, formant values were extracted for all intra-lexical vowels in the three languages, and average formant values computed for each vowel.

Figure 2 shows the vocalic system of French using the average intra-lexical vowel formant values in a F1/F2 space. The vocalic hesitation is represented with "&" in the F1/F2 space along with the other French oral vowels. The resulting representation allows the distance between hesitation vowels and the phonemic vowels to be visualized. Similarly Figure 3 shows hesitation vowels in the vocalic systems for English and Spanish.

The comparison between vocalic hesitations and the vocalic systems reveals interesting differences across the 3 languages [12]. For European Spanish and to a lesser extent for French, vocalic hesitations exhibit timbres coinciding with another vowel in the system. In French the timbre is very close to a central /œ/ vowel, but slightly more

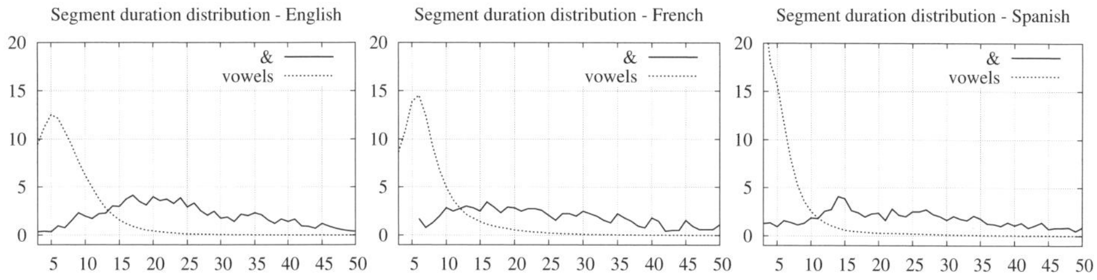

Figure 4. English, French and Spanish (from left to right) segment duration distributions for intra-lexical (vowels) and hesitation (&) vowels. X-axis : segment duration in centiseconds; Y-axis : percentage of segments in population.

open. Differences in the degree of opening might be correlated with durational aspects: as shown in [10], acoustic vowel spaces are more centralized, as the duration of the segments decreases. In contrast, for American English, the average timbre of the vocalic hesitation does not directly correspond to a vowel of the language inventory, as measured in our corpora (Figure 3). The average timbre of the hesitation vowel is closest to /ʌ/ and thus to a central position, thus it is more open than the vocalic hesitation in French.

As can be observed in Figures 2 and 3 the relation between the vocalic hesitation and language inventory varies across languages. Among the languages analyzed here, French and European Spanish vocalic hesitations exhibit timbres which are very close to existing vowels in the vocalic system. The average timbre of vocalic hesitations in American English is rather close to a central realization, however it does not coincide with an existing vowel. The average timbres of vocalic hesitations, close to central, mid-open realizations in French and American English and closer to a front mid-closed vowel in European Spanish, suggest that vocalic hesitations do probably not correspond to a universal speech "rest" position but that they are language-dependent. These results bring new acoustic evidence to the question of language-dependent articulatory settings supporting the hypothesis made by Gick et al. [11]. This also suggests that vocalic hesitations might be considered as a speech item carrying language-dependent information for automatic language identification. Consequently it may be interesting to use language-specific acoustic models which take into account timbre variability across languages.

3. Duration

Vocalic hesitations are widely described as being perceptually and objectively longer than intra-lexical vocalic segments [5,7,15]. Phonemic alignment of large speech corpora in different languages allows the comparison of segment durations of intra-lexical vowels and vocalic hesitations.

Figure 4 shows duration distributions of vocalic hesitations and intra-lexical vowels. Whereas the vowels have a modal value around 60 ms (even shorter for Spanish) vocalic hesitations exhibit rather flat distributions with a large proportion in the range of 150 to 250 ms. Intra-lexical segments are rarely lasting for more than 150ms.

In a recent study we have shown that duration can furthermore be correlated with speaking style: under stress vocalic hesitations tend to be even longer than in more con-

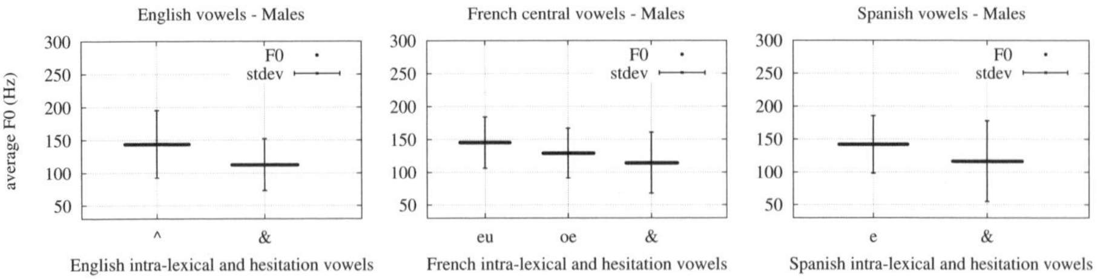

Figure 5. English, French and Spanish (from left to right) average F0 measures and standard deviations for intra-lexical vowels and hesitation vowel (&). For each language the selected intra-lexical vowels are closest neighbors to the hesitation vowel: for French /ø/ and /œ/; for Spanish /e/; for English /æ/ and /ʌ/.

trolled conditions such as semi-prepared journalistic speech [16]. These factors have been neglected here.

4. Fundamental frequency (pitch)

As for duration, fundamental frequency exhibits common patterns across languages, i.e. relatively stable, flat, slightly descending contours with low average values. Average F0 measures are shown in Figure 5 for the hesitation vowel and the closest intra-lexical vowels in each language. In order to give an idea, average pitch values for French male speakers were computed using 27k segments of the central mid-open vowel /œ/ [1], 3k segments for the central mid-closed vowel /ø/ and 1k for the hesitation vowel. Whereas significantly less data were used for female speakers the observed tendencies are globally the same. The average fundamental frequency of hesitation vowels is consistently lower than for intra-lexical vowels.

As for duration pitch can also vary with speaking style, for example speech produced under stress and language proficiency, i.e. the expression in native vs. foreign language [16].

5. Conclusion & Perspectives

This paper has studied the acoustic and prosodic characteristics of vocalic hesitations in a cross-language perspective. We aimed at establishing whether some language universals as for instance the concept of speech "rest" position can be related to the hesitation phenomena. From another point of view, the relation between the vocalic hesitations and the languages' vowel inventories was explored. The gained insights may also interest the acoustic modelling research community for application to automatic speech and language recognition.

Vocalic hesitations in French, American English and European Spanish were automatically extracted from large journalistic broadcast and parliamentary debate corpora. Duration, fundamental frequency and formant values were measured and compared both from an intra- and inter-language perspective. The results on timbre quality show that

[1] The average F0 measure for vowel /œ/ is computed using segments from both realized schwas (by far the most frequent ones) and [œ] segments.

vocalic hesitations are realized differently across languages. This suggests that, firstly, hesitations do not necessarily result from neutral realizations close to a rest position of speech, and secondly, this position, if salient, is language-dependent. Furthermore, the vocalic hesitations exhibit similarities with some intra-lexical vowels but the degree of similarity varies across languages. In European Spanish the vocalic hesitation almost coincides with the mid-closed front-vowel /e/ whereas in American English and French, it is central slightly more open than /œ/ (in French) and mid-open central between /ʌ/ and /æ/ (in American English). Finally, in the framework of language identification approaches, the results suggest that vocalic hesitations require language-specific acoustic models.

Fundamental frequency and duration exhibit common patterns across languages. The average fundamental frequency of the intra-lexical segments is consistently higher than that of vocalic hesitations, whereas the duration of vocalic hesitations is usually significatively longer than intra-lexical vowels.

Many more detailed studies can be carried out in the future, including analyses depending on different hesitation functions. Extensions to other languages, regional accents and to more spontaneous speech collected in various interaction contexts, are certainly promising future research directions which can contribute to an extensive description and more in-depth understanding of hesitation phenomena in speech.

6. Acknowledgements

This work has been partially financed by the French CNRS MIDL and European EC TC-STAR and CHIL projects. The authors would like to thank Maria Candea (Paris 3 - Sorbonne Nouvelle), for her collaboration to parts of this work; and Rena Nemoto for her work comparing the hesitation vowels to the vocalic systems in French, American English and European Spanish.

References

[1] M. Adda-Decker and L. Lamel, "Pronunciation variants across system configuration, language and speaking style", Speech Communication, vol. 29, pp. 83-98, 1999.

[2] M. Adda-Decker and al., "A Disfluency study for cleaning spontaneous automatic transcripts and improving speech language models", In DiSS-2003, Papers in Theoretical Linguistics, vol. 90, pp. 67-70, 2003.

[3] P. Boersma and D. Weenink, "Praat, a system for doing phonetics by computer", Institute of Phonetic Sciences of the University of Amsterdam", pp. 132-182, 1999.

[4] M. Candea, Contribution à l'étude des pauses silencieuses et des phénomènes dits d'hésitation en français oral spontané. Phd dissertation, University of Paris 3,2000.

[5] M. Candea, I. Vasilescu and M. Adda-Decker, "Inter- and intra-language acoustic analysis of autonomous fillers", In DiSS-2005, pp. 47-51, 2005.

[6] H.H. Clark and J.E. Fox Tree, "Using uh and um in spontaneous speaking", Cognition, vol. 84, pp. 73-111, 2002.

[7] Duez, D., "Caractéristiques acoustiques et phonétiques des pauses remplies dans la conversation en français", Travaux Interdisciplinaires du Laboratoire Parole et Langage, vol. 20, pp. 31-48, 2001.

[8] D. Duez, "Modelling Aspects of Reduction and Assimilation in Spontaneous French Speech", In Proc. of IEEE-ISCA Workshop on Spontaneous Speech Processing and Recognition, Tokyo, 2003.

[9] J.L. Gauvain, L.F. Lamel, G. Adda, "The LIMSI Broadcast News Transcription System", Speech Communication, vol.37(1-2), pp. 89-108, 2002.

[10] C. Gendrot, M. Adda-Decker, "Impact of duration on F1/F2 formant values of oral vowels: an automatic analysis of large broadcast news corpora in French and German", Proc. of Eurospeech-Interspeech, Lisboa, Portugal, 2005.

[11] B. Gick and al., "Language specific articulatory settings: evidence from inter-utterance rest position", Phonetica, vol. 61(4), pp. 220-233, 2004.

[12] R. Nemoto, "Hésitation vocalique autonome vs. système phonétique de la langue : étude acoustique en plusieurs langues", Mémoire de DESS (Master dissertation), LIMSI-CNRS, 2006.

[13] H. Maclay and C.E. Osgood,"Hesitation phenomena in spontaneous English speech", Word, vol. 15, pp. 19-44, 1959.

[14] E. E. Shriberg, Preliminaries to a Theory of Speech Disfluencies. PhD thesis, University of California at Berkeley,1994.

[15] E.E. Shriberg, "The 'errrr' is human: ecology and acoustics of speech disfluencies", Journal of the International Phonetic Association, vol. 31(1), 2001.

[16] I. Vasilescu and M. Adda-Decker, "Language, gender, speaking style and language proficiency as factors characterizing autonomous vocalic fillers in spontaneous speech", Proc. of ICSLP 2006, Pittsburgh, USA, 2006.

[17] M. Watanabe,Y. Den, K. Hirose and N. Minematsu, "The effects of filled pauses on native and non-native listeners speech processing", In DiSS-2005, pp.169-172, 2005.

[18] Y. Zhao and D. Jurafsky, "A preliminary study of Mandarin filled pauses", In DiSS-2005, pp.179-182, 2005.

A. Esposito et al. (Eds.)
IOS Press, 2007

Intonation, Accent and Personal Traits

Michelina SAVINO[1]
Dept. of Psychology, University of Bari, Italy

Abstract. Intonation contributes to conveying information on speakers' regional accent personal trait, as accent identifies a person as belonging to a specific linguistic sub-community. In this paper a discussion on this topic is presented, taking Italian as an example of descriptive aspects of regional variation in intonation, and on German language as a case study on the perceptual identification of varieties by relying on intonation.

Keywords. Intonation, regional varieties , accent identification, personal traits

Introduction

Pitch conveys a great deal of information on speakers' personal traits, such as age and gender or emotional state. These can be considered as non-linguistic or paralinguistic aspects of pitch use (for discussions and more explanations on this topic, see Ruth Bahr's paper "Age as a disguise in a voice identification task", this volume). However, there are also linguistic aspects of pitch use relating to dialectal and sociolectal features. One could be, for example, the use of intonation for the identification of regional accent, henceforth 'accent'. Accent can also be considered a personal trait, since it gives information about the geolinguistic background of the interlocutor, i.e. it identifies a speaker as belonging to a specific linguistic (sub)community.

In recent years, a large amount of scientific work has been devoted to intonational variation in a relevant number of languages (see for example [1], [2]), as is also testified by the increasing number of international conferences and specialised workshops directly or indirectly focussing on this topic (also in relation to technological applications). A number of works has dealt with the problem of ascertaining the specific role of intonation in accent (or, more generally, in language) identification (see section 3 for references). Results obtained so far have not given satisfactory answers to this question and are still calling for further research, the main problem consisting in the intrinsic difficulty of separating suprasegmental and segmental information in a reliable way when presenting speech materials to listeners, as well as in determining the role of segmental features in the accent identification process, and whether, as if so how, they relate to intonational features. Another relevant parameter to be controlled is the role of listeners' linguistic background in the accent identification task, for example whether native accent competence plays a crucial role in recognising native accent or not.

[1] Corresponding Author: Dept. of Psychology, University of Bari, Italy, piazza Umberto I, 1 – 70121 Bari, Italy; e-mail: m.savino@psico.uniba.it .

On the other hand, descriptive analysis of how language varieties differ intonationally is of fundamental help in accent identification task studies. In this respect, some intonation contours can be considered as more crucial than others during the accent identification process.

The aim of this paper is to present some aspects of this topic, taking Italian as an example of descriptive aspects of regional variation in intonation (section 2), and on German language as a case study of the issue relating to the perceptual identification of varieties by intonation (section 3). These will be preceded by an introductory section on intonation analysis.

1. What is Intonation?

Giving a comprehensive overview on intonation is beyond the scope of this section, the topic being notoriously complex[2]. This brief introduction simply aims at providing few basic concepts useful for readers who are less acquainted with the subject, in order to facilitate the comprehension of the intonational issues discussed in the following sections.

Intonation can be generally defined as the "organised" use of some phonetic features (mainly fundamental frequency, intensity and duration) to convey linguistic (and paralinguistic) meanings at the utterance level. When listening to people talking, we can perceive that

a) some words are more prominent than others

b) words are grouped into chuncks or units (prosodic phrasing)

Observation in a) is due to the fact that the stressed syllables of more prominent words are realised with stronger intensity, longer duration and with some pitch obtrusion on or around them with respect to the remaining syllables.

The marking of the boundary between two prosodic phrases (b) also has some phonetic correlates, such as change in pitch, lengthening of the phrase-final syllable, or pause insertion.

Marking the relative prominence of syllables and determining prosodic edges is a basic function of intonation. At the phonological level, pitch excursions occurring on or around prominent syllables and at prosodic edges are identified as discrete tonal events. According to the autosegmental-metrical approach to intonation analysis (the most currently widespread framework, starting with the seminal work by Pierrehumbert [7], see [5] for a complete review), these tonal events can be described in terms of High (H) and Low (L) F0 targets: tones associated with prominent syllables are referred to as pitch accents, whreas those associated with prosodic edges are identified as boundary or edge tones. For describing pitch accent types the star symbol '*' following the tone is used, for example H* designates a F0 peak on the prominent syllable, L* a F0 valley, etc.. Pitch accents can also consist of two tones, generally describing pitch movements like rising or falling ones: in these cases, the symbol '+' is used to join the two tones, where the starred tone indicates the tone associated with the prominent syllable. For example, L+H* normally describes a rising movement starting somewhere before or at the prominent syllable and reaching the peak somewhere within that syllable, whereas H+L* tends to identify a falling movement on that syllable (the details of the phonetic

---

[2] for a more comprehensive introduction to the topic and more suggestions for further readings, see [3], [4], also [5] and [6].

realisation of a pitch accent depending on the language or language variety). The last pitch accent within a prosodic phrase is called a nuclear pitch accent and it is normally the most important one from the pragmatic point of view.

Different combinations of pitch accents and boundary tone types result in different intonational sequences or *tunes*[3], which can be related to different pragmatic meanings. For example, the realisations of the Bari Italian expression "Mangia una banana" [(s/he) eats a banana] as a declarative or as interrogative depend on the intonation contour, especially around the nuclear syllable. The two relating text-tune representations are the following (prominent syllables are capitalised, the nuclear syllable is additionally in boldface):

Declarative	Interrogative
MANgia una ba**NA**na	MANgia una ba**NA**na
H* **H+L*** L-L%	H* L+**H*** L-L%

In these examples, tones accompanied by the "-" and the "%" symbols refer to their associations with prosodic edges. The transcription used here is inspired by the ToBI (Tone and Break Indices) intonation labelling system[4], which accounts for two levels of phrasing, an intermediate one ("-") and an intonational one ("%"): in this case they both coincide. The F0 traces along with the text-tune alignment/associations for both Italian renditions are shown in Figure 1a (declarative) and Figure 1b (interrogative), respectively.

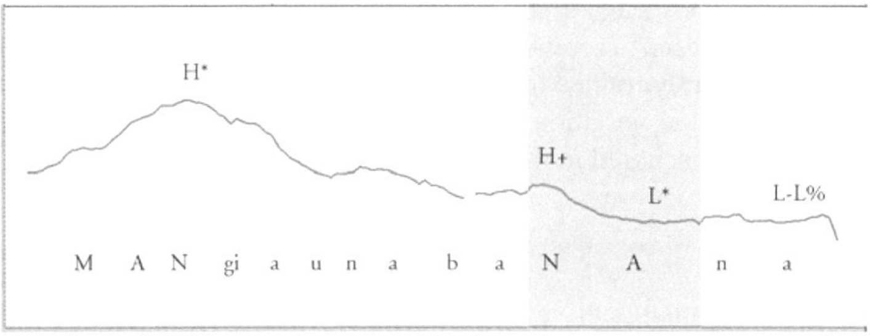

Figure 1a. F0 trace and relating text-tune association/alignment of the Bari Italian declarative utterance "MANgia una baNAna" [(s/he) eats a banana]. Prominent syllables bearing a pitch accent are capitalised, nuclear pitch accented syllable is highlighted in grey.

Configurations of H and L targets for pitch accents and boundary tones, and intonational sequences can vary from language to language, and most of the differences among varieties can also be captured by such phonological descriptions.

---

[3] In this model, F0 traces between one target and the following are dealt with some interpolation rules, as F0 minor variations are considered as not affecting the pragmatic meaning of a given utterance.

[4] ToBI is a widespread intonation labelling system within the autosegmental-metrical framework, initially developed for American English [8], and currently also developed or under development for several languages (see for example, [9], [2])

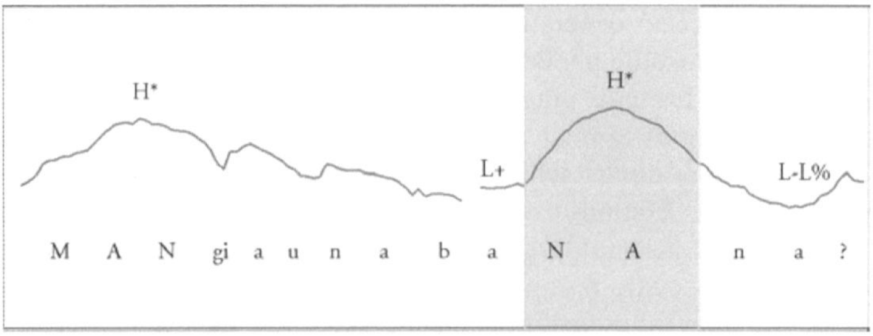

Figure 1b. F0 trace and relating text-tune association/alignment of the Bari Italian interrogative utterance “MANgia una baNAna?” [(does s/he) eat a banana?]. Prominent syllables bearing a pitch accent are capitalised, nuclear pitch accented syllable is highlighted in grey.

2. How Can Language Varieties Differ Intonationally? Some Examples in Italian

For Italian speakers, spoken language represents a reliable way for identifying the geolinguistic trait of their interlocutors in verbal interactions, as they always speak with an accent, in both formal and informal situations. This is a consequence of the particular status of Italian as compared with other languages: for historical reasons, in fact, the process of standardisation has been successfully achieved for the written form but not for the spoken language, which is presently characterised by quite strong regional accents [10], [11]. As a matter of fact, standard Italian has never been taught at any level of the national education system, and its use has been restricted to a small number of professional speakers and actors, and therefore in very specific contexts. Such contexts have been currently more restricted, since even newsreaders on television are not currently trained to use the standard pronunciation.

There is considerable variation between Italian accents, some examples of which will be presented and discussed in the following sub-sections.

2.1 Yes-no Questions

A clear example of intonational variation across Italian accents is yes-no question. In traditional literature on “standard Italian” intonation, it is claimed that the basic distinction between polar question and statement is marked by a terminal rising movement in questions vs a terminal fall in statements ([12], [13], [14], [15], [16]). Yet regional varieties present a large number of different intonational realisations of polar questions, whereas statements are generally characterised by a (low) falling final movement. Some examples of such variability among varieties are presented in the following, and the kind of questions described are more specifically information-seeking yes-no questions, i.e. questions asking information about something which is completely unknown to the locutor.

For example, Bari and Turin varieties have completely different melodic shapes: Bari questions (as mentioned in section 1) are characterised by a rising-falling movement on the nuclear syllable, staying low until the end of the utterance ([17], [18], [19], [20]), whereas typical Turin Italian questions have a falling-rising movement on the nuclear syllable followed by a slightly falling boundary tone [21]. Such differences can be phonologically described as L+H* L-L% (Bari) vs H+L* H-L% (Turin) as illustrated in the examples in Figure 2 (in this and all forthcoming Figures, nuclear syllables are highlighed in grey).

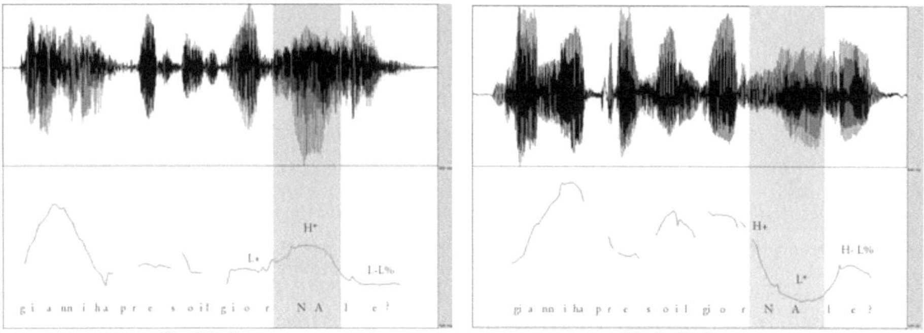

Bari Italian Turin Italian

Figure 2. Different melodic shapes in Bari and Turin Italian polar question "Gianni ha preso il giorNAle?" (*Did John buy the newspaper?*). Turin Italian original audio file taken from [21]

Another case is represented by a different tonal alignment of the same nuclear pitch accent movement, an example of which is found in Bari and Neapolitan question intonation. Both varieties are characterised by a rising-falling intonation contour, but such movement is aligned with the nuclear syllable in a different way. In fact, in Bari Italian the peak is reached in the middle of the nuclear syllable, and the falling part of the movement starts already within that syllable. In Neapolitan Italian, on the other hand, the peak is reached at the nuclear syllable offset, so that the fall is realised afterwards [22]. See Figure 3 for an example of the same question spoken by a Bari and a Neapolitan speaker. In addition to a difference in alignment there is also a different degree in steepness of the rising part of the movement in the two varieties. These nuclear accent differences can be phonologically captured: L+H* in Bari vs L*+H in Neapolitan. A similar difference can also be found between Bari and the variety spoken in Palermo [23].

Another kind of variability is that found for example between the varieties spoken in Rome and in Lecce: preliminary observations on questions have shown that both varieties are characterised by a F0 peak on the nuclear syllable followed by a terminal fall, but the peak is realised differently, namely as a medial peak in Rome and as a late peak in Lecce, as it is shown in Figure 4. In this case the difference can be phonetically described and marked by using H*L-L% for Rome vs H*[late peak] L-L% for Lecce. However, it is worth noting that intonational analysis for these regional accents are still at a very preliminary stage and that more work is needed in order to ascertain the specific characteristics of their intonational systems.

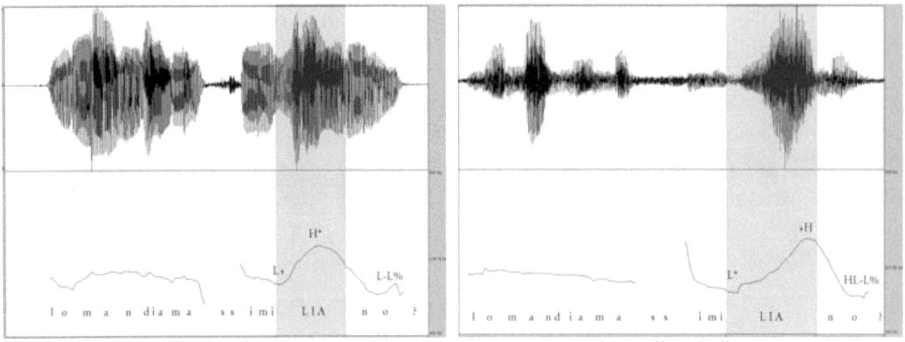

Bari Italian Neapolitan Italian

Figure 3. Different nuclear tonal alignment in Bari and Neapolitan Italian polar question "lo mandi a MassimiLIAno?" (*Will you send it to Maximilian?*) (original audio files taken from [23])

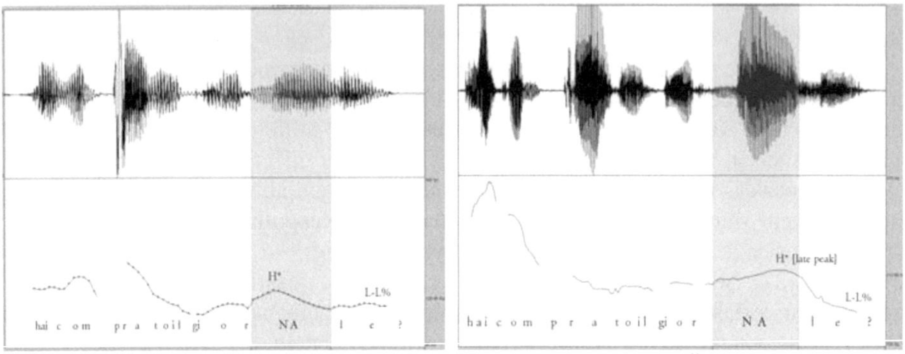

Rome Italian Lecce Italian

Figure 4. Different phonetic realisation of the nuclear peak accent in Rome and Lecce Italian polar question "hai comprato il giorNAle?" (*Did you buy the newspaper?*) Note erroneous automatic F0 values extraction on the beginning part of the Rome Italian utterance ("hai comp"), resulting in halved pitch.

From these examples (and from the research they are based on) it clearly appears that intonation does play an important role in characterising Italian local varieties, and also that, at least for this language, the yes-no question intonation contour might be considered as relevant for identifying different regional accents.

A finer distinctions between questions asking for new information (i.e. information interlocutors believe is not shared, henceforth QUERIES) and those asking for confirmation about given information (i.e. information locutors believe is shared, henceforth CHECKS) could also be useful for investigating intonational variation. Intonation analysis of Bari Italian Map Task [24] dialogues has shown that in this variety such a distinction is marked by using a different pitch accent [17] [18] [20]. In fact, QUERIES are characterised by the rising pitch accent L+H* as described in Figure 1, whereas CHECKS asking confirmation about completely given information are always realised by a falling pitch accent, which could be either a high-falling H*+L – conveying also a certain degree of contrast – or by a fall in reduced range !H+L* as in neutral statements. On the other hand, CHECKS asking for confirmation about information which is textually, situationally or inferentially accessible (i.e. only partially given) to the interlocutor can be realised by all of the three accents. The choice in this case depends on the degree of interlocutor's confidence as to the correctness of the inference made in the question [25] [26]. The rising L+H* pitch accent is used when s/he is totally unconfident (as in QUERIES, in fact), a falling H*+L or !H+L* in cases of more confidence, as schematised in Table 1.

Table 1. Intonational sequences of polar questions asking for confirmation about accessible information according to different speaker's attitudes as to correctness of inferred information in Bari Italian (adapted from [18])

BI intonational sequences of polar questions asking for confirmation about accessible information		
SPEAKER ATTITUDE	PITCH ACCENT	BOUNDARY TONE
Not Confident	L+H*	L-L% / L-H%
Confident	H*+L	L-L%
Very Confident	!H+L*	L-L%

Other Italian varieties appear to share only partially these features, or do not share them at all. For example, a preliminary analysis on Pisa Italian Map Task dialogues has shown that in this case speakers use the same falling-rising H+L* H-L% intonational sequence, even though it appears that a difference between QUERIES and CHECKS can rely on some different phonetic realisations (for example, in the height of the leading H tone of the pitch accent) [27] [28]. It is worth noting that in the same study, some examples of more confident CHECKS with a rising-falling melodic shape, intonationally described as L+H*L-L% were also found: this pitch accent conveys contrastiveness, as its Bari confident CHECK counterpart which has a falling H*+L pitch accent instead (see Table 1). These preliminary results on Pisa hints at the hypothesis that intonational strategies used by Pisa Italian speakers might be more complex than these first results seem to indicate.

In Neapolitan, on the other hand, another preliminary study based on the same type of speech materials has claimed that in this variety the difference between information-seeking and confirmation-seeking questions is not marked intonationally at all (not even phonetically), both types of questions being characterised by a rising-falling L*+H HL-L% contour [29] (see the Neapolitan question example in Figure 2).

2.2 Non-Final and Pre-Final Utterances

Intonational characterisation of Italian varieties can also be found in relation to melodic strategies used by speakers for signalling subordination and/or coordination relationships within discourse segments, and also for signalling upcoming discourse boundaries. It is well known, in fact, that intonation can be used for cueing utterance position within a discourse segment: it can signal that the current utterance is the non-final one in a sequence (i.e., that more utterances are coming, typically expressed intonationally by a terminal rising "continuation rise" contour), or that it is the final one (i.e., that no more utterances are expected to be produced for the current discourse unit, typically conveyed by a falling intonation). In some languages and language varieties, it is even possible to signal intonationally that the end of a discourse unit is *approaching*, i.e. that the current utterance is the penultimate – or "pre-final" – one in a sequence.

In Italian, studies conducted on Bari and Pisa similar speech materials, involving Map Task dialogues and lists readings, have shown they both have a quite rich inventory of intonational sequences for signalling the boundaries of discourse units [30] [31] [32] [33], as it is summarised in Table 2.

Table 2 also shows that in Bari Italian the difference between generic non-finality and "pre-finality" functions is captured phonologically, as it has a specific set of intonation contours for signalling penultimate position of an utterance in a sequence, whereas it seems that Pisa Italia speakers cannot rely on such a phonological distinction in intonation [33].

Table 2. Distribution of intonation contours signalling non-final, pre-final and final utterance position in Bari and Pisa Italian (adapted from [30], [33])

BARI ITALIAN		**PISA ITALIAN**	
NON-FINAL	L* L-H%	NON-FINAL	H+L* H-H%
	H* H-L%	PRE-FINAL	H* H-H%
	H* L-L%		H+L* L-L%
			H* L-L%
PRE-FINAL	H* H-H%		
	H+L* L-H%		
	H+L* L-L%		
FINAL	!H+L* L-L%	FINAL	H+L* L-L%

Therefore, it can be argued that intonational strategies used by speakers for cueing utterance position within a sequence can also be useful for discriminating among different varieties of Italian.

3. Can Regional Varieties Be Perceptually Identified by Intonation? A Review of Some Results on German

It is a widely shared belief that listeners can identify a language or a language variety by relying solely on pitch information, and a number of perceptual experiments have been carried out in the past aiming at giving experimental evidence for this. In overviewing the traditional studies on this issue, Peters and collegues [34], [35] report that in those experiments stimuli were created by filtering the original signal using different techniques, all having the scope of eliminating or minimising segmental information ([36] [37] [38], more recently [39]). Results in these experiments did not show very high success rate, and some authors argued that this might be due to the fact that such filtering techniques may eliminate crucial prosodic information such as word and syllable boundaries which are relevant for language identification by intonation. Since it seems that information about the relationship between intonation contours and segmental structure is crucial for listeners, Peters and collegues [34] performed two set of identification experiments with varieties of German for testing the role of intonation where they retained such "higher-level information" in the stimuli[5]. Since they wanted also test the role of listeners' linguistic competence in the identification, in both sets of

[5] In a successive work Peters and collegues compared the two approaches in another identification task of German varieties – one in which the stimuli were presented without segmental information (by using speech signal filtering techniques), and the other in which such linguistic information were preserved and different contours were superimposed on the same set of original utterances. Resultes showed that subjects performed better with the second set of stimuli [35]

experiments they subdivided the subjects into native speakers of the variety to be identified and having lived in their respective town since birth ("native residents") and non-native speakers of that variety but having lived there for a long time ("non-native residents").

In the first experiment subjects were asked to discriminate between Hamburg Urban Vernacular (HUV) and Northern Standard German (NSG) utterances based on three different but functional equivalent types of intonation contours in the two varieties. The phonological representations of the contour types (along the GToBI guidelines, i.e. the ToBI labelling system developed for German as described in [40]) used in this experiment for both HUV nd NSG, along with their function are summarised in Table 3.

Stimuli were obtained in two steps: in the first one, original utterances extracted from some HUV natural conversations were inserted into a carrier text and read by a NSG speaker; the second step consisted in superimposing HUV and NSG F0 contours on such read sentences with PSOLA pitch resynthesis [41] (available in the speech analysis software package Praat, [42]). This technique permits manipulation of F0 values without altering the spectral quality of the speech signal, thus preserving the segmental information content of the utterances.

Table 3. Intonation contour types in HUV and NSG used in the first identification perceptual experiment by Peters and collegues [31] (adapted)

Contour	Function	Hamburg Urban Vernacular	Northern Standard German
1	continuation	H* L-H%	L*H-%
2	final	%^H L* (H*) H* L-%	H* (!H*) !H* L-%
3	yes-no question	^H* L-H%	H* L-H%

The main results of this experiment are that HUV listeners can identify HUV contours in a pairwise identification test (HUV vs SNG), and that non-native residents listeners perform better than native ones. Peters and collegues explained the latter apparently counter-intuitive result with two different hypotheses: 1) non-native residents perfomed better as they had been more exposed to German varieties other than HUV, and have therefore developed more awareness or/and sensitivity to the differences between their own native variety and the one spoken in Hamburg; 2) non-native residents may have used a sort of "elimination strategy" i.e. excluding the NSG (as being "non HUV") more easily that natives, as they are more familiar with NSG than HUV native speakers (for the same reason of having been "more exposed to other varieties" as explained above). In any case, such results gave evidence that listeners' linguistic competence play a very important role in the identification task.

A second experiment was carried out in order to verify the elimination strategy hypothesis and therefore the role of a listener's linguistic competence. In this experiment native, non-native resident and non-native non-resident Berlin Urban Vernacular (BUV) speakers were asked to identify the BUV variety in comparison – this time – with two other different German varieties, namely SNG and Low Alemannic German (LAG). Stimuli consisted again in utterances reproducing three different but functional equivalent types of intonation contours in the relating German varieties under examination, obtained by the same method described above for HUV, except for an additional element. Since in the second experiment the authors wanted also to verify the influence of segmental parameters, the original natural BUV utterances were read by both a BUV and a NSG native speakers. The hypothesis to be checked was that BUV stimuli taken from original BUV read utterances would be identified more successfully than BUV stimuli originating from NSG read utterances. Again, target utterances were obtained by superimposing BUV, NSG and LAG F0 contours on BUV and NSG read sentences by using PSOLA resynthesis.

Table 4. Intonation contour types in HUV, NSG and LAG used in the second identification perceptual experiment by Peters and collegues [34] (adapted). The symbol '<' in contour type 2 for BUV indicates peak delay

Contour	Function	Berlin Urban Vernacular	Northern Standard German	Low Alemanic German
1	contrast	L*+H L- %	H*+L L- %	H+L(*) L- %
2	declarative	H* < L- %	H* L- %	H*+L H(*)+L L-%
3	non-final	H* H- %	H* L- H%	L*+H L- %

Results of the second experiments seem to confirm the main hypotheses tested in the first one, namely that subjects are able to identify BUV contours when compared with NSG and LAG ones, and that non-native residents perform better than natives, since they are also familiar with other varieties (even though in this case the possible role of the "elimination strategy" was not clearly ascertained, nor was the behaviour of the non-native non-residents always consistent in the identification). Results seem also to indicate that success rate increases when BUV contours are superimposed on BUV original read utterances, i.e. that segmental informational also play a role in the identification, even though this result is not always very clearly determined by the statistics. Moreover, it appears that the type of intonational contour has an influence on the identification rate, i.e. some contours are more "salient" than others.

4. Concluding Remarks

In this paper some indications on the role of intonation in identifying different varieties of a language have been presented. From the descriptive viewpoint, a number of examples where differences among regional accents are marked by intonation have been presented for Italian In some cases it is the tune used which determines the difference (as is the case for information-seeking polar questions), whereas in other cases it can be the presence vs absence of an intonational strategy for cueing a particular communicative intention which marks the difference (as in signalling general non-final and penultimate position, which are two separate intonational categories in Bari variety, but just one category in Pisa). A third difference is the phonetic realisation of particular contours, for example differences in tonal alignment.

Description of intonational variation is crucial for studying the other side of the accent identification issue, namely the perceptual aspect. Some perceptual experiments carried out on German language have been presented, whose results seem to confirm the specific role of intonation in accent identification task (among two and among three different varieties of German), and that success rates also depends on the linguistic competence of the listeners, even though in a way which might have at first appeared counter-intuitive. In fact, it might have be assumend that native speakers of a variety can perform better than non-native ones in identifying native accents, whereas results have shown that non-native listeners had better performance (but it is still unclear whether this is due to a truly better linguistic competence – native plus non-native accents – or rather to the application of some "elimination strategy").

As expected, the contribution of segmental features increases the percentage of accent identification, even though results reported on German are not completely clear-cut. Also, it would be worth investigating whether and, if so, in which way segmental and suprasegmental feature interact in the identification process. Since German results have also shown that some contours are more "salient" than others in the identification task, it may hypothesised that intonational features may be more used by speakers when listening to utterances marked by such contours, whereas segmental features may predominate when listening to utterances characterised by non salient melodic patterns. In Italian, the first case could be represented by accent identification listening to questions (characterised by strong intonational differences), the second by listening to statements (whose intonational contour is similar among the Italian varieties). These hypotheses need to be verified, calling for further investigation on the topic.

Also, including the gestural feature in modelling the identification of the regional accent personal trait would be useful, as it well know that some gestures are to some extent related to specific languages and cultures (see for example [42], [43]).

Acknowledgments

I would like to thank Martine Grice for helpful comments on an earlier version of this paper.

References

[1] Gilles P & Peters J (eds), *Regional variation in intonation*, Tuebingen: Niemeyer Verlag, 2004.
[2] Jun Sun-Ah (ed.), *Prosodic Typology. The phonology of Intonation and phrasing*, NY: Oxford Univ. Press, 2005.
[3] Grice M. *Intonation*, in K. Brown (ed.), Encyclopedia of Language and Linguistics, 2nd ed, Oxford: Elsevier, vol.5, pp.778-788, 2006.
[4] Grice M & Baumann S, *An introduction to intonation – functions and models*, in J. Trouvain & U. Gut (eds.), Non-native prosody: phonetic descriptions and teaching practice, "Trends in Linguistics", Berlin: Mouton de Gruyter, to appear.
[5] Ladd R, *Intonational Phonology*. Cambridge: Cambridge Univ. Press, 1996.
[6] Gussenhoven C., *The Phonology of Tone and Intonation*, Cambridge: Cambridge Univesity Press, 2004.
[7] Pierrehumbert J, *The phonetics and phonology of English Intonation*. MIT, doctoral dissertation, Bloomington: Indiana Univ. Linguistics Club, 1980.
[8] Beckman M & Ayers G, *Guidelines for ToBI labelling*, Ohio State University, 1994.
[9] Beckman M. e J. Pierrehumbert, Intonational structure in English and Japanese, *Phonology Yearbook* **3** (1986), 255-310.
[10] De Mauro T., *Storia linguistica dell'Italia unita*, Bari: Laterza, 1963.
[11] Tosi, A., *Language and society in a changing Italy*, Multilingual Matters, Clevedon, UK, 2001.
[12] Agard F. & Di Pietro R. J., *The sounds of English and Italian*, Chicago: Univ. of Chicago Press, 1965.
[13] Fiorelli P., *Corso di pronunzia italiana*, Padova: Radar, 1965.
[14] Chapallaz M., *The Pronunciation of Italian. A Practical Introduction*, London: Bell & Hyman, 1979.
[15] D'Eugenio A, *Major problems of English phonology. With special reference to Italian-speaking learners*, Foggia: Atlantica, 1982.
[16] Canepari L., *Italiano standard e pronunce regionali*, Padova: CLEUP, 1980.
[17] Savino M., *Il ruolo dell'intonazione nell'interazione comunicativa. Analisi strumentale delle domande polari in un corpus di dialoghi spontanei (varieta' di Bari)*, doctoral dissertation, Università & Politecnico di Bari, Italy, 1997.
[18] Savino M, *Descrizione autosegmentale-metrica di alcune tipologie intonative dell'italiano di Bari*, in E. Burr (a cura di), Tradizione & Innovazione – Atti del VI Convegno della Società di Linguistica e Filologia Italiana (Duisburg 28 giugno-2 luglio 2000), Firenze: Franco Cesati Editore, pp. 163-178, 2005.

[19] Grice M & Savino M., *Low tone versus 'sag' in Bari Italian intonation; a perceptual experiment*, Proceedings of the XIII International Congress of Phonetic Sciences, Stockholm, vol. 4, pp. 658-661, 1995.

[20] Grice M & Savino M., *Can pitch accent type convey information-status in yes-no questions?*, Proceedings of the ACL Workshop on Concept-to-Speech Generation Systems, Madrid, pp. 29-38, 1997.

[21] Interlandi G, *L'intonazione delle interrogative polari nell'italiano parlato a Torino: tra varietà regionale e nuova koiné*, doctoral dissertation, Università di Pavia, Italy, 2003.

[22] Grice M, D'Imperio M, Savino M, Avesani C, *Strategies for intonation labelling across varieties of Italian*, in: Sun-Ah Jun (Ed.), Prosodic Typology: the Phonology of Intonation and Phrasing, New York: Oxford University Press, pp. 362-389, 2005.

[23] Grice M., *The intonation of interrogation in Palermo Italian; implication for intonation theory*, Tuebingen: Niemeyer, 1995.

[24] Carletta J, Isard A, Isard S, Kowtko J, Doherty-Sneddon G, Anderson A, The Reliability of a Dialogue Structure Coding Scheme, *Computational Linguistics* **23 (1)** (1997), 13-32.

[25] Grice M & Savino M, Map Tasks in Italian: asking questions about given, accessible and new information, invited paper, *Catalan Journal of Linguistics*, special issue on Romance Intonation (P.Prieto, ed.) **2** (2003) 153-180.

[26] Grice M & Savino M, *Information Structure and Questions – Evidence from Task-Oriented Dialogues in a Variety of Italian*, in P. Gilles & J. Peters (eds.), Regional Variation in Intonation, Niemeyer: Tuebingen, pp.161-187, 2004.

[27] Gili Fivela B., *The Phonetics and Phonology of Intonation: the Case of Pisa Italian*, doctoral dissertation, Scuola Normale Superiore di Pisa, Italy, 2004.

[28] Gili Fivela B., *L'intonazione nella varietà pisana di italiano: analisi delle caratteristiche principali*, in Atti del Gruppo di Fonetica Sperimentale dell'Associazione Italiana di Acustica, Macerata 2-4 Dic. 2001, pp.103-111, 2002.

[29] Crocco C., *Prosodic and informational aspects of polar questions in Neapolitan Italian*, Proceedings of Speech Prosody 2006, Dresden May 2-6, 2006 (on CD-ROM).

[30] Savino M, *Non-finality and Pre-finality in Bari Italian Intonation: a Preliminary Account*, Proceedings of the VII European Conference on Speech Communication and Technology, Aalborg 3-7 Sept. 2001, vol.2, pp.939-942, 2001.

[31] Savino M, *Sul ruolo dell'intonazione nel (pre)segnalare i confini di unità del discorso: un esperimento percettivo*, in Atti delle XV Giornate del Gruppo di Fonetica Sperimentale dell'AIA (Associazione Italiana di Acustica), Viterbo, 4-6 Dec. 2003, pp.203-208, 2004.

[32] Savino M, *Intonational Cues to Discourse Structure in a Variety of Italian*, in P. Gilles & J. Peters (eds.), Regional Variation in Intonation, Niemeyer: Tuebingen, pp.145-159, 2004.

[33] Savino M, Grice M, Gili Fivela B, Marotta G, *Intonational cues to discourse structure in Bari and Pisa Italian: perceptual evidence*, Proc. Speech Prosody, Dresden May 2-5, 2006 (on CD-ROM).

[34] Peters J, Gilles P, Auer P, Selting M, Identification of regional varieties by intonational cues. An experimental study on Hamburg and Berlin German, *Language and Speech*, **45(2)** (2002), 115-139.

[35] Peters J, Gilles P, Auer P, Selting M, *Identifying regional varieties by pitch information: a comparison of two approaches*, Proceedings of ICPhS 2003, Barcelona, pp.1065-1068, 2003.

[36] Atkinsons K, Language identification from non-segmental cues, *Working Papers in Phonetics (UCLA)* **10**, (1968), 85-89.

[37] Ohala J & Gilbert JB, *Listeners' ability to identify languages by their prosody*, in PR Lèon & M Rossi (eds), Problèmes de Prosodie II: Expérimentations, modèles et fonctions, Ottawa: Didier, 1981.

[38] Maidment J A, Language recognition and prosody: further evidence, *Speech, Hearing and Language: Work in Progress UCL* **1** (1983), 133-141.

[39] Schaeffler F & Summers R, *Recognizing German dialects by prosodic features alone*, Proc. of ICPhS (International Conference of Phonetic Science) 99, San Francisco, August 1-7, pp.2311-2314, 1999.

[40] Grice M & Baumann S, Deutsche Intonation und GToBI, *Linguistische Berichte* **191**(2002), 267-298.

[41] Moulines E & Charpentier F., Pitch Synchronous Waveform Processing Techniques for Text-to-Speech Synthesis using Diphones, *Speeech Communication* **9** (1990), 453-467.

[42] Boersma P. & Weenink D., Praat. *A system for doing phonetics by computer*, http://www.fon.hum.uva.nl/praat/, 1999.

[43] Kendon A., Gesture as illocutionary and discourse structure markers in Southern Italian conversation, *Journal of Pragmatics* **23** (1995), 247-279.

[44] McNeill D., *Hand and mind*, Chicago: Chicago University Press, 1992.

Prosodic Cues for Automatic Word Boundary Detection in ASR

Klara VICSI[1], György SZASZÁK
Budapest University of Technology and Economics
Department for Telecommunication and Mediainformatics

Abstract: This article presents a cross-lingual study for agglutinative, fixed stressed languages, like Hungarian and Finnish, about the segmentation of continuous speech on word level by examination of supra-segmental parameters.
We have developed different algorithms based either on a rule based or a data-driven approach. The best results were obtained by data-driven algorithms (HMM-based methods) using the time series of fundamental frequency and energy together. This HMM based method will be described in this article.
Word boundaries were marked with acceptable accuracy, even if we were unable to find all of them. On the base of this study a word level segmentationer has been developed which can indicate the word boundaries with acceptable precision for both languages.
The evaluated method is easily adaptable to other fixed-stress languages.

Keywords: Automatic Speech Recognizers (ASR), word and phrase boundaries detection, prosodic recognizer, data-driven (HMM) approach, Finish language, Hungarian language

Introduction

Supra-segmental features are an integral part of every spoken language utterance. These can provide cues to the linguistic structure of the speaker's message, emotional state or communicative intent. These features are intonation, stress, rhythm, etc. They can signal the structuring of utterances into larger discourse segments and provide additional information for human speech processing.
Using supra-segmental features in automatic speech recognition to increase its robustness is a tendency again [3,4,8,7]. Some trials were conducted in the mid-eighties, but it has not yet been possible to exploit such knowledge in an automatic speech recognition system [1,2,6]. This relative failure, according to Philippe Langlais [2], is mainly due to three types of difficulties:
- significant contextual variability of prosodic knowledge (type of speech, speaker, structure and content of sentences, nature of the environment, etc);
- complexity of relations between prosodic information and various linguistic organization levels of a message;
- problems encountered with accurate measurement of prosodic parameters, and their possible integration on a perceptual level.

[1] Corresponding Authors: Budapest University of Technology and Economics Department for Telecommunication and Mediainformatics, vicsi@tmit.bme.hu.

Speech production is a continuous movement of the articulating organs, producing a continuous acoustic signal. Linguistic content and phonological rules help the human speech processing to separate syntactic units, such as words, phrases (sections between two intakes of breath), or sentences. In our experiments we have examined how words can be automatically separated in continuous speech in fixed stress languages such as Hungarian and Finnish (both of these languages belong to the Finno-Ugrian language family).These languages are highly agglutinative, so they are characterized with longer average word length than English, for example, and also with a relatively free word order. Stress is generally on the first syllable of the word, except in the case of conjunctions or articles. This means that the word-level intonation units we have dealt with during our experiment are composed of a word, stressed on the first syllable, together with unstressed conjunctions or articles, if any exist. This word level intonation unit is shorter than the phrases. Henceforward in this article we will strictly use the expression "word unit", and the boundaries of these units will be called "word boundaries".

A word level segmentationer has been developed capable to indicate the word boundaries with acceptable precision for both languages. The ultimate goal is to increase the robustness of speech recognition on the language modeling level by the detection of word boundaries, thus decreasing significantly the searching space during the decoding process. Searching space reduction is highly important in the case of agglutinative languages.

1. Methodology

In our experiment we have measured the fundamental frequency, the energy level and the time course. These parameters (all or some of them) are necessary for the realisation and perception of stress in the Finno-Ugrian language family. The peaks of energy and fundamental frequency clearly represent the first syllables of the words. Syllable length, it was found, is not greatly influenced by the stress.

If stressed syllables can be detected, the word boundary can be marked by finding the minimum of the energy and the fundamental frequency just before the stressed point. Hence the stressed vowels may be situated in the onset or even in the nucleus of the syllable, by searching the minimum point, the absolute minimum within an interval just before the stress should be found. The length of this interval is typically a half-syllable to avoid overlap with previous word. This value was set in empiric way, 100 ms interval length was found to be suitable for normal Hungarian read speech rate. We have developed different algorithms based either on a rule based or a data-driven approach [11]. The best results were obtained by data-driven algorithm for the determination of word units and their boundaries, therefore this approach will be described in this article.

Hungarian BABEL [5] and Finnish [10] continuous read speech databases were used for the examination. The databases were labelled and segmented on word level by an expert using audio visual method. For Hungarian 1600 sentences of 32 speakers and for Finnish 250 sentences of 4 speakers were used.

1.1. Acoustic Pre-processing

Fundamental frequency (Hz) and energy level (dB) were measured and their first and second order deltas. Syllable duration was found to be problematic to measure exactly without using speech recognizer output, hence it was discarded. For the detection of fundamental frequency, autocorrelation method was used.
The autocorrelation function for x(n) discrete signal is:

$$R(k) = \sum_{n=1}^{N-k} x(n)x(n+k) \tag{1}$$

F_0 at the ith frame $F_0(i)$ was obtained after median filtering:

$$F_0(i) = \mathbf{med}\{ F_0(i-3), F_0(i-2), F_0(i-1), F_0(i), F_0(i+1), F_0(i+2), F_0(i+3) \} \tag{2}$$

The energy E(i) was calculated with an integration time of 100 ms:

$$E(i) = \frac{1}{M+1} \sum_{n=i-\frac{M}{2}}^{i+\frac{M}{2}} x^2(n) \tag{3}$$

where M is the number of samples pro 100 ms.
The fundamental frequencies and the energy levels are given in every 25.6 ms.

1.2. Data-driven Approach

In data-driven approach the common HMM method [9] was used for the determination of word units and their boundaries. This type of examination needs a speech database segmented on word level to train the prosodic pattern HMM models. The word boundaries and the type of word units were marked in the speech database by an expert relying on fundamental frequency and energy cues, using audio-visual segmentation method. The segment boundaries of the word level prosodic curves were marked so that they overlap with word boundaries. 5 different word level prosodic curves, its boundaries and the silent periods were marked. Thus by the training a set of 6 different Hidden Markov Models were constructed. These models represent 5 types of word level prosodic curves which are descending, rising, floating, jumping and rise-fall while the 6th model is a silence model. Training examples for different curves are presented in *Figure 1* between cursors. The sentence in *Figure 1* was selected from the database in such a manner that it presents all of the used models in one example. However in the database 80% was descending, 10 % was silent and another 10 % was rising, floating, jumping and rise-fall together of all word units.

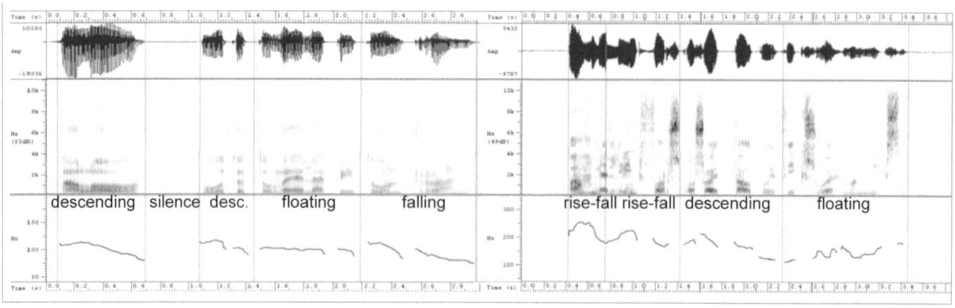

Figure 1. Examples for the trained prosodic curve types. Time function, spectrogram, F_0 contour on Hungarian speech data.

The prosodic model itself may be of interest in the future, but for the moment, we use only the boundaries of these word units for the evaluation.

We have also adapted this data-driven approach for Finnish language using the same 6 prosodic models as for Hungarian. Training strategy for Finnish was the same as for Hungarian. A cross- and multilingual experiment with Hungarian and Finnish was performed. In this case, we used prosodic HMM models trained on one language to segment speech on the other language. Moreover a hybrid was also trained to segment Hungarian and Finish speech.

1.3. Evaluation

To evaluate results the obtained prosodic segmentation is compared with the original one. Two measures are used to present our results. First one is the **correctness** which denotes whether a unit boundary predicted by our algorithm was correct or not (in %).

$$Corr\ [\%] = \frac{\#\_of\_correctly\_marked\_word\_boundaries}{\#\_of\_all\_marked\_word\_boundaries} * 100$$

Our second measure is the **effectiveness** which says how many *word* boundaries were found at all (in %):

$$Eff[\%] = \frac{\#\_of\_correctly\_marked\_word\_boundaries}{\#\_of\_all\_real\_word\_boundaries\_(in\_reference\_transcription)} * 100$$

Note that this second measure is expected to be less than 100 %, since not the all of the words in speech are emphasized. In this way articles and conjunctions can never be separated because they are not emphasized.

Since for speech recognition tasks we require that whenever a word boundary is predicted it should be detected correctly the correctness is more critical then the effectiveness. (We have required 80% correctness at least.) Of course, the higher effectiveness the more robust the system, but we cannot permit this at the expense of lower correctness.

The predicted segment boundary of word units is regarded as correct if it deviates no more than 2 frames forward or backward from the real word boundaries.

1.4. Optimization of main HMM parameters

By training of the HMM prosodic recognizer the first task was the optimization of the main HMM parameters. This means mainly to find the optimal number of states: since word level prosodic units are longer than phonemes – modeled usually by 5 state models at 10 ms frame rate – the number of states for prosodic models can be changed between 9 and 15. We have applied a frame rate of 25.6 ms so a 9 state model requires a minimal phrase length of ~230 ms while minimal phrase length for a 15 state model is ~380 ms which is according to our examinations a typical value for Hungarian. The optimal number of states was found to be 11 as it can be seen on *Figure 2*.

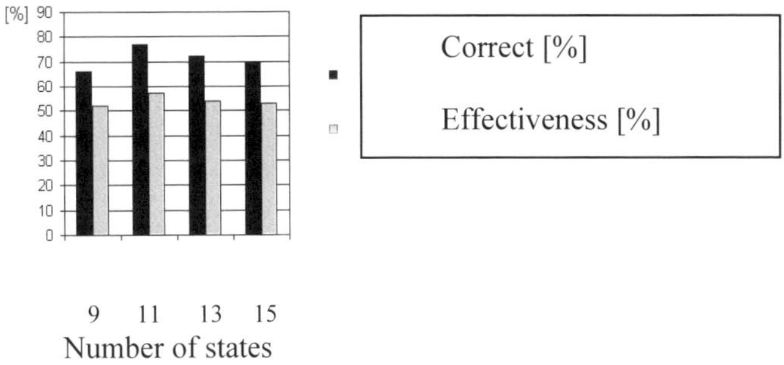

Figure 2. Correctness and effectiveness depending on the number of HMM states (trained by 4 speakers, using 2 Gauss, Hungarian)

Only 6 dimensional observation sequences (F0, energy, first and second order deltas) were used and 2 or at most 4 Gaussian components were sufficed to describe output distribution of each state. The first and the last states were non-emitting ones [9].

Two acoustic pre-processing alternatives were examined with the above mentioned fixed HMM parameters: In the first case we have used either F_0 or Energy data only with appended first and second order deltas while in the second case both F_0 and Energy data have been applied as input with their first and second order deltas appended.

Table 1. Correctness and effectiveness of word unit's boundary determination with different acoustical preprocessing

Used prosodic parameter(s)	Language	Training corpus	Correctness/Effectiveness [% / %]
$F_0+\Delta F_0+\Delta^2 F_0$	Hungarian	14 persons	67.4 / 58.4
$E+\Delta E+\Delta^2 E$			67.7 / 66.6
(F0+ΔF0+Δ2F0 +(E+ΔE+Δ2E)			**76.5 / 53.0**

In *Table 1* the results are presented as a function of the acoustic pre-processing parameters. The best results were obtained when fundamental frequency-type parameters were used together with energy-type parameters.

We have also tested several training strategies for the constructed HMMs. During the examination a 14 speaker data set(14x50 sentences) was used for training and 18 speakers (18x50 sentences) for testing. First the size of the training material was changed. The training set was reduced to 4 persons (4x50 sentences) and finally to one person (50 sentences), while the test set consisted of the same 18 speakers (18x50 sentences) in all cases.

Table 2 Correctness and effectiveness of boundary determination of word unit with HMM for different training settings.

Used prosodic parameter(s)	Language	Training corpus	Correctness/Effectiveness [% / %]
(F0+ΔF0+Δ2F0 +(E+ΔE+Δ2E)	Hungarian	1 person	77.1 / 49.6
		4 persons	**77.4 / 57.2**
		14 persons	76.5 / 53.0

If the HMMs were trained on few speakers these speakers were carefully selected in order to ensure a relatively accurate training corpus. Results are shown in *Table 2*.

Surprisingly, there is no relevant difference in correctness if fewer speakers are involved in the training corpus. Effectiveness, however, depends very much on the number of speakers, and to achieve effectiveness over 50% at least four speakers' data should be used for training. In comparison with the results obtained by the rule-based approach effectiveness is much higher with an acceptable correctness. If we use only F_0 or energy patterns, effectiveness is excellent, but we have ~10% reduction in correctness. The overall best result was 77.4% correctness with 57.2% effectiveness, obtained with HMMs trained on 4 speakers' F_0 and energy data.

For further experiments the optimized parameters were applied: frequency and energy level values together, the speech examples of 4 persons and the state of the models were 11.

2. Results

The developed system can be convenient for automatic segmentation of word units. An example is presented in Figure 3 of how the developed segmentation method works on word level. Time function of the speech signal, F_0 and energy level contour are visible on the screen. The first row at the bottom contains an expert-made hand segmentation taken as reference while the second row illustrates the output obtained by the automatic prosodic segmentation. Segmentation accuracy means the correctness of the word unit boundaries' determination.

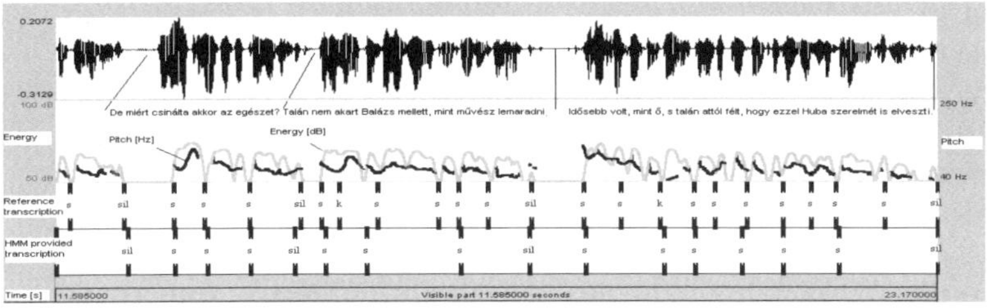

Figure 3 HMM provided word level segmentation versus expert-made hand segmentation on a passage of 3 Hungarian sentences: [s] and [k] denote word boundaries, [sil] silence

Results obtained for Finnish with the same data-driven method as for Hungarian are presented in Table 3. We may note a small fall in correctness while effectiveness is high. The reason for this may be that Finnish speech is much slower than Hungarian, and Finnish words often contain long plosive sounds where F_0 and energy contour show a very similar behavior to the one they have on real word boundaries. As a result, many more word boundaries can be found than in

Table 3 Correctness and effectiveness of boundary determination of word unit with HMM for Hungarian and Finnish

Used prosodic parameter(s)	Language	Training corpus	Correctness/Effectiveness [% / %]
$F_0+\Delta F_0+\Delta^2 F_0$ + $E+\Delta E+\Delta^2 E$	Hungarian	4 persons	77.4 / 57.2
	Finnish	4 persons	69.2 / 76.8

Hungarian, but we can also detect some in-word secondary emphasis which explains the lower correctness.

Cross-lingual prosodic word boundary segmentation results both for Finnish and Hungarian are presented in Table 4. It can be seen that segmentation of Finnish speech with models trained on a Hungarian database gives nearly the same result as the

Table 4. Results of direct cross-lingual word boundary segmentation for Finnish and Hungarian

HMM models	Test data	Corr/Eff [%/%]
Hungarian (4 speakers)	Hungarian	77.4 / 57.2
Hungarian (4 speakers)	Finnish	67.1 / 52.1
Finnish (4 speakers)	Finnish	69.2 / 76.8
Finnish (4 speakers)	Hungarian	70.7 / 52.3
Finnish (4 speakers) + Hungarian(4 speakers)	Hungarian	75.0 / 68.2
Finnish (4 speakers) + Hungarian(4 speakers)	Finish	69.7 / 83.7

Finnish models. On the other hand, segmented Hungarian data with Finnish models yields 70.7% correctness which is somewhat better than on Finnish data (67.1%). This is probably due to the sparseness of Finnish data.

Generally this means that a prosodic word boundary segmentator well trained with a Hungarian database can be applied for automatic segmentation of unknown Finnish speech material and vice versa. Naturally hand-made correction is necessary.

3. Conclusion

Our word level prosodic segmentation method based on measuring fundamental frequency and energy level functions gave promising results. Word boundaries can be marked with acceptable correctness, even if we are not able to find all of them. Two measurements, correctness and effectiveness, were used to describe the behavior of this prosodic segmentation system.

Word boundaries are found with acceptable correctness and effectiveness for fixed-stress languages like Hungarian and Finnish. In case of Finnish we obtained results comparable with Hungarian with lower correctness and higher effectiveness which may be the result of the difference between the two languages and also of data sparseness in the Finnish database.

Moreover, these results ensure that the integration of a prosodic recognizer into a CSR system can help reduce the searching space and thus improve speech recognition performance. The importance of this searching space reduction is great in recognition of agglutinative languages such as Hungarian and Finnish where the possible number of word forms is more than hundreds of millions. In this domain further investigations are needed.

Summarizing the results of our experiments it is clearly worth continuing research in this field. The evaluated method is easily adaptable to fixed-stress languages other than Finnish and Hungarian. The examination of such languages would be useful.

We have presented a single example how it is possible to use prosodic information, but there may be several solutions.

A practical result emerged from these experiments: this prosodic recognizer can be used as a word-level automatic segmentationer for Hungarian and Finnish languages.

4. Acknowledgement

We would like to express our thanks to Toomas Altosaar (Helsinki University of Technology, Laboratory of Acoustics and Audio Signal Processing) for his kind help and his contribution to the use of the Finnish speech database. The work has been supported by the Hungarian Scientific Research Foundations OTKA T 046487 ELE and IKTA 00056.

References

[1] Di Cristo: Aspects Phonétiques et Phonologiques des Éléments Prosodiques. In Proceedings of *Modèles linguistiques Tome III*, 2, 24-83, 1981.

[2] P. Langlais, H. Méloni: Integration of a Prosodic Component in an Automatic Speech Recognition System. In Proceedings of 3rd European Conference on Speech Communication and Technology, Berlin, 2007-2010, 1993.
[3] S. Mandal, A.K. Datta, B. Gupta: Word Boundary Detection of Continuous Speech Signal for Standard Colloquial Bengali (SCB) Using Suprasegmental Features. In Proceedings of FRSM., IIT, Kanpur, 2003.
[4] B. Peters: Multiple Cues for Phonetic Phrase Boundaries in German Spontaneous Speech. In Proceedings of 15th ICPhS, Barcelona, 1795-1798, 2003.
[5] P. Roach: BABEL: An Eastern European Multi-Language Database. In Proceedings of International Conference on Speech and Language Processing, Philadelphia, 1996.
[6] M. Rossi: A Model for Predicting the Prosody of Spontaneous Speech (PPSS model). *Speech Communication,* 13, .87-107, 1993.
[7] A. Salomon, C.Y. Espy-Wilson, O. Deshmukh: Detection of Speech Landmarks: Use of Temporal Information. *Journal of Acoustic Society of America*, 115, 1296-1305, 2004.
[8] L. Yang: Duration and Pauses as Phrase and Boundary Marking Indicators in Speech. In Proceedings of 15th ICPhS, Barcelona, 1791-1794, 2003.
[9] S. Young, G. Evermann, D. Kershaw, G. Moore, J. Odell, D. Ollason et al.: *The HTK Book.* Cambridge University, 22-130, 2002.
[10] M. Vainio, T. Altosaar, M. Karjalainen, R. Aulanko, S. Werner: Neural Network Models for Finnish Prosody. In Proceedings of 14th ICPhS 1999. San Francisco, 2347-2350, 1999.
[11] K. Vicsi, G. Szaszák: Automatic Segmentation of Continuous Speech on Word Level Based on Supra-segmental Features. *International Journal of Speech Technology*, 14, 55-64, 2006.

A. Esposito et al. (Eds.)
IOS Press, 2007

Non-Speech Activity Pause Detection in Noisy and Clean Speech Conditions

Vojtěch STEJSKAL[1a], Zdenek SMÉKAL[a], Anna ESPOSITO[b]
[a]*Brno University of Technology, Dept. of Telecommunications, Czech Republic.*
[b]*Department of Psychology, Second University of Naples and International Institute for Advanced Scientific Studies, Vietri sul Mare, Italy.*

Abstract. Nowadays, successful pause detection plays an important role not only in the process of speech recognition and speech coding but also in the biometrical field for detecting stress in the speaker's emotional state due to uncomfortable situations or in interactive dialog systems for making more natural the human-machine interaction more natural. Most of the recordings exploited in practical applications are made under adverse conditions and few algorithms have been proposed to handle noisy conditions. This paper proposes two methods for non-speech activity pause detection in spontaneous speech recordings made in noisy environments. The input signal is transformed into log spectral energy and is divided into specific frequency bands. Each band is smoothed and tracked by dynamically adjusted thresholds based on noise energy estimation. Thresholds are adapted taking into account the dynamic changes of the speech signal under environmental noise. The proposed methods run in real time and do not require a priori knowledge of the SNR and a priori threshold values. Experimental results show that their performance is comparable with standard VADs.

Keywords. Non-speech activity pause detection, energy, noise, sub-band based segmentation.

Introduction

Spoken utterances can be divided into segments with high information value (speech) and non-speech segments consisting either in noise from the background environment or non-lexical items produced by the speaker. Accurate detection of the second kind of segment, which we will designate as *Non-Speech Activity* pauses (NSA), is crucial for most of today's speech processing methods. For example, in *speech recognition*, the correct detection of word boundaries has a significant impact on the final recognizer efficiency [1], and NSA pauses often correspond to word boundaries. The detection of non-speech regions is also very important in *speech enhancement* especially for noise reduction [2], and in *speech coding* [3] methods since their detection will allow the reduction of the average bit rate by using a variable transmission mode. Practical use can be seen in the definitions of the European Telecommunication Standards Institute (ETSI) or in the International Telecommunication Union (ITU) standards.

Moreover, the interest in developing better methods for NSA pause detection also arises from the need to develop better interactive dialog systems and intelligent avatars

[1] Corresponding Author: Vojtěch Stejskal, Brno University of Technology, Dept. of Telecommunications, Czech Republic, e-mail: v.stejskal@phd.feec.vutbr.cz , v.stejskal@gmail.com .

that are able to engage a natural interaction with the user. In this context speech pauses appear to play several functions that guide the flow of the interaction, among those that of signalling a turn change in which someone else might attempt to speak, or indicate internal speaker processes such as anxiety, cognitive difficulty, or problems in encoding the utterance, as well as the absence of the hearer attentiveness [21]-[23]. Furthermore, the frequency and the length of NSA pauses can be an useful biometrical information on the emotional state of the speakers, since it has been shown that when the situation is uncomfortable (for example, when telling a lie) speakers tend to include in their speech a higher number of NSA pauses of longer duration than in non stressed emotional states [9]. A machine communicator, among others, can exploit NSA pause detection to be able to detect the register the speaker is using, or when he is looking for the gaze of his recipients.

The speech/non-speech segment discrimination is not as trivial as it appears and usually most of the detection algorithms fail due to the combination of background noise and speakers' wrong articulation effects. In this context researchers paid more attention in finding new algorithms with special emphasis on the derivation and study of noise robust features and decision rules able to reduce these drawbacks. Earlier algorithms are frequently based on energy thresholding [10], pitch detection [11], zero-crossing rate [7], periodicity measure [12], cepstral features [13], spectrum analysis [14], and Linear Prediction Coding (LPC) [15] or their combinations. The efforts to enhance detection performance have led to the implementation of statistical models with decision rules derived from the Likelihood Ratio Test (LRT) applied to a set of hypotheses [8]. Recently, the Gaussian statistical model improved with the incorporation of an effective hang-over scheme based on the Hidden Markov Model (HMM) was applied in order to achieve more reliable results [16]-[17]. A different approach exploits a set of fuzzy rules implemented into the detection algorithm [18]. Beyond that, it has been shown that algorithm robustness can be improved by using the signal-to-noise ratio (SNR) parameter and long-term information about the speech/non-speech signal performed separately on each filtered spectral band to formulate the appropriate decision rule for the problem under examination [4] [19].

Each of the above proposed solutions has proved to give satisfactory performance when tested on standard databases, such as TIMIT [20], or NTIMIT [21], or Aurora framework [22] where noise has constant attributes that do not change from one record to another or during record. However, the above algorithms performance decreases when the environmental noise changes due to variations of the recording environment.

To overcome this problem, we propose an approach that exploits a different NSA detection strategy based on adaptive energy thresholding, where the detection of NSA pauses is made exploiting along time speech signal energy values measured in different frequency regions. The rationale is that high energy regions can be assigned to speech segments, whereas, low energy regions as a part of a bounded pause. The main problem for a direct detection of NSA pauses is caused by local energy fluctuations, not only due to transient consonants, but also to the presence of environmental noise. Fixed threshold methods are of low efficiency [10] because of high-level energy variations across the speech signal. In our approach, to solve this problem an adaptation of the threshold values is implemented in real-time. Threshold values are continuously adjusted on the basis of long-term speech and noise energy information as well as additional parameters that will be described in the following.

A block diagram of the proposed NSA pause detection system is shown in Figure 1. The raw speech signal is taken as the input and the system output discriminates speech from NSA pauses via identifying the borders of the non-speech regions. Individual diagram elements are described in the following text.

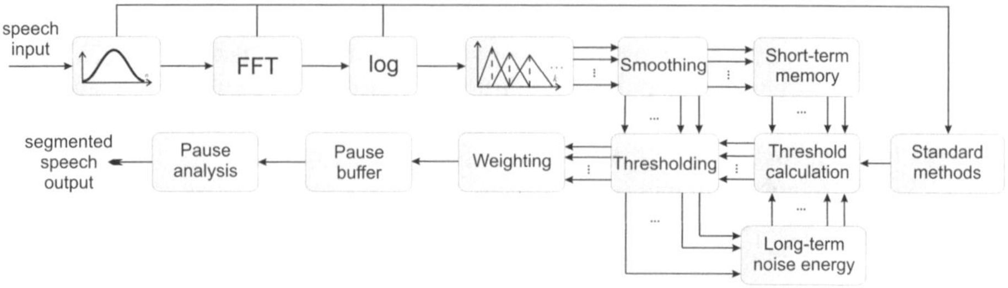

Figure 1 Diagram of the speech pause segmentation system.

1. Feature extraction

The appropriate extraction of an adequate feature vector for NSA pause detection and the development of a robust decision rule are challenging problems that affect the performance of the entire detection algorithm particularly when it operates under noisy conditions.

An input speech signal is divided into frames using a 20 ms Hamming window with 10 ms overlap. The frame is purposely chosen to be as small as possible to ensure detection of rapid fluctuations in the speech waveform. The log-energy values are then computed for each window by the Fast Fourier Transform (FFT). A logarithm function is applied to improve signal resolution in energy. Next, the computed spectrum is divided into several sub-bands by a filter bank. The band edges of the filters are chosen to follow two frequency scales [5] that match psycho-acoustically-based filters related to the human ability to perceptually resolve sounds with respect to frequency. The *Mel-Scale* was chosen in preference to the *Bark scale* because of the better performance we experimentally obtain when this band subdivision is used. On the Mel-scale, the band edges are linear for low frequencies and increase logarithmically with higher frequency, imitating the auditory critical bands resolution. Filters are normalized according to their varying bandwidth. A Mel-frequency m is computed from a linear frequency f as:

$$m = 2595 \log_{10}(1 + f / 700)\,. \tag{1}$$

Instead of a standard filter distribution, our approach employs a small number of filters with boundaries on critical or Mel frequencies. A proper selection of these filters is critical to the nsa pause detection process. We observed that signal parameters of interest for our task can be successfully described by as few as four filters distributed in the range of [0-4 kHz] because this frequency range contains an adequate amount of information for vocal activity and vocal tract articulatory features (see Figure 2). The frequency region from 0 to 1 kHz is largely covered by one or two filters that approximately specify non-speech segment locations. The rest of the filters serve as a correction for the presence of noise and consonant closures. The pre-processing of the speech signal then results in a set of four features each obtained from the four filter bands.

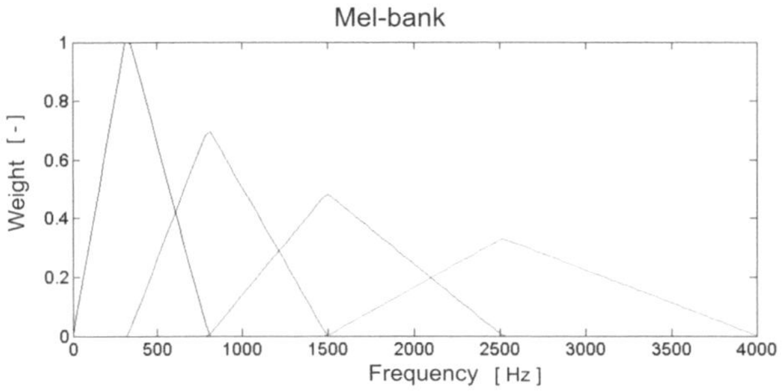

Figure 2 Applied Mel-bank filters.

The high energy feature fluctuations that result from the small window size are then simply suppressed by the recursive smoothing algorithm on each filter channel described in Eq. (2), where E_c is a feature computed from k^{th} filter in n^{th} time window and α is a forgetting factor whose value is experimentally set to 0.9:

$$E_c(k,n) = \alpha E(k,n-1) + (1-\alpha)E(k,n). \qquad (2)$$

2. Adaptive threshold algorithms

The decision whether the processed frame is a speech or a non-speech segment, is made by applying a thresholding algorithm to the output of each filter band, which are simultaneously thresholded. The threshold computation takes into account not only the features provided by the filter-bank but also, information describing backward signal characteristics, i.e. acoustic features computed on the previous frames. The activity of the thresholding algorithm in each sub-band can be in general summarized as follow:

- Two different threshold variables, T_s and T_p, are computed. A speech frame is detected if the energy distribution value of the boundary regions exceeds the threshold value T_s. On the contrary, if the energy distribution value falls below the threshold T_p, a non-speech frame is detected. This separation partly approximates the energy distribution value of the boundary regions.
- Both thresholds are recomputed any time the previous threshold value T_p allow us to assume that a non-speech frame is detected, i.e. both thresholds T_s and T_p are recomputed only during the non-speech frames. Thresholds then partly approximate the time-variant noise levels in speech.
- A primary threshold value T is computed for the first band, when the algorithm begins to operate, as a mean of all first-frame energy features independently of whether the first frame is a non-speech or a speech segment. The first NSA segment is detected when all energy features fall below the threshold T in any of the following frames. The primary value Ts is set as equal to T at the beginning of the first NSA segment. The following values of T_s and T_p are computed directly from each NSA frame as will be described in the following. This initial threshold value set-up may result in a minor acceptable misclassification since the first detected NSA frame position may show a slight offset with respect to its manual detected position.

- The primary T energy values for the remaining filter bands are taken in the same frame position as where the first NSA segment in the first band was detected.
- The time-varying signal acoustic features exploited by the proposed approach are: the mean energy in the detected pauses, the mean energy of the speech signal, the maximum noise energy in non-speech regions, the minimum noise energy in the current detected pause and the none-speech energy slope vector which serve for decision whether the minimum energy in non-speech segment is reached or not. These features are used in threshold equations described in sections 2.1 and 2.2.
- Several algorithms for threshold calculation and adaptation were designed and experimentally tested due to difficulty to identify a procedure capable of preventing threshold value fluctuations that might arise from either a random noise or long-term silent segments. Among them two were selected because of their robustness to adapt to signal fluctuations and are described in the following text. They are called the Min/Max and the Spectral Flatness method.

2.1. Min/max method

The *min/max* algorithm is based on the ratio of predicted minimal noise energy in a detected non-speech region to the maximal noise energy computed on recently detected non-speech regions. Since the entire detection system runs in real-time, the processing of minimum noise value must be predicted on-line in concurrence with the detection of NSA pauses. The min/max ratio allows an adaptation of the generalized threshold level in response to changes of the noise level. The computation of the generalized threshold is described by Eq. (3), where T_k is the threshold value computed for k^{th} band and n^{th} frame, $\overline{N}$ is the mean of the noise energy computed on recently detected pauses, $N_{\max}$ is the maximum noise energy, $N_{\min}^{p}$ is the minimum noise energy in the current detected NSA pause p:

$$T_k(n) = \overline{N} + \overline{N} \cdot \left(1 - \frac{N_{\max}}{N_{\min}^{p}}\right) \tag{3}$$

Thresholds T_s and T_p are then equal to $T_k(n) = T_{ks}(n) = 1.2 \cdot T_{kp}(n)$. They are also protected against overflow and underflow of appropriate energy levels, where the upper level is represented with a long-term mean signal energy value and the lower level with a long-term mean noise value. The long-term mean signal energy is computed from the last previous 10 seconds of input signal and the long-term mean noise value from the last previous 2 seconds of detected NSA segments.

2.2. Spectral flatness method

An alternative to the min/max method is to introduce a correction factor into the adaptive algorithm and this modification is described by Eq. (4), where F_c is the threshold correction obtained from the spectral flatness function described in the Eq.

(5) below, λ is constant signal-to-noise ratio (SNR) correction which is estimated once during the first detected non-speech segment and then it remains constant over the following computations. Instead of using the threshold computation procedure proposed in the above section, here an SNR is computed as the ratio of the average speech energy and the noise energy in the first detected NSA pause. The SNR reflects the speakers' and background energy variations in each utterance. Another parameter involved in the generalized threshold computation is the constant γ, which can take two different values (identified after trial and error processes) according to the nature of the processed frame; γ is set to $\gamma_{T_s} = 1$ if the processed frame is a speech frame, otherwise is set to $\gamma_{T_p} = 1.6$ for non-speech frames

$$T_k(n) = \lambda\overline{N} + \frac{\gamma}{4}\left(1 - F_c(n)\right). \tag{4}$$

The introduction of a correction factor revealed useful for emphasizing word boundaries and intra-pause noise. After several trial and error processes, and comparisons of the obtained experimental performance, the most appropriate procedure identified to compute the correction factor was the *spectral flatness* method (see Eq. (5)), where f is a spectral log-energy and n is the time window:

$$F_c(n) = \int \log(S(f,n))\,\mathrm{d}f - \log\left(\int S(f,n)\,\mathrm{d}f\right). \tag{5}$$

Figure 3 illustrates with an example how the NSA pause detection algorithm exploits spectral flatness information to compute the different thresholds involved in the identification of speech and non-speech segments. The parameter T_{lp} represents the pause threshold, T_{ls} the speech threshold, and $\overline{N}$ is the mean of noise energy of the first band.

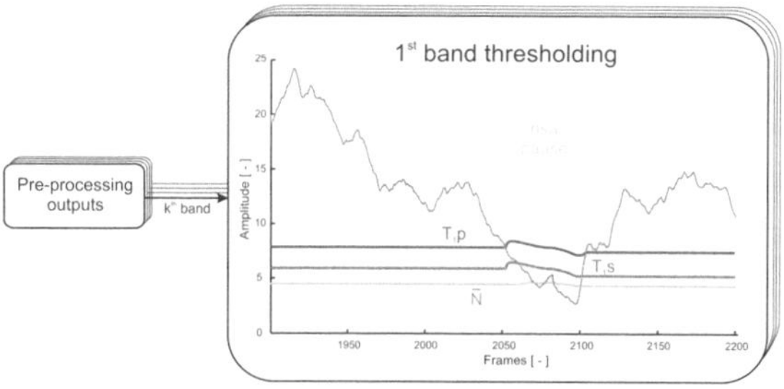

Figure 3 Spectral flatness thresholding.

3. Post-processing

The output of the detection algorithm is a vector of $J=4$ binary thresholded outputs, one for each sub-band, that is then processed by a mapping algorithm indicating if the frame under examination belongs or not to NSA pauses. The mapping algorithm exploits the vector component values, where a value of 1 hypothesizes a possible NSA pause, and 0 otherwise (see Figure 4). A frame is assumed to belong to an NSA pause when almost all the components of the binary vector outputted by the filter bank (only the output of the first band filter can take a value of 0) will take a value of 1.

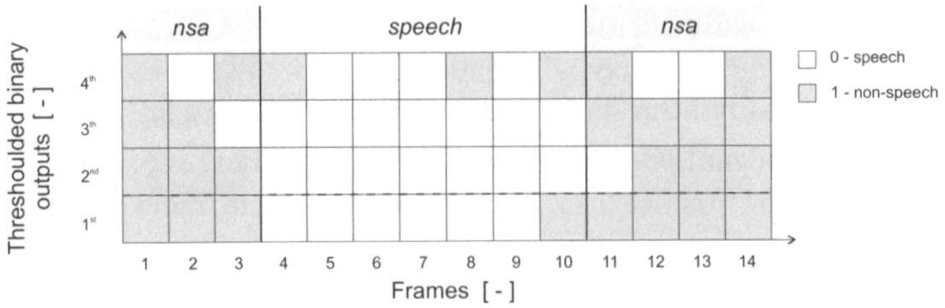

Figure 4 Mapping algorithm.

Backward analysis is performed after each new detected NSA pause to avoid detection errors due to small speech segments (e. g. weak fricatives or reduced vowels) whose energy does not exceed the computed threshold value. The purpose of the analysis is to avoid including in the computation of the energy and the slope vector associated to the NSA pauses also acoustic features belonging to these speech segments.

4. Comparative Tests

The system efficiency was tested on the narration recordings of twelve Italian and 1 American English female speakers (9 Italian males and 2 females). The narrations were all referred to an episodic cartoon that the speakers had seen and were reporting to a friend. For this purpose the speakers were videotaped at different environmental noise levels with a digital video-camera and the audio to test our proposed algorithms was extracted directly from the video. The average narration length was 5 minutes ± 2.5 (standard deviation). The audio was sampled at 32 kHz and quantized at 16 bits. Noise sources included color noise, babble noise, echo and environmental noise. The minimum average SNR was 3dB. For reference the signal was manually labeled for speech and NSA pause frames using the Speechstation2™ from Sensimetrics. A cleaned version of the audio was obtained using the Adobe Audition™ noise cancellation techniques [6] and the proposed algorithms were also tested on this cleaned speech.

Detection performance was assessed in terms of the non-speech hit-rate (HR0) and the speech hit-rate (HR1) defined in Eq. (6) as a fraction of all present non-speech or speech frames that are correctly detected as pause or speech frames, while $N_{0,0}$ and $N_{1,1}$ are the numbers of correctly classified non-speech and speech frames respectively and N_0^{ref} and N_1^{ref} are the number of real non-speech and speech frames respectively, coming out from the manual labeling.

$$HR0 = \frac{N_{0,0}}{N_0^{\mathrm{ref}}}, \quad HR1 = \frac{N_{1,1}}{N_1^{\mathrm{ref}}} \tag{6}$$

The results are summarized in Table 1 where are reported the performance of two proposed algorithms both on noisy and cleaned speech. The average accuracy of speech pause detection with 100 ms tolerance for the spectral flatness algorithm was 55% and 57% for noisy and clean speech respectively. For the min/max method it was 36% in noisy and 39% in clean speech. We can notice relatively high values of non-speech frame hit-rate (HR0) above all in the result of the spectral flatness algorithm. An explanation can be given on the basis of several mutually dependent methods that are used in the detection system. The SNR computation allows us to adapt the threshold with respect to varying environmental and speaker characteristics. The main advantage of exploiting the spectral flatness in the detection algorithm is that with this procedure pause boundaries are more appropriately detected, especially pause onsets, which represents one of the main difficulties in accurate pause detection and significantly increases the HR0 value.

It is worth noting that pause detection performance does not vary significantly in noisy or clean conditions. This implies possible difficulties with NSA pause detection even if the speech is without noise and supports the robustness of our algorithms as regards noise presence.

Table 1 Average speech/non-speech hit-rates.

Input files	Detectors	Min/max		Spectral flatness	
		Male	Female	Male	Female
Original noisy speech	HR1 (%)	95.95	96.59	98.91	98.34
	HR0 (%)	62.19	54.78	74.21	67.85
Cleaned speech	HR1 (%)	96.60	96.42	97.91	98.47
	HR0 (%)	62.37	60.27	77.49	68.90

5. Conclusions

A software system for NSA speech pause detection in noisy and cleaned conditions, which exploits two adaptive log-energy based thresholding methods is proposed. Both the methods show similar performance and their efficacy is comparable with standard VADs, [7]-[8]. No significant performance differences are noticed in the cleaned and noisy speech conditions suggesting that the reliability of the proposed methods do not depend on the signal quality even though, according to the results obtained, NSA pause detection remains a difficult task that should be investigated more.

The advantage of the proposed methods is that they are robust with respect to high energy noise localized randomly in sub-bands and the low energy wideband noise. They do not require a priori knowledge of the SNR and a priori threshold values. However, they present a disadvantage with regards to the high energy noise spread widely over the band. In our future work, the methods will be embedded in a multimodal pause detection system using both speech and gestures to improve the naturalness of intelligent avatars and interactive dialog systems.

Acknowledgements

The paper has been supported by the National Research Project "Information Society" No 1ET301710509, by the Ministry of Education, Youth and Sports, project number F2101/2006/G1, and by the **COST Action**

2102: "Cross Modal Analysis of Verbal and Nonverbal Communication (CAVeNC), http://www.cost.esf.org/index.php?id=110&action_number=2102."

References

[1] X. Huang, A. Acero, H. Hon: Spoken Language Processing: a Guide to Theory, Algorithm, and System development. New Jersey: Prentice-Hall, 2001, 375-463.
[2] L. Rabiner, B. Juang: *Fundamentals of Speech Recognition*. New Jersey: Prentice-Hall, 1993.
[3] R. Goldberg, L. Riek: *A Practical Handbook of Speech Coders*. Boca Raton: CRC Press, 2000.
[4] J. Ramírez, J. Segura, C. Benítez, A. Torre, A. Rubio: Efficient Voice Activity Detection Algorithms Using Long-term Speech Information, *Speech Communication*, **42** (2004), n.3-4, 271-287.
[5] T. Quatieri: *Discrete Time Speech Signal Processing*. New Jersey: Prentice Hall, 2002.
[6] Adobe Audition 2, Adobe Audition Team, Adobe System Incorporated, San Jose, United States, [Computer Program].
[7] ITU, A Silence Compression Scheme for G.729 Optimized for Terminals Conforming to Recommendation V.70. In *ITU-T Recommendation* G.729-Annex B, 1996.
[8] ETSI, Voice Activity Detector (VAD) for Adaptive Multi-Rate (AMR) Speech Traffic Channels. In *ETSI EN 301 708 Recommendation,* 1999.
[9] P. Ekman: *Telling lies*, W.W. Norton & Company, Inc, NY, 2001
[10] S. Gerven, F. Xie: A Comparative Study of Speech Detection Methods. In *Proceedings of 5th European Conference on Speech Communication and Technology EUROSPEECH*, Rhodos, **3**, 1095-1098, 1997.
[11] R. Chengalvarayan: Robust Energy Normalization Using Speech/Nonspeech Discriminator for German Connected Digit Recognition. In *Proceedings of EUROSPEECH 1999*, Budapest, Hungary, 61–64, 1999.
[12] R. Tucker: Voice Activity Detection Using a Periodicity Measure. In *Proceedings of Inst. Electr. Eng.*, **139**, 377–380, 1992.
[13] F. Stouten, J. Martens: A Feature-Based Filled Pause Detection System for Dutch. In *Proceedings of IEEE Automatic Speech Recognition and Understanding Workshop*, St. Thomas, U.S.A., 309-314, 2003.
[14] M. Marzinzik, B. Kollmeier: Speech Pause Detection for Noise Spektrum Estimation by Tracking Power Envelope Dynamics. *IEEE Trans. Speech Audio Process.*, **10**, no. 6, 341–351, 2002.
[15] L. R. Rabiner, M. R.Sambur: Voiced-Unvoiced-Silence Detection Using Itakura LPC Distance Measure. In *Proceedings of IEEE International Conference in Acoustics, Speech, and Signal Processing*, 323–326, 1977.
[16] J. Sohn, N. S. Kim, W. Sung: A Statistical Model-Based Voice Activity Detection. *IEEE Signal Processing Letters*, **6**, no. 1, 1–3, 1999.
[17] J. Stadermann, V. Stahl, G. Rose: Voice Activity Detection in Noisy Environments. In *Proceedings of European Conference on Speech Communication and Technology*, Aalborg, Denmark, 1851–1854, 2001.
[18] F. Beritelli, S. Casale, A.Cavallaro: Adaptive Voice Activity Detection for Wireless Communications Based on Hybrid Fuzzy Learning. In *Proceedings of IEEE Global Telecommunications Conference*, **3**, 1729-1734, 1998.
[19] J. Ramírez, J. C. Segura, C. Benítez, A. Torre, A. Rubio: A New Voice Activity Detector Using Sub-Band Order-Statistics Filters for Robust Speech Recognition. In *Proceedings of ICASSP*, Montreal, **1**, 849-852, 2004.
[20] Garofolo, J., et al.: *DARPA TIMIT Acoustic-Phonetic Continuous Speech Corpus*, Gaithersburg, National Institute of Standards and Technology, NTIS Order No. PB91-505065, CD-ROM, 1990.
[21] Jankowski et al.: NTIMIT: A Phonetically Balanced, Continuous Speech, Telephone Bandwidth Speech Database, in *Proc. of ICASSP*, 1990.
[22] H. G. Hirsch, D. Pearce: The AURORA Experimental Framework for the Performance Evaluation of Speech Recognition Systems Under Noise Conditions. In *Proceedings of ISCA ITRW ASR2000*, 2000.
[23] G. Aversano, A. Esposito, A.M. Esposito, M. Marinaro: A New Text-Independent Method for Phoneme Segmentation. In R. L.Ewing et al. (Eds), *Proceedings of the IEEE International Workshop on Circuits and Systems*, **2**, 516-519, 2001.
[24] C. Goodwin: Restarts, Pauses, and the Achievement of a State of Mutal gaze at the Turn Beginning. *Sociological Inquiry*, 50, 272-302, 1984.
[25] E. Goldman-Eisler: Pauses, Clauses, Sentence. *Language and Speech*, 15, 103-113, 1972.
[26] S. Jr. Duncan: Some Signals and Rules for Taking Speaking Turn in Conversation. In S. Weitz (ed.) *Nonverbal Communication,* New York, Oxford University Press, 1974.

Fundamentals of Verbal and Nonverbal Communication and the Biometric Issue
A. Esposito et al. (Eds.)
IOS Press, 2007

On the Analysis of F0 Control Characteristics of Nonverbal Utterances and its Application to Communicative Prosody Generation

Ke LI[1a], Yoko GREENBERG[a], Nagisa SHIBUYA[a], Nick CAMPBELL[b], Yoshinori SAGISAKA[a]
[a]*GITI Waseda University,*
[b]*ATR Spoken Language Communication Research Laboratory*

Abstract. Aiming at communicative prosody control, fundamental frequency (F0) control characteristics of nonverbal utterance "n" were modeled. For the input to F0 control, a three-dimensional vector is employed to quantify its perceptual impressions, since MDS (Multi-Dimensional Scaling) analysis can efficiently reduce impression vector space dimension to three-dimensional as shown in our previous studies. For the output, F0 generation model parameters were employed to efficiently reduce control freedoms only by the magnitude of long-range control (phrase component) and short-range control (accent component) and their timing. From the analysis of nonverbal utterances using an F0 generation model, their control characteristics were well understood and directly linked to impression vectors. Based on the training of a non linear mapping from an impression vector to F0 generation model parameters, it was experimentally confirmed that F0 contours were generated directly from input impression vectors.

Keywords. Prosody control, communicative speech synthesis, paralinguistic information

Introduction

In the study of prosody control, up to now, linguistic control correlates have mainly been studied. Though it is quite natural to formulate prosody control from linguistic viewpoints, this tendency may restrict the understanding of prosody control only to written language. As it is very difficult to specify all input factors needed to generate communicative prosody of spoken language and there is little studies on the description of communicative prosodic patterns, there is no quantitative modeling for prosody in real human communications. For better understanding of spoken language and the generation of communicative speech synthesis, we have been studying the F0 characteristics of nonverbal utterances of "n" [1]-[3]. The utterance "n" is, acoustic-phonetically, a sustained utterance of either [m], [n], or [eng]. In Japanese, it can be an interjection, rejoinder or filler depending on the context or situation. Most importantly, it does not have any particular lexical meaning, default intonation or accent, so that its prosody conveys so-called *paralinguistic* information.

Through these studies, we have tried to specify the input and the output for communicative prosody generation. By analyzing F0 of nonverbal utterances "n" in conversational situations, we found that they could be classified in terms of their heights and dynamic patterns. The MDS (Multi-Dimensional Scaling) analysis on

[1] Corresponding Author: Ke Li, GITI Waseda University, rikoku@akane.waseda.jp.

subjective listening tests asking their perceptual impressions revealed the good correspondences between perceptual impressions and their F0 characteristics [1]. Moreover, strong correlations between utterance contents and their prosody suggested a possibility of communicative prosody generation for general single-phrase utterances [2][3].

In this paper, we tried to control prosody of "n" by its perceptual impressions using a neural network. For the input to this control model, we employed a vector reflecting multi-dimensional impressions. For the output, we employed F0 generation model parameters to efficiently reduce control freedoms only by the magnitude of long-range control (phrase component) and short-range control (accent component) and their timing. In the following sections, first, we analyze F0 characteristics of "n" based on F0 generation model proposed by Fujisaki in Section2. In Section 3, a computational model is proposed for F0 generation parameters from an impression vector. Section 4 shows the experimental results on the prediction of F0 generation parameters. Finally, we summarize all findings and discuss further steps towards communicative speech synthesis.

1. Prosodic analysis of nonverbal utterances "n" based on an F0 generation model

1.1 Parametric description of F0 using a generation model

We adopted parametric description of an F0 contour by F0 generation model [4]. As well known, this model decomposes an F0 contour into phrase components and accent components obtained as an output of second order critical damping system for impulsive inputs. A phrase component *Gp(t)* and a n accent component *Ga(t)* are given by the following equations.

$$G_{pi}(t)=\begin{cases}\alpha_i^2 t\exp(-\alpha_i t), & t\geq 0,\\ 0, & t<0,\end{cases} \quad (1)$$

$$G_{tj}(t)=\begin{cases}\min[1-(1+\beta_j t)\exp(-\beta_j t),\gamma], & t\geq 0,\\ 0, & t<0,\end{cases} \quad (2)$$

where α and β are the time constants for the phrase and accent control, respectively. Based on the former results, they were fixed at $3.0\,s^{-1}$ and $20.0\,s^{-1}$, respectively. The ceiling parameter γ was also fixed at 0.9. An F0 contour *F0(t)* is given by the sum of these components described the following equation.

$$\ln F_0(t)=\ln F_{\min}+\sum_{i=1}^{I}A_{pi}G_{pi}(t-T_{0i})+\sum_{j=1}^{J}A_{aj}\{G_a(t-T_{1j})-G_a(t-T_{2j})\} \quad (3)$$

where, lnF_{min} is the bias level, i and j are the number of phrase and accent commands, respectively. A_{pi} and A_{aj} are the magnitude of the ith phrase command and the jth accent command respectively. T_{0i} is the time of the ith phrase command. T_{1j} and T_{2j} are the onset and the reset time of the jth accent command respectively. Since we apply this model to a single utterance of "n", only six parameters (A_p, A_a, T_0, T_1, T_2, lnF_{min}) are needed to represent one F0 contour.

1.2 F0 generation parameter distributions of nonverbal utterance "n"

Randomly selected 1,076 samples from 23,648 nonverbal utterances of "n" collected in JCT/CREST ESP project [5] were used for the analysis. As shown in our previous studies [1]-[3], we can classify these F0 patterns by their dynamic patterns and heights.

Table1. The averages and the standard deviations (σ) of F0 generation parameters for four dynamic patterns of nonverbal utterances "n"

F0 Patterns		Ap	Aa	T0	T1	T2	lnFmin
Rise	Average	0.550	0.400	-0.180	0.160	0.960	4.70
	σ	0.128	0.128	0.550	0.242	0.073	0.169
Flat	Average	0.560	0.210	-0.270	-0.150	0.260	4.55
	σ	0.186	0.082	0.036	0.066	0.344	0.214
Fall	Average	0.650	0.630	-0.290	-0.170	0.240	4.38
	σ	0.149	0.158	0.003	0.031	0.152	0.180
Rise&Fall	Average	0.630	0.510	-0.250	-0.050	0.570	4.56
	σ	0.154	0.208	0.050	0.087	0.153	0.220
All	Average	0.590	0.390	-0.250	-0.070	0.460	4.55
	σ	0.168	0.216	0.055	0.171	0.363	0.225

Table 1 shows the analysis results classified by four dynamic patterns (rise, fall, flat and rise&fall). As shown in the Table, four dynamic patterns have different distributions. Though four patterns have overlapped in each parameter, their combinations specify the dynamics of each pattern. The perceptual listening tests using synthetic speech with F0 recalculated from these average values confirmed the strong correlation to their perceptual impressions. As shown in the previous studies, we can approximate perceptual impressions in three dimensions of *positive-negative*, *confident-doubtful* and *allowable-unacceptable*. Through the analysis, we could find that four types of F0 dynamics (rise, flat, fall, rise&fall) systematically corresponds to the perceptual impressions. On the other hand, F0 average height gives the distinction between high group (confident, allowable and positive) and low group (doubtful, unacceptable and negative).

2. A computational model of F0 generation parameters from an impression vector

2.1 A perceptual impression vector

To specify prosody, we adopted a perceptual impression vector as input information. It consists of multi-dimensional eight level scores, 0(not at all) - 7(very much), to measure how much each speech sample related to 26 impression words (doubt, ambivalence, understanding, approve, deny, objection, agreement, dark, weakly, not interested, bad mood, heavy, bothering, audacious, anger, annoying, cheerful, delight, gentle, good mood, excited, happy, light, interested and bright). Multi-Dimensional Scaling (MDS) analysis showed that 26 dimensions could be effectively reduced to three dimensions that can nicely correlate to the variation of impressions with F0 dynamics and heights [1]. We use this reduced three-dimensional vector as an input impression vector.

2.2 A computational model of F0 generation parameters using neural nets

F0 generation parameters (A_p, A_a, T_0, T_1, T_2, lnF_{min}) were calculated using three-layered feed-forward neural networks. To check the dependence between F0 generation parameters, we trained the following three types of neural network (NN)s.

(1) Each: An independent NN for each F0 generation parameter, i.e. six NNs in total

(2) Group: Two types of NNs for (A_p, A_a, lnF_{min}) and (T_0, T_1, T_2) to constrain parameter dependencies in frequency and time domains independently

(3) All: One NN for all six F0 generation parameters to constrain all parameter dependencies

3. Experiments on the prediction of F0 generation parameters

3.1 Experimental setup

For the experiments, we asked five subjects to score their perceptual impressions as 26-dimensional vectors. 43 samples of "n" utterances were extracted from the VQ centroids of 1,076 samples shown in Section2. These vectors were reduced to three dimensional impression vectors using MDS analysis results and used as input information.

Three types of three-layered NNs; (1) six 3×3×1, (2) two3×3×3 and (3) one 3×3×6, were trained using the training data. Furthermore, to check the validity of MDS analysis, 26-dimensions impression vectors were directly input to (26×12×6) NN. To compare the performance to open data, we measured not only the performance for the training data but also for 20% held out open data.

3.2 Experimental results

Table 2 shows the RMS errors between predicted F0 generation parameters and observed ones of open data. As shown in the Table, the errors of (2) Group and (3) All turned out to be smaller than (1) Each for open data. This result suggests the parametric dependencies among F0 generation parameters. Though some of the RMS errors for the open data are still big, we could have confirmed that the predicted F0 generation parameters could reproduce F0 contours reasonably well.

As for 26-dimensions impression vector input, the errors turned out to be smaller than (1) Each, but bigger than (2) Group and (3) All for open data. Though NN has a capability of reducing the dimensions, the reduction using MDS gave better performance than automatic reduction by NN. This may be caused by the difference in dimension reduction criteria. The dimension reduction using MDS is carried out to extract dominant factors from redundant input to explain the relationship between data, i.e. distances between data. On the other hand, the reduction using NN can only reduce the dimension simply by reducing total prediction error.

Table2. RMS errors of F0 generation parameters for the training data and open data

	Training data				**Open data**			
F0 parameter	**Each**	**Group**	**All**	**26-vectors**	**Each**	**Group**	**All**	**26-vectors**
Ap	0.098	0.137	0.067	0.006	0.382	0.259	0.232	0.313
Aa	0.109	0.116	0.112	0.008	0.356	0.212	0.203	0.303
T0	0.018	0.063	0.025	0.019	0.052	0.045	0.054	0.061
T1	0.100	0.145	0.094	0.018	0.418	0.182	0.218	0.276
T2	0.060	0.101	0.078	0.008	0.411	0.229	0.281	0.348
lnFmin	0.146	0.107	0.094	0.008	0.383	0.341	0.355	0.358

4. Conclusion and future works

This paper introduced experimental results on the F0 control modeling of nonverbal utterance “n”. Three-layered feed-forward neural networks were trained using three-dimensional input vectors representing perceptual impressions of the corresponding utterance and six-dimensional outputs of F0 generation parameters. The prediction experiments of F0 generation parameters showed reasonable performance for open data. Through errors between predicted parameters and the observed ones were not small enough, the regenerated F0 patterns from input impression vectors showed the same characteristics both in their shapes and perceptual affects. Since the proposed modeling framework is easy to expand to general utterances, this computational modeling has an advantage to generalize to ordinary communicative prosody generation as already confirmed in some pilot studies using general single phrases [3]. We would like to complete further quantitative evaluations and apply this modeling to more general utterances phrases.

Acknowledgements

A part of this work is supported in part by Waseda Univ. RISE research project of "Analysis and modeling of human mechanism in speech and language processing" and Grant-in-Aid for and Scientific Research B-2, No. 18300063 of JSPS.

References

[1] Kokenawa, Y., Tsuzaki, M., Kato, K., and Sagisaka, Y., “F0 control characterization by perceptual impressions on speaking attitudes using Multiple Dimensional Scaling analysis”, Proc. ICASSP, pp.273-276, Mar.2005

[2] Greenberg, Y., Tsuzaki, M., Kato, K., and Sagisaka, Y., “Communicative speech synthesis using constituent word attributes”, Proc. INTERSPEECH2005, pp.517-520, Sep.2005

[3] Greenberg, Y., Tsuzaki, M., Kato, K., and Sagisaka, Y., “A trial of communicative prosody generation based on control characteristic of one word utterance observed in real conversational speech”, Proc. Speech Prosody 2006, pp.37-40, May 2006

[4] H. Fujisaki, K. Hirose, “Analysis of voice fundamental frequency contours for declarative sentences of Japanese,” J. Acoust. Soc. Japan (E),Vol.5, No.4, pp.233-242, 1984.

[5] JST/CREST Expressive Speech Processing project. http://feast.atr.jp/esp/esp-web/

Fundamentals of Verbal and Nonverbal Communication and the Biometric Issue
A. Esposito et al. (Eds.)
IOS Press, 2007

Effectiveness of Short-Term Prosodic Features for Speaker Verification

Iker LUENGO[1], Eva NAVAS, Inmaculada HERNÁEZ
University of the Basque Country, Spain[2]

Abstract. In this work prosodic information is considered in order to improve current state-of-the-art Automatic Speaker Verification (ASV) systems. A system based on pitch and energy curves is combined with a traditional MFCC based one to determine whether simple short-term prosodic information is useful. Experiments carried out over read speech show that there is not significant improvement in using short-term prosody. Results are presented and discussed.

Keywords. Speaker verification, prosody

Introduction

In the last few years, the interest on biometrics based automatic identification systems has grown considerably. This fact is reflected in the increase of biometric verification system contests, such as the Fingerprint Verification Competition [1] or the NIST Speaker Recognition Evaluations [2] where new algorithms and methods are being proposed in order to improve existing results. Automatic Speaker Verification (ASV) is not an exception. Most of the current implementations use features related to the spectral envelope to parameterize the voice (MFCC, LPCC, etc.) achieving very good performance (Equal Error Rate (EER) close to 3% for telephonic channel) [3] [4] [5]. However, recent research considers prosodic information in the system in order to reduce error rates.

Speech prosody refers to the intonation, energy and rate of the speech. It is well known that these features are related to the speaker's identity. Furthermore, prosody is related to the voice source in the source-vocal tract model and it can be assumed that it is uncorrelated with the spectral envelope shape. Therefore, adding prosodic features to the already used spectral features may lead to an improvement in the system's performance.

Most of the works in this area focus on the use of some kind of long-term prosodic information [6] [7] [8], together with the fusion with a state-of-the-art system. Some others use short term intonation and power values [9] [10]. This last approach is very appealing, because the new features can be easily combined with the traditional cepstral coefficients.

The work described in this paper focuses on determining whether simple short-term prosodic information can be used to improve current state-of-the-art ASV. The

[1] Corresponding Author: Departamento de Electronica y Telecomunicaciones, Escuela Técnica Superior de Ingeniería, Alda. Urquijo s/n. 48013 Bilbao, Spain. e-mail: ikerl@bips.bi.ehu.es.
[2] This work has been partially financed by Spanish Ministry through TIC200308382C0503

article is organized as follows: First, the new ASV system using both spectral and prosodic features is presented. Then, a description of the database used to carry out the experiments is given, followed by a report of the performed experiments and the results. Finally, these results are commented.

1. Verification System Description

1.1 Baseline System

The baseline system consists of a traditional GMM-UBM system [11] with Mel frequency cepstral coefficients (MFCC) parameterization. More precisely, an 18 MFCC feature vector was extracted every 10 milliseconds. These vectors were augmented with the first and second order derivatives. Cepstral Mean Subtraction (CMS) was applied in order to reduce channel effects [12].

Speaker models were trained by Maximum a Posteriori (MAP) adaptation of the previously trained Universal Background Model (UBM) [3]. Only the means were adapted, leaving the variances and weights unaltered. As it is usual in this kind of systems, the scores were normalized using UBM and HNorm normalization [13].

1.2 Prosodic System

In the prosody based system, energy and intonation related information was used. Separate models were created for prosody modelling in voiced and unvoiced regions, in order to handle the discontinuities of the intonation curve. The signal's power was estimated every 10 milliseconds using Hamming windowing 30 milliseconds length. F0 was also estimated every 10 milliseconds, using a method based on cepstrum transformation followed by a Viterbi algorithm. This method not only gives the estimated value of F0, but also decides whether a frame is voiced or not. Once the power and intonation curves were estimated, their first and second order derivatives were calculated, in order to take into account the dynamics of the features.

The voiced frames were separated from the unvoiced frames to generate two streams of feature vectors. Voiced frames were parameterized with five features (instantaneous F0, its first and second derivative, and the first and second derivative of the power), whereas unvoiced frames were parameterized with only two features (the first and second derivatives of the power). The instantaneous power was discarded in both cases, because this value is more related to the channel gain than to the speaker's identity.

Using these two vector streams, two different models were trained for each speaker, using a traditional GMM-UBM scheme. First, two UBM models were developed, one for voiced and another one for unvoiced frames. Then, the speaker models were created from these by means of MAP adaptation.

During the test phase, and for each recording, two different scores were calculated, one for the voiced and another one for the unvoiced frame streams. These two scores were normalized using UBM normalization before being fused using the product rule. i.e., the final score for the prosodic information is calculated as the product of the scores of the voiced and the unvoiced frame streams.

1.3 Fusion of the Systems

In order to combine the results of both classifiers, a late fusion scheme was chosen. First, the scores using both traditional and prosody-based systems are calculated; next, a final score is obtained by combination of those scores. For this last operation, a Linear Discriminant Algorithm (LDA) [14] was selected. The LDA is capable of finding the linear combination of scores that best separates the scores of the user and the impostors. However, it must be trained over a validation set of speakers before being applied.

2. Database Description

The Ahumada database [15] was used for the experiments carried out in this work. This database consists of voice recordings from 103 male Spanish speakers and was specifically recorded for the training and testing of automatic speaker recognition systems. In fact, it was used during the NIST speaker recognition evaluation campaigns of 2000 and 2001 [2], together with an extension including female speakers.

Although the complete database contains high quality microphone recordings as well as telephone speech recordings, only the later have been used for our experiments. This allows capturing the effects that channel distortion has on the system's behaviour. These telephonic recordings were carried out in three different sessions, and in each of them the speakers used a different telephone handset.

The first session (named T1) was recorded through an internal call with a flat frequency response. As a result there is no spectral distortion in the signals. In the second session (T2) the recording was done through the speaker's home telephone, so the handset type remains unknown. In the third session (T3) each of the speakers used one of the nine handsets that were available in the recording laboratory. Hence, in session T3 the handset type is known for each speaker. In T3 the telephone microphones were either of type carbon button or electret.

From the items in the database only two items were selected for each speaker and session. The first item selected is a common text that was read aloud by each speaker, i.e. one common text for all speakers and sessions. This item will be referred to as *task C* (for Common). The second item is a text, different for each of the speakers and sessions. This second item will be referred to as *task S* (for Specific). All these recordings were sampled at 8 kHz and 16 bit per sample. The mean length of the selected items is about 65 seconds.

In summary, the set of signals used in our experiments consists of 618 items. (103 male speakers * 3 telephone sessions * 2 items selected per session = 618 items) This set was divided into 3 groups as follows: 51 speakers were reserved to train the UBM models, 26 were used for the validation tests, and the remaining 26 were kept as users of the system. Every user was treated as an impostor for the rest of the users which is equivalent to having 25 impostors. The speakers were randomly assigned to one of the three groups in order to assure unbiased results.

Speaker model training was done using both tasks C and S from session T1. This gives about 130 seconds of training material for each speaker. As the T1 session was recorded without spectral distortion, the resulting models are channel independent. Tasks C and S from session T3 were used for the development tests because knowing the handset type used makes it possible to determine the HNorm coefficients for handset normalization. Finally, task S from session T2 was reserved for the final test phase. This way, the final tests were carried out with a handset that was unknown to the identification/verification system as it has not been used before, neither in the training nor in the development phase. Furthermore, using only task S also the contents of the recordings were new to the system.

For the tests (both for development tests as well as for the final tests), nine speech segments of 10 seconds each were extracted from each item allowing a 50% overlap between two consecutive segments. Hence for each speaker nine validation test segments were available. As a result each user was tested against 25*9=225 impostor segments.

All this makes the design of our experiments very realistic. It simulates an environment in which the users make the recordings for training and development in the laboratory, but then try to gain access from their own homes or offices using different phones.

3. Results

In order to estimate the best number of mixtures for the models, different order GMMs were developed, from 2 mixtures up to 1024, and both for the MFCC and the prosody-based systems. The final number of mixtures was selected in order to minimize the EER. Table 1 shows the EER values for the trained systems. As expected, the baseline system achieves better performance than the prosody-based system. It can also be seen that energy information of unvoiced frames does not help to raise the final result of the prosody-based verification system.

According to the achieved EER values, GMMs of 1024 and 128 mixtures were selected for the baseline and prosody-based systems respectively. After applying the LDA and fusing the scores, a final EER of 3.84% was achieved. Figure 1 shows the DET curves [16] of the systems before and after the fusion. It can be seen that the DET curves for the baseline and the combined systems are very close together.

Table 1: %EER values for the baseline and prosody based systems

# mix.	*2*	*4*	*8*	*16*	*32*	*64*	*128*	*256*	*512*	*1024*
Baseline	29.13	20.29	17.18	13.25	10.36	8.12	6.75	5.71	5.21	3.85
Voiced	35.47	33.21	29.88	28.96	27.78	25.32	23.93	24.03	24.74	23.93
Unvoiced	46.53	49.45	47.86	45.86	44.44	43.16	42.74	43.18	43.59	44.44
V+UV	35.27	32.48	29.49	28.77	27.33	25.64	23.93	24.89	24.53	24.14

4. Conclusions

Several reasons can be found to explain the lack of improvement when fusing the prosody and the baseline systems. First, the prosody-based system shows a poor performance. The analysis of the data shows that this is due to the large inter-session variability of the features, which suggests that some kind of normalization should be applied to reduce it. Some tests have been made using simple mean subtraction, but with no success. Second, the designed baseline system performs very well with the data used. This leaves a short margin for improvement. Preliminary experiments using spontaneous speech (from the same database) have shown worse results with the baseline system and a larger improvement when using also prosody.

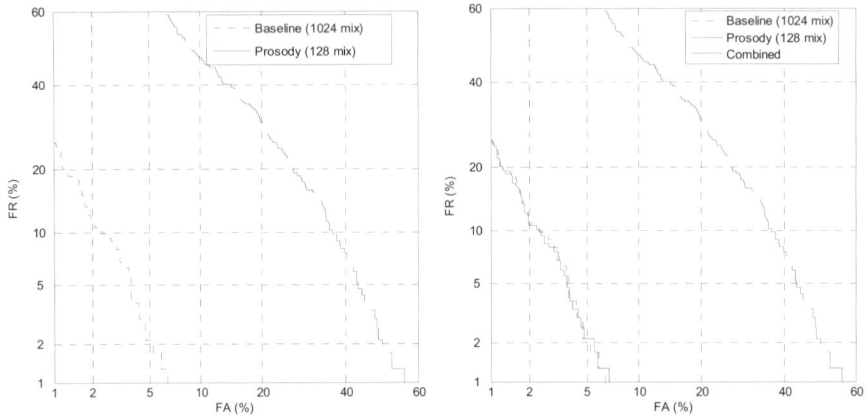

Figure 1: DET curves for the baseline and prosody based systems (left) and the final combined system (right).

References

[1] FVC2006 web site: http://bias.csr.unibo.it/fvc2006/

[2] NIST Speaker Recognition Evaluations' web site: http://www.nist.gov/speech/tests/spk/

[3] D. A. Reynolds, T. F. Quatieri, and R. B. Dunn, "Speaker Verification Using Adapted Gaussian Mixture Models," Digital Signal Processing, vol. 10, pp. 19-41, 2000.

[4] M. A. Przybocki and A. F. Martin, "NIST Speaker Recognition Evaluation Chronicles," presented at Odyssey, Toledo, España, 2004.

[5] A. F. Martin and M. A. Przybocki, "The NIST Speaker Recognition Evaluations: 1996-2001," presented at Speaker Odyssey 2001, Creta, Grecia, 2001.

[6] M. J. Carey, E. S. Parris, H. Lloyd-Thomas, and S. Bennett, "Robust Prosodic Features for Speaker Identification," presented at ICSLP, Philadelphia, EEUU, 1996.

[7] A. G. Adami, R. Michaescu, D. A. Reynolds, and J. J. Godfrey, "Modeling Prosodic Dynamics for Speaker Recognition," presented at ICASSP, Hong Kong, 2003.

[8] B. Peskin, J. Navratil, J. Abramson, D. Jones, D. Klusacek, D. A. Reynolds, and B. Xiang, "Using Prosodic and Conversational Features for High Performance Speaker Recognition," presented at ICASSP, Hong Kong, 2003.

[9] D. A. Reynolds et al., "SuperSID project final report," SuperSID project (http://www.clsp.jhu.edu/ws2002/groups/supersid/) 2002.

[10] M. Arcienega and A. Drygajlo, "A Bayesian Network Approach for Combining Pitch and Spectral Envelope Features for Speaker Verification," presented at COST275 workshop, Roma, Italia, 2002.

[11] F. Bimbot, J. F. Bonastre, C. Fredouille, G. Gravier, I. Magrin-Chagnolleau, S. Meignier, T. Merlin, J. Ortega-García, D. Petrovska-Delacrétaz, and D. A. Reynolds, "A Tutorial on Text-Independent Speaker Verification," EURASIP Journal on Applied Signal Processing, vol. 4, pp. 430-451, 2004.

[12] C. Barras and J. L. Gauvain, "Feature And Score Normalization for Speaker Verification of Cellular Data," presented at ICASSP, Hong Kong, 2003.

[13] R. Auckenthaler, M. J. Carey, and H. Lloyd-Thomas, "Score Normalization for Text-Independent Speaker Verification Systems," Digital Signal Processing, vol. 10, pp. 42-54, 2000.

[14] J. B. Kennedy and A. M. Neville, Basic Statistical Methods for Engineers and Scientists. New York: Harper & Row, 1986.

[15] J. Ortega-García, J. González-Rodríguez, and V. Marrero-Aguiar, "AHUMADA: A Large Speech Corpus in Spanish for Speaker Characterization and Identification," Speech Communication, vol. 31, pp. 255-264, 2000.

[16] A. F. Martin, G. R. Doddington, T. Kamm, M. Ordowski, and M. A. Przybocki, "The DET Curve in Assessment of Detection Task Performance," presented at Eurospeech, Rhodes, Grecia, 1997.

Fundamentals of Verbal and Nonverbal Communication and the Biometric Issue
A. Esposito et al. (Eds.)
IOS Press, 2007

N-gram Distributions in Texts as Proxy for Textual Fingerprints

Carl VOGEL
Centre for Computing and Language Studies, O'Reilly Institute, School of Computer Science and Statistics, Trinity College Dublin, Dublin 2 Ireland (vogel@tcd.ie)

Abstract. Recent experiments using mainly character unigram distributions in authorship attribution tasks are discussed. Results so far indicate efficacy in similarity judgements seemingly good enough for 'balance of probabilities' standards, but not yet for proof 'beyond reasonable doubt'.

Keywords. Authorship attribution, text classification, forensic linguistics

Introduction

This paper describes an area of text classification applied to the problem of very fine-grained categories, authorship attribution, in particular. The connection to verbal communication is fairly obvious in most applications of statistically driven text classification techniques. The possibility of authorship as a category that can be reliably classified on the basis of corpora of texts is where the issue of nonverbal communication inheres. Texts can be reliably classified to the level of granularity required by authorship attribution to the extent that authors encode in their texts, consciously or unconsciously, details that make the tests automatically identifiable as such.

The work here is driven by hypotheses within forensic linguistics that there are valid and reliable methods which can be applied to authorship attribution tasks. Behind these hypotheses is the claim that individuals essentially unconsciously fingerprint themselves in the texts that they produce. Many metrics are proposed in the literature. Research analyzes consciously manipulated author style, and other aspects of text that are extremely difficult for an author to manipulate. Orthography is one such aspect of texts. An author may deliberate a great deal on selection of lexical items from open-class categories, somewhat less reflection tends to be involved in closed class categories, and therefore it is interesting to explore distributions of words in closed class categories used by an author across texts and genre, and in comparison with other authors as a complement to any analysis of lexical richness or quirkiness. However, orthographic analysis crosses both categories,[1] and is basically driven by the fact that while one can choose one's words, one does not generally choose how words are spelled.

Along these lines I and my research group have been experimenting with character n-gram analyses of texts and using similarity of character n-gram distributions to guide

[1] That orthography is open-class requires a moment's reflection since it is usually dealt with in terms of a finite alphabet of anticipated characters. However, new characters (and fonts) are always on the horizon.

the assessment of similarity among texts.[2] The research involves comparative analysis, varying both the value of n and the scale of tokenization—that is, character n-grams, word n-grams, n-grams of part of speech tags—as well as accounting for stop-lists, etc. Some of the experiments have been on closed systems of texts in which authorship is actually known, others involve partly open systems of texts in which the claim of single authorship is disputed, and still more open ended explorations in sentiment analysis (§2).

An assumption is that through the sequences of words that constitute them, and aside from physical evidence like handwriting or signatures, texts communicate information about authorship that is unintended by the author, yet which can potentially be used to identify the author. It must be acknowledged that some authors do directly manipulate orthography for effect—*lipograms* provide a relevant example: [20] is a 50,000 word English novel written without a single instance of the letter E, quite a task given the frequency in English of that letter. The next section describes the method we have been exploring; §2 details some of the results so far; and, §3 outlines ongoing work.

1. The Method

The first step is to prepare the electronic corpus with an initial classification of texts, ideally balanced for size, concatenating or splitting files as appropriate, and balancing numbers of files in each category. Next is choice of a unit to count and its size. The unit measure could be letters, alphanumeric characters including spaces and punctuation, words, or part-of-speech tags assigned to words. For size of n and token choice, the experiments reported here will address mainly letter unigrams, following suggestion from forensic linguistics [2,3,4]. The perl scripts that implement the method are parameterized for both choices, but letter unigrams have proven most reliable in experiments so far.

Obviously, any number of other properties of texts can and have been examined, and word level analyses are quite popular (for example [1,9]). It is also possible to take a number of linguistic 'habits' into account at once: average word length, average sentence length, proportion of open class and closed class categories, and so on [5]. Many of these other possibilities are more clearly stylistic than letter unigrams. However, the interesting thing about letter n-gram distributions inherent in text is that they are so difficult for an author to consciously manipulate as they are only a consequence of the word choices that an author makes, as noted above. This is of separate interest from being objectively individuated in a way that 'average sentence complexity' cannot. Note that even reproducible measures such as Sampson's [18] average depth of words depend on theoretically unresolved issues: node depth depends on the constituent analysis assumed, for example, for coordination. The research reported here follows the suggestion of Chaski [2] that letter n-grams are actually the most reliable thing to count if forensic analysis of texts using corpus linguistic techniques is to satisfy the Daubert test of admissibility of expert testimony in criminal court. Certainly, letter n-gram statistics are reproducible. They also have validity in that they depend on the words selected, and exhibit zipfian distributions inherent in other aspects of language use.

---

[2]When I use "we" in this paper, it refers to myself and past and present members of my research group involved in this research who are named in the acknowledgements section, and to a some extent folks who will join this activity within the group in the near future.

For each file, a list of n-grams and their absolute and relative frequencies is extracted. Ultimately the relative-frequency distributions of particular n-grams in texts will be compared among files to identify the files with most similar distributions. The idea is simple and can be tested using any number of statistical tests in order to associate confidence intervals with judgements of similarity. In work designed to locate idiosyncrasies in non-native English produced by Finnish emigrants to Australia, [15] use a vector of relative frequencies of POS trigrams and calculation of cosine to gauge similarity, with permutation tests to assess statistical significance (this amounts to assessing the differences between texts using some measure δ and then estimating the proportion of times random samples from the compared texts behave the same with respect to δ). We combine χ^2 testing and the Mann-Whitney rank ordering test for confidence intervals.

Using statistics for hypothesis testing in this context requires comment. Kilgarriff, among others, has pointed out the risks of using hypothesis testing in corpus linguistics [7,8]. The null hypothesis, that the tested samples are randomly drawn from the same population, is only loosely applicable to corpus linguistics since the underlying phenomena are not at all random. Texts are structured, and random sampling from them will reveal distributions of letters, words, parts of speech, that are far from random, and the structure will be revealed increasingly with the total number of n-grams sampled and the value of n. With large enough sample sizes, the χ^2 critical value will be exceeded. However, the test is still interesting for the relative magnitude of the χ^2 value. If one wanted to spot locations of specific differences between two texts one would examine large values of the χ^2, inspecting the exact n-grams for which large values obtain. We calculate the χ^2 value for each n-gram presented by two compared files on the basis of the different observed and expected frequencies in each. Highly influenced by [7], the cumulative sum of the χ^2 value for each of the n-gram tests is then divided by the number of the n-grams tested, yielding a symmetric measure of similarity between the two files. Normally each cell in the contingency table for an individual test must contain at least five observed instances; thus, in allowing a zero (an n-gram not even shared by the files), the analysis we have been exploring amplifies the differences between files that don't have tokens in common. It seems best to make it as hard as possible for the tests to conclude that the files are similar enough to have been written by the same author.

Any authorship attribution task involves in principle comparing more than two texts. The task typically compares a text of known provenance and a questioned text. The metric described above provides a measure of similarity between them. However, there is typically more than one candidate author, and comparisons must be made with known texts of the contending authors. All of the pairwise comparisons of files are made to obtain the cumulative χ^2 value as a similarity ranking index in which smaller values indicate greater similarity, as described above. The resulting rank list is valuable for direct inspection of similarity of categories of files. It makes sense to proceed in two ways simultaneously. The first way is to consider the texts as each constituting a unit category,[3] identifying the pairs of files that are close to each other according to the similarity score. This effects a sense of clustering since it is done without reference to the actual categories the individual files might be members of. All files about one theme might well cluster together more than all files by one author. Thus, where clustering is by author, it is quite interesting. The second method is to gather the texts into categories according to their

[3]A unit category is a category with one instance.

provenance and compare the questioned texts to the categories corresponding to the texts of each of the authors. This brings more external knowledge to bear on the question than considering the entire corpus of texts as each constituting a category.[4] In either method one has rankings of a file according to its similarity to some file(s) and a set of other files. One can then use the Mann-Whitney test to assess whether there is statistical significance associated with the consequent categorization. Where one has external knowledge about the categories the files belong to naturally (e.g. "all files written by author X" or "not spam") it is best to use it, thus preferring the assessments provided by the second method, but making note of the ranked similarity of the first method as well.

For each file, the method examines it with respect to each of the possible categories in turn. This means examining the file's similarity with respect to each file of some category as opposed to each of the files in the complement of that category. The Mann-Whitney test is used independently for each pair of category checks. The null hypothesis is that the similarity with the category considered and its complement are indistinguishable in similarity rank. So, the null hypothesis is rejected when the file is more like one category than it is like the complement, subject to the confidence intervals supplied by the Mann-Whitney. Thus, when assigning a unique category to each file as in the clustering method one is essentially asking what other file is the file most similar to (which one already knows from the χ^2 divided by the number of degrees of freedom) and whether, taking ties into account, that is statistically significant. The method using unit categories is defined in a way that makes it arithmetically difficult to obtain actual statistical significance (by construction, a file is compared to a category with one file in it in relation to its similarity with the complement category, which is constituted by the remainder of the files). For the first method, the actual similarity score is the most interesting statistic. The second method allows one to examine each file within a category to consider its proximity to that category in relation to each other possible category. This means that one can test the homogeneity of a category.

Because the Mann-Whitney test is conducted independently for matching a file between all of the files in one category and the complement of that category, one is not guaranteed a unique match even at a strict significance threshold. A particular file might be like the undisputed works of Shakespeare (as opposed to all other files) with confidence $p \leq .02$ and also like the works of Bacon with the same confidence. However, if, for example, a letter unigram analysis and an analysis based on part of speech tags agree, then one might have increased confidence in the homogeneity of the questioned file with respect to its attributed category, but false positives are not eliminated. Nonetheless, it is clear that given the sorts of questions that one approaches texts with in authorship attribution problems (i.e. many texts and many candidate authors, where each author constitutes a category), a test which admits several categories at once rather than independent tests of a category vs. complement would also be worth adapting and exploring.[5]

2. Representative Experiments

The method has been applied in a number of experiments in search of conditions that invalidate the approach. Using binary categories, spam vs. nonspam, the letter unigram

[4] Balancing the file sizes and amalgamating and splitting the files has analogous impact.

[5] The Mann-Whitney test is a special case of the Kruskal-Wallis test where there are just two categories.

method scored perfectly in an evaluation as a spam filter, against a Bayesian method based on words [16]. Admittedly, this is not a hard test given a fixed training set, since the problem with spam email is that the training set requires constant updating. The methods have also been explored in the context of aligning political parties on a left-right spectrum using only political manifesto texts [19] (this is not an invented problem but one that is topical within political science research [11,10]).

It has also been tested on speeches by prominent politicians of this century and the last [6]. Using the actual authorship identification criterion, results that were relievingly noisy emerged. Texts of Eamon de Valera, Franklin Roosevelt, Gerry Adams, George W. Bush, Huey P. Long and Margaret Thatcher were all correctly assigned using letter unigrams, but only Dick Cheney and de Velara had that status using word unigrams. Bertie Aherne's texts were assigned entirely incorrectly to John Hume, and the texts of George H.W. Bush were assigned to Dick Cheney, Bill Clinton and G.W. Bush. Clearly the tests are noisy on political speeches. This is good because politicians rarely author their own speeches and many have multiple speech writers.

A study intended to determine conditions of applicability of the method are reported by [13]. For example, in work on the Federalist papers, blind classification using the 'genre' category (e.g. letter vs. essay, etc.) were more successful than actual authorship attribution. In another experiment, midi recordings of Lennon and McCartney music were textualized into ABC encoding.[6] There were 141 resulting Beatles files, and a control of 72 files with ABC encodings of Mozart. Again, mixed results obtained: each of the Beatles was correctly discriminated from Mozart, but not from each other.

Another set of experiments used literature archived by Project Gutenberg [14]. Authors sampled were Yeats, Wilde, Shaw, Edgeworth, Gregory. Letter unigrams and bigrams proved best for clustering the texts correctly by their authorship. In many cases where correct categorization was made, the statistical significance reported by the Mann-Whitney test was insufficient to credit the method with providing a correct assessment. In a related test the textual contributions of each character from four plays extracted into files, and clustered characters by author. Results here were quite noisy, except that Yeats' characters highly significantly ($p < .001$) classified as Yeats' using letter unigrams and word unigrams. Some authors evidently have deeper fingerprints than others.

3. Reflections

The experiments demonstrate mixed results. However, when the classifications come up correct, one cannot help but feel there is something real. It is unclear as yet how to merge tests using different units. A sort of intersection may be inappropriate. Also temporal dimension has yet to be thoroughly explored. All of the tests we have conducted effectively assume that the texts were authored at the same time. We have done no classifications that study diachronic effects. Certainly, this adds noise to any synchronic analysis: the texts of early Wittgenstein may be very different from later Wittgenstein not only in their philosophical content. Thus, assessing language change over time within individuals is urgent to explore further. It is important also because if a path of normal change can be established, it may prove useful in research on aging and early detection of neuro-degenerative disorders which impinge on language production [17].

[6]This was done using freely available software by James Allwright of University of Westminster.

Acknowledgements

Support of the NATO Advanced Study Institute and Science Foundation Ireland funding to RFP 05/RF/CMS002. This research has benefitted from collaboration with Sofie Van Gijsel, Lucy Hogan, Cormac O'Brien, Niamh McCombe, Julia Medori, Myriam Mencke and Mary Ronner. The initial perl scripts were implemented by McCombe [12] and subsequently developed further by both me and Medori [13].

References

[1] John Burrows. 'Delta': a measure of stylistic difference and a guide to likely authorship. *Literary and Linguistic Computing* **17** (2002), 267–287.

[2] Carole Chaski. Who wrote it? Steps toward a science of authorship identification. *National Institute of Justice Journal* **233** (1997), 15–22.

[3] Carole Chaski. Linguistic authentication and reliability. In *Proceedings of National Conference on Science and the Law* (1999), 97–148.

[4] Carole Chaski. Empirical evaluations of language-based author identification techniques. *Forensic Linguistics* **8** (2001), 1–65.

[5] Jill M. Farringdon. *Analysing for Authorship*. With contributions by Morton, A.Q., M.G. Farringdon and M.D. Baker. Cardiff: University of Wales Press, 1996.

[6] Lucy Hogan. A corpus linguistic analysis of American, British and Irish political speeches. Master's thesis, Centre for Language and Communication Studies, Trinity College, University of Dublin, 2005.

[7] Adam Kilgarriff. Comparing corpora. *International Journal of Corpus Linguistics* **6** (2001), 97–133.

[8] Adam Kilgarriff. Language is never, ever, ever random. *Corpus Linguistics and Linguistic Theory* **1-2** (2005), 263–275.

[9] Moshe Koppel, Shlomo Argamon, and Anat Rachel Shimoni. Automatically categorizing written texts by author gender. *Literary and Linguistic Computing* **17** (2002), 401–412.

[10] Michael Laver, editor. *Estimating the Policy Position of Political Actors*. Routledge, 2001.

[11] Michael Laver, Kenneth Benoit, and John Garry. Extracting policy positions from political texts using words as data. *American Political Science Review* **97** (2003), 311-331.

[12] Niambh McCombe. Methods of author identification. B.A. (Mod) CSLL Final Year Project, TCD, 2002.

[13] Julia Medori. Experiments testing the reliability and validity of author identification methods. Master's thesis, Computational Linguistics Group, Trinity College Dublin, 2005.

[14] Myriam Mencke. Experiments to validate scientifically reliable author identification techniques. MPhil in Linguistics, Centre for Language and Communication Studies, Trinity College, University of Dublin. 2004.

[15] John Nerbonne and Wybo Wiersma. A measure of aggregate syntactic distance. In John Nerbonne and Erhard Hinrichs, editors, *Linguistic Distances Workshop*. COLING/ACL, 2006.

[16] Cormac O'Brien and Carl Vogel. Spam filters: Bayes vs. chi-squared; letters vs. words. In Markus Alesky, et al. (eds), *Proceedings of the International Symposium on Information and Communication Technologies*, (2003), 298–303.

[17] Mary Ronner and Carl Vogel. Iris Murdoch: Writing over a lifetime—a study in form. Presented at the 2006 Meeting of the Iris Murdoch Society, Kingston, UK. September 15-16, 2006.

[18] Geoffrey Sampson. *Empirical Linguistics*. New York: Continuum, 2001.

[19] Sofie Van Gijsel and Carl Vogel. Inducing a cline from corpora of political manifestos. In Markus Aleksy, et al. (eds), *Proceedings of the International Symposium on Information and Communication Technologies*, (2003), 304–310.

[20] Ernest Vincent Wright. *Gadsby*. Wetzel Publishing Co., 1939.

A. Esposito et al. (Eds.)
IOS Press, 2007

An MPEG-7 Architecture with a Blind Signal Processing Based Front-End for Spoken Document Retrieval

Fausto IACCHELLI, Giovanni TUMMARELLO[1], Stefano SQUARTINI, Francesco PIAZZA
D.E.I.T., Università Politecnica delle Marche, Ancona, Italy

Abstract. Metadata extracted from Multimedia or live sensoring is set to play a major role in any intelligent and multimodal interactions between humans and computers. Furthermore, it is generally required that such metadata is structured and encoded according to well agreed standards. This is fundamental to enable interoperability and create complex applications as a mesh of heterogeneous services and components. On purpose, the MPEG-7 standard for dealing with multimedia metadata and the tools developed within the Semantic Web initiative are providing today the basic framework. Their application to real world problems, however, is made problematic by the fact that the data are often captured from difficult live conditions. It is therefore of primary importance to enhance the quality of the observable signals before the metadata extraction algorithms are employed. In particular, for the case of audio signals, it is important to perform separation and deconvolution of audio signals captured in real environments and in blind conditions. In this work a real world multimedia metadata assisted living scenario is addressed using a combination of Blind Signal Processing and MPEG-7 based metadata techniques. In such example, an array of microphones captures speech signals and thanks to MPEG-7 technologies the user can select multimedia content to be played.

Keykwords. Spoken Document Retrieval, MPEG-7, Blind Source Separation.

Introduction

Multimedia Metadata has been long recognized as a key to cope with the immense increase in production and availability of content in the Internet era.

Initiatives such as the W3C semantic web and the MPEG-7 [1] standard provide tools which address such distributed multimedia metadata scenarios. Despite the availability of such tools since many years now, the adoption has been particularly slow. In particular, the real world applicability on complex scenarios it is probably still to be demonstrated. In this work we address one such scenario, and in particular the combination of an MPEG-7 metadata based voice clip search engine and a preprocessing engine based on Blind Signal Processing techniques.

[1] Corresponding Author: Giovanni Tummariello, DEIT, Università Politecnica delle Marche, Ancona, Italy, e-mail: g.tummarello@gmail.com.

The voice clip search engine that we present works on the principle of phonetic matches and in particular displays characteristics which make it applicable to a variety of scenarios:

- Voice annotations can be added as quick description of other multimedia clips
- Can be a document itself, called sometimes Spoken Document
  Can be a query of the user toward a service on the web

All such three scenarios involve a task known as *Spoken Document Retrieval* (SDR) [1], where an *Automatic Speech Recognition* (ASR) engine is merged with an *Information Retrieval* (IR) engine. A number of SDR have been proposed in literature based on the principle of speech to text followed by textual search on a database of automatic or manually extracted audio transcripts [3]-[5] The limit of this approach is however evident, even though the Word Error Rate (WER) of the speech to text is relatively low, the vocabulary is nevertheless limited a priori, thus penalizing the fidelity of transcription and search, especially with respect to names of persons, cities etc.

A sub-lessical and or phonetic approach has been studied in [6] and partially solves this specific issue. The high WER however penalize the IR phase, which is usually performed by a classic *Vector Space Model* (VSM) engine [7], [8]. In later years, hybrid alternatives have been proposed and in particular some which exploit the tools provided by MPEG-7 specifically for scenarios of SDR. The results have been particularly interesting, with document *recall* rates beyond 70%[9], [10].

This paper deals with a SDR similar to the latter; it uses a phonetic approach and MPEG-7 structures. We named it the *Speak 'n' Browse* framework. In particular it is realized within a general framework of MPEG-7 extraction and processing called MPEG7AudioDB[1], which, on top of providing tools to extract and process MPEG-7 metadata, makes it easy to use and create annotations in W3C Semantic Web formats (RDF, OWL etc).

The main novel characteristic of the system we propose is the adoption of a pattern matching algorithm directly applied to the ASR lattices of the Spoken Documents of the *query* and of the *resource.* The algorithm falls in the category of *Inexact Sub-Graph-Matching* which starts from the assumption that the lattice can be treated as an acyclic graph and that the search operation can be conducted as a search of identical or quasi identical sub-graphs. The algorithm works directly on the MPEG-7 Spoken Content descriptions and draws from principles of *String-edit Distance* in the context of graph alignment: the approach is described in all its parts in sections I-IV. Section IV is about the set of trials and simulations performed in order to validate the efficacy of what implemented.

To move from an ideal scenario to a more practical setting, input disturbances coming from live environments cannot be ignored. In section VI we implement and evaluate a Blind Signal Processing (BSP) based front-end which enhances the quality of the audio signals feeding the MPEG-7 SDR architecture. We describe there the main characteristics of the involved BSP procedure and the benefits in terms of retrieval performances, as confirmed by further computer simulations.

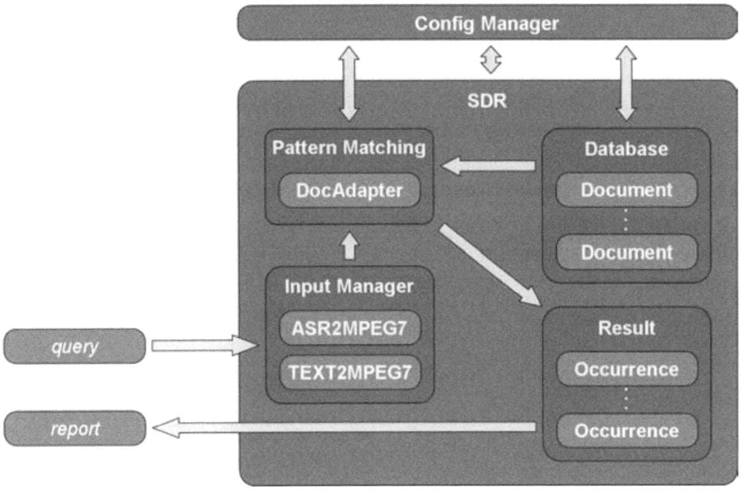

Figure 1. Structure of the Speak 'n' Browse framework.

1. The Speak 'n' Browse SDR framework

The *Speak 'n' Browse* framework has been developed as an extension of the *MPEG7AudioDB* [12] project, an advanced handling tooking for MPEG-7 metadata.

In such system, a *Document* is the MPEG-7 Spoken Content representation of an audio stream, while the *Database* is a structure which is fed by such XML documents and stores them in an optimized and indexed way.

As illustrated in **Figure 1**, the *InputManager* block handles the user queries, which can be both textual and vocal; the *TEXT2MPEG7* and *ASR2MPEG7* handle respectively text, via a simplified speech synthesis like approach, or Automatic speech recognition output and turn them into the MPEG-7 Spoken Content encoding.

The *PatternMatching* block is at the heart of the system; it handles the matching between the content of the database and the incoming query. The list of matches forms a query *Result*, which is then presented to the user in an appropriate, rank sorted list.

The system itself is configured in XML and can be tuned at runtime trough the *ConfigManager* block.

2. The first block of the ASR2MPEG7 component: *Sphinx-4*

The ASR2MPEG7 block uses an Open Source ASR framework known as *Sphinx-4* [11]. Sphinx-4 is a state of the art ASR; release 4 has been implemented in Java and extends the HMM approach to continuous and semi continuous models. The system is particularly flexibly especially with respect to its ability of working both stand alone and embedded. Aspects of the ASR processing (Sampling, signal processing and filtering, features extractions) can be handled separately. For our purposes we configured Sphinx to operate based on features such as MFCC, Δ e ΔΔ. The dictionary has been the 45 phonemes of the English idiom which constitute a continuous HMM trained from a set of 5000 words from the Wall Street Journal (WSJ5K). The recognition is at phonetic level, but does not happen in a brute force single decision fashion: the intelligent pruning of the recognition lattice is based on a tri-gram lexicon derived from the WSJ2K set itself. This effectively filters incoherent phoneme recognition and limits the size of the lattice itself. The configuration we have used in our experiments is based on a batch recognition, with processing times limited to 160% of the segment duration on a Intel Centrino 1.6MHz with 512MB RAM.

3. Transcription into MPEG-7 Spoken Content format

The ISO/IEC MPEG-7 standard, formally *Multimedia Content Description Interface*, defines instruments and structures to describe at metadata level the content of images, sounds, videos and combination of these. MPEG-7 [1] consists in basic *Descriptors* (D), which represent the main visual and acoustic characteristics at different abstraction level. These contain features encoded as *Datatypes* which extend those of XML schema with complex types such as vectors and matrices. Descriptors are often grouped in hierarchical structures said *Description Schemes* (DS). The D elements and the rules which compose DS follow a syntax which is defined in the *Description Definition Language* (DDL), a vocabulary which enable the extension of the existing D and DS as well as the creation of new elements. DDL, DS, D and the Datatypes themselves are defined in *XML Schema* and coherently, the metadata serialization and archival happens in XML format.

The *Spoken Content* section of the MPEG-7 standard defines D e DS to represent various aspects which deal with different use cases in speech recognition. Those that are of most interest for the use case we present are:

- *PhoneLexicon* and *WordLexicon*: describe the kind of dictionary used in the transcription; the earlier indicates a dictionary composed by sublexical elements such as, e.g. Syllables and phonemes. The latter indicates a dictionary composed by words.
- *ConfusionInfo*: define the structures which report statistical information on the performance of the recognizer. These are written in terms of the typical values of *Insertion*, *Deletion* e *Substitution* which are generated by the regression-tests of most ASR systems.
- *SpeakerInfo*: handles the profiling of the speaker (name, sex, nationality etc..). This is used to indicate in the vocal stream the onset of a new speaker and handles the confusion information specific to the ASR/speaker coupling.
- *Block*: is the main MPEG-7 Spoken Content container. Its task is to collect and organize the acoustic characteristics of the stream, the textual tags of the vocal segment, the temporal index of the segment the probability/dependability of the recognition process, the links between speakers and other elements in the transcription.
- *Node*, *PhoneLink* e *WordLink*: are elements which are also in the *Block* hierarchy and enable the transcription in form of complex lattice along with a series of attributes and DS which can interconnect the components.

4. Pattern matching based on Inexact sub-Graph-Matching via String-edit Alignment extension

Defining an IR engine involves the definition of a pattern matching technique which can evaluate the best matches for a given query. The quality result of such operation is said *Retrieval Status Value* (RSV), that is the global evaluation value of the informative cross-relation [13]-[14]. In our work we opted for a graph based implementation, in which the entire MPEG-7 transcription lattice is evaluated for similarities with the query, itself transformed into MPEG-7 Spoken Content format.

Given the nature of both transcriptions, which often display a WER higher than 30%, it is necessary to have a flexible and tolerant matching algorithm. In our case the algorithm performs a homomorphic sub-graph-matching based on the String Edit Aliment [14]-[16] of the lattices. This procedure seeks to evaluate the edit distance between the two. The edit distance measure compares strings using simple, local operations called edits. The distance is evaluated as the minimum number of such basic changes which would turn on string into the other.

While the classic techniques determines the distance between two character sequences, in our case there is need to alight two lattices. These can however be seen as a compact version of a series of N sequences of symbols related to a single object.

To illustrate how the algorithm operates we show some example graphs G1 e G2 which represent two generic strings (which are however phonemes sequences) as a starting hypothesis we consider:

- Each character in the string (phoneme) is an arc in the graph;
- Graph nodes work as character delimiters (temporization of the phonemes);
- The similarity matrix will have as many rows and columns as the number of nodes in both graphs;

Such matrices are constructed as follows:

$$DG(i,j) = \textit{Global Distance} \text{ at position } (i,j) \tag{1}$$

$$DL(i,j) = \textit{Local Distance} \text{ at position } (i,j) \tag{2}$$

$$\begin{aligned} &DL(0,0) = 0 \quad DL(0,j) = j \\ &DL(i,j) = \begin{Bmatrix} 0 \text{ if } ch_i = ch_j \\ 1 \text{ if } ch_i \neq ch_j \end{Bmatrix} \end{aligned} \tag{3}$$

$$DG(i,j) = \min \begin{Bmatrix} DG(i,j-1) + DL_{\max} \\ DG(i-1,j) + DL_{\max} \\ DG(i-1,j-1) + DL(i,j) \end{Bmatrix} \tag{4}$$

The example in **Figure 2** shows the situation in which the ASR engine matches two different symbols to the same audio segment (confidence values omitted).

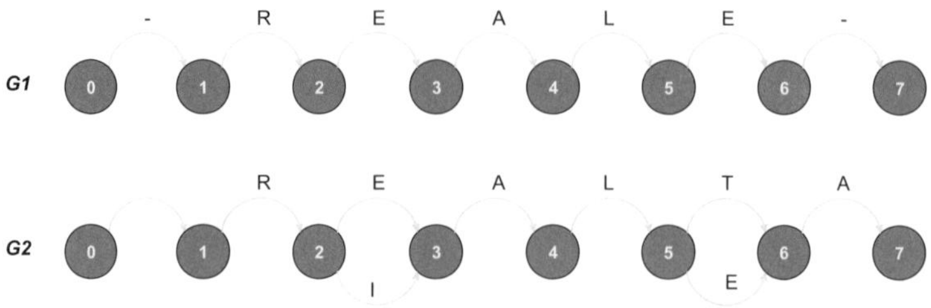

Figure 2. A stored audio segment (G1) and two different symbols are matched to the same query audio segment (G2).

Table 1. Local Distance matrix

	G 1	0 -	1 R	2 E	3 A	4 L	5 E	6 -	7 /
G 2	0	1	1	1	1	1	1	1	1
0 -	1	0	1	1	1	1	1	0	1
1 R	1	1	0	1	1	1	1	1	1
2 E I	1	1	1	0	1	1	0	1	1
3 A	1	1	1	1	0	1	1	1	1
4 L	1	1	1	1	1	0	1	1	1
5 T E	1	1	1	1 0	1	1	1 0	1	1
6 A	1	1	1	1	0	1	1	1	1
7 /	1	1	1	1	1	1	1	1	0

Table 2. Global Distance matrix

	G 1	0 -	1 R	2 E	3 A	4 L	5 E	6 -	7 /
G 2	0	1	2	3	4	5	6	7	8
0 -	1	0	1	2	3	4	5	6	7
1 R	2	1	0	1	2	3	4	5	6
2 E I	3	2	1	0	1	2	3	4	5
3 A	4	3	2	1	0	1	2	3	4
4 L	5	4	3	2	1	0	1	2	3
5 T E	6	5	4	3	2	1	0	1	2
6 A	7	6	5	4	3	2	1	1	2
7 /	8	7	6	5	4	3	2	2	1

The local similarity matrix shown in **Table 1** is used in (4) to compute the global distance matrix, shown in **Table 1**: when a double link is involved, the evaluation of the red cells happens for both local distance paths; the path with higher weight is then discarded.

The example shows how the double link between nodes 2 and 3 in the end produces the same local distance components: the internal similarities contribute to the simplification of the search.

Figura 3 examples show a situation where a G2 arc links two nodes by skipping others: the greater extension of the outgoing star causes a structural alternative which corresponds to a contraction in the length of the sequence. In the classical technique, this would lead to a similarity matrix of lesser dimension than the initial one.

Given that such variable geometry cannot be handled in the current spatial realization, the algorithm proceeds by propagating the value of the similarity induced by the arcs to all the nodes that have been skipped by the arc itself. The final effect is the "stretching" of the similarity matrix to the point of covering the surface of the original global matrix: **Table 3** shows a line-values stretching along all columns.

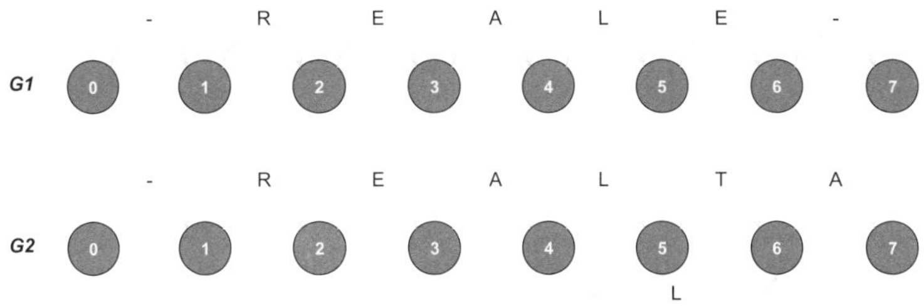

Figura 3. A link in G2 skips a node.

Table 3. Local Distance matrix

	G 1	0 -	1 R	2 E	3 A	4 L	5 E	6 -	7 /
G 2	0	1	1	1	1	1	1	1	1
0 -	1	0	1	1	1	1	1	0	1
1 R	1	1	0	1	1	1	1	1	1
2 E I	1	1	1	0	1	1	0	1	1
3 A	1	1	1	1	0	1	1	1	1
4 L L	1	1	1	1	1	0 0	1	1	1
5 T L	1	1	1	1	1	1 0	1	1	1
6 A	1	1	1	1	0	1	1	1	1
7 /	1	1	1	1	1	1	1	1	0

Table 4. Global Distance matrix

	G 1	0 -	1 R	2 E	3 A	4 L	5 E	6 -	7 /
G 2	0	1	2	3	4	5	6	7	8
0 -	1	0	1	2	3	4	5	6	7
1 R	2	1	0	1	2	3	4	5	6
2 E I	3	2	1	0	1	2	3	4	5
3 A	4	3	2	1	0	1	2	3	4
4 L L	5	4	3	2	1	0	1	2	3
5 T L	5	4	3	2	1	0	1	2	3
6 A	6	5	4	3	2	1	1	2	3
7 /	7	6	5	4	3	2	2	2	2

This is the same also in cases of greater complexity (see **Figure 4**), but at implementation level there are further algorithmic simplification that make it so that the global distance matrix is constructed at the same time as the identification of the local distances, that is, without the need to actually construct the local similarity matrices.

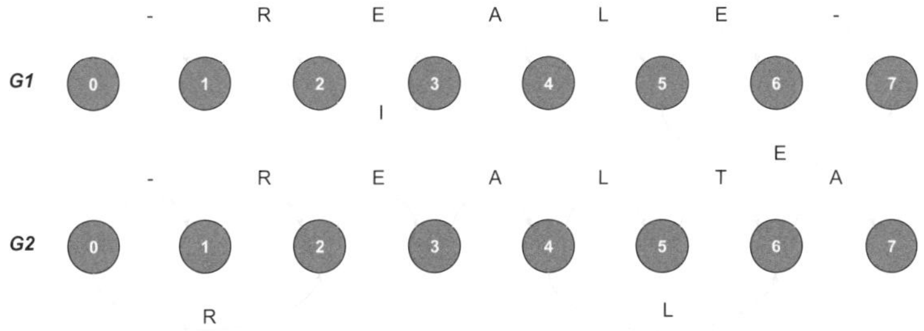

Figure 4. A more complex example.

5. Tests and Results

The framework has been verified by constructing an archive of 270 documents in XML MPEG-7 format, extracted from 80 sentences in English language with an average length of 3s, pronounced by a inhomogeneous set of 115 speakers of different sex and provenience (taken from the *DARPA TIMIT* Database [17]).Two tests have been performed:

- Batch: six queries with sentences from the archive itself and difference relevance threshold.
- Real-Time: 30 queries via microphone.

The system performance is evaluated in terms of the following metrics:

$$precision = \frac{Valid\ documents\ found}{Total\ documents\ found} \quad (5)$$

$$recall = \frac{Valid\ documents\ found}{Total\ valid\ documents} \quad (6)$$

$$mAP = \frac{1}{L}\sum_{i=1}^{L}\left(\frac{1}{N}\right)\sum_{k=1}^{N_i}\left(\frac{k}{rank_{ik}}\right) \quad (7)$$

Where L is the total query number, N_i is the number of documents that is returned form the q_i query and $rank_{ik}$ is a similarity score for the q_i occurrence and the d_k document, that is, the degree of relationship that exists between the document and the query, also known as *retrieval status value* (RSV).

From the trials it is possible to evaluate a Recall/Precision graph in which the area comprised by the curve is the global evaluation parameter mAP. In an ideal system, such parameter would be 1.

Figure 5 shows the behaviour of the system during the Batch trials, resulting in an estimated mAP around 88%.

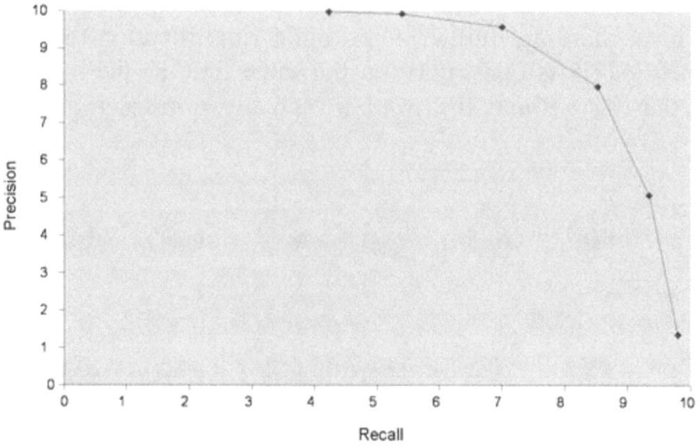

Figure 5. Recall/Precision graph during the batch trials.

As expected, the Real-Time test has shown lower performance than those of the batch test. It was shown however that the performance greatly varied according to the speaker. In particular, as our test speakers were not English mother tongue, the recall has been higher than 70% for the first 30 documents returned by the query for those with a good knowledge of the English language while dropped to 30% for those speaking a beginner to lower medium quality English.

Expectedly, the search shows to benefit from query of longer duration: queries longer than 1s consistently produced more precise matches.

About operational parameters influencing the search, the most important property is the SEARCH_SPAN of the PM module. This is an expected behaviour due to the nature of the parameter: a value of 30% indicates that the sliding of the query graph on the document graph proceeds in steps of 30% of the query graph nodes. To lower this parameter means to make the search more accurate and increase precision but can requires consistently more processing power.

6. Application of the BSP front-end to the SDR architecture

As aforementioned, the MPEG-7 standard defines the rules on how to deal with metadata coming from available signals. It is an open standard, i.e. it does not give strict limitations on the algorithms to use in the different application we are involved.

This leaves free to the implementer the choice of specific algorithms and preprocessing blocks to use on the signals. In order to advance our study into a more realistic scenario we here study the effect of a DSP front-end (based on Blind Signal Processing [18]-[19]) in order to make the live captured input signals more tractable for our purpose. In an assisted living scenario, the MPEG-7 database is driven by queries captured in live environments, e.g. Spoken out loud by a person while others might also be conversing. In this case, the data captured by fixed ambient microphones are in general convolutive noisy mixtures of the single audio sources. This scenario is usually referred to as the Cocktail Party problem [18]-[19]. We need to recover the original sources in order to feed our MPEG-7 based structure with "clean" audio sources. On purpose, we believe that BSP (Blind Signal Processing) techniques represent a general and effective pre-processing tool set in various operating conditions. In details, the situation is the one depicted in Fig.6: we have audio mixtures, coming from common microphones recordings and the SDR architecture composed by the SPHINX-IV plus MPEG-7 metadata processing algorithms; in between we place the Separation Process, a Blind Signal Processing algorithm suitable for the task. Furthermore, considering the modularity of the SDR architecture described above, such a BSP based front-end can be implemented as a plug-in of the *Speak 'n' Browse* framework.

The proper choice of the separation procedure strictly depends on the operating conditions (number of sources and sensors, nature of the acoustic environment, typology of the useful signals and noises present). Here, we intend to highlight the feasibility of the general approach, showing that the application of the BSP front-end allows the MPEG-7 based SDR architecture to improve its performances when clean recordings are not available. For the purpose of this task we consider the following assumptions: the reverberation effect of the room is neglected and the sensors outnumber the sources. The latter assumption lets us employ the INFOMAX approach for ICA as blind separation algorithm [18]-[19].

For our computer simulations we used synthetic mixture of speech and music signals (3sec recordings, 16kHz sampling frequency, 16 bit quantization, PCM, TIMIT database, English with Italian accent). Two are the scenarios addressed, namely:

1. *Call Center.* Three persons give vocal commands to the system that is supposed to be able to process them at the same time; the sentence spoken by each person is disturbed by the others due to the vicinity of the colleagues and so separation is needed by the system to get all sentences and execute the commands therein
2. *Classic Speech Enhancement.* The vocal command is spoken in a noisy environment; a music source is here used as disturbing signal. Also other approaches could be employed (like AEC, ANC, Spectral Subtraction), in dependence on the nature of the disturbance. The separation allows us to keep track of the information contained in the non-speech signals (music), that could be maybe useful for other applications.

Results are reported in the following tables. Three operating conditions have been considered in both scenarios: clean sources, mixed sources and recovered original sources. It must be underlined that, in order to make the results more readable, we considered the following assumptions regarding the definition of the recall and precision quality indexes:

- RSV threshold = 70%;
- Total number of sentences Retrieved = 10 queries with highest RSVs

The algorithm performs very satisfactory, to the point that the first and the third case result identical. Original sources, on the other hand, do not operate as well or in some case do not work at all. This depends on the nature of the mixing matrix. Concerning the Scenario 1, if the contribution of the sources to the mixture signal is energetically equivalent (matrix $A1$ chosen with entries uniformly distributed) then the performances are very poor; otherwise, if one source is predominant (the realistic case where each call center speaker has a direct microphone in close vicinity) within a certain mixture signal (matrices $A2$ and $A3$, generated by $A=I+cR$, where I is the identity matrix R is a uniformly distributed random matrix and c is equal to 0.3 and 0.15 respectively) then the performances gradually increase. The same occurred in the Scenario 2 case study, where $A1$ is random and $A2$ is defined as above with c=0.4. Further experiments, not reported for the lack of space, showed that the performance in term of recovering recall and precision in the MPEG-7 database does not depend on the chosen sentences and on the speaker involved.

Table 5. Scenario 1 – Recall and Precision for the three sentences retrieved in three different operating conditions: clean sources, mixed sources and recovered original sources. Mixing Matrix: *A1*.

Recall %	**Source**	**mix1**	**mix2**	**mix3**	**Rec**
Sen1	66	0	0	33	66
Sen2	50	0	0	0	50
Sen3	100	0	0	0	100
Precision %	**Source**	**mix1**	**mix2**	**mix3**	**Rec**

Sen1	70	0	0	7	70
Sen2	70	0	0	0	70
Sen3	100	0	0	0	100

Table 6. Scenario 1 – Recall and Precision for the three sentences retrieved in three different operating conditions: clean sources, mixed sources and recovered original sources. Mixing Matrix: *A2*.

Recall %	Source	mix1	mix2	mix3	Rec
Sen1	66	66	0	0	66
Sen2	50	0	50	0	50
Sen3	100	0	0	100	100
Precision %	**Source**	**mix1**	**mix2**	**mix3**	**Rec**
Sen1	70	40	0	7	70
Sen2	70	0	66	0	70
Sen3	100	0	0	100	100

Table 7. Scenario 1 – Recall and Precision for the three sentences retrieved in three different operating conditions: clean sources, mixed sources and recovered original sources. Mixing Matrix: *A3*.

Recall %	Source	mix1	mix2	mix3	Rec
Sen1	66	66	0	0	66
Sen2	50	25	25	0	50
Sen3	100	0	0	66	100
Precision %	**Source**	**mix1**	**mix2**	**mix3**	**Rec**
Sen1	70	37	0	0	70
Sen2	70	7	37	0	70
Sen3	100	0	0	70	100

Table 8. Scenario 2 – Recall and Precision for the noisy sentence retrieved and in three operating conditions: clean sources, mixed sources and recovered original sources. Mixing Matrices: *A1* and *A2*.

%	Source	mix1	mix2	Rec
Recall	100	0	0	100
Precision	100	0	0	100
Recall	100	0	50	100
Precision	100	0	52	100

Figure 6. BSP front-end and MPEG-7 based SDR architecture: conceptual block scheme.

7. Conclusions

In this work we developed a framework for the Spoken Document Retrieval. The approach is phoneme based and in particular deals maximally with the tools and formats defined in the MPEG-7 standard. The main characteristics of the frameworks areImplemented in Java language;

- Modular, it is possible to redefine each stage of the process
- it handles external audio streams, XML MPEG-7 streams, text and direct sampling via microphone

As the system enables the the use of DSP front-end plugins, we investigated on how a BSP algorithm could help recover the loss of performance deriving from real world signal capturing conditions (e.g. Microphones in a call center). The results show that the system performs well, with numerical results very similar to those obtained in more more favorable conditions.

As real world recording situations also include convolutive mixing processes and additive noise, future development should take into considerations algorithms such as [18]-[19]. As far as the SDR engine is concerned, the performance can be said to be in line and comparable with those of state of the art systems [3]-[5]. There is however room for improvement. Literature shows, the results that are given by a single search algorithm can be improved as much as 15% by techniques such as retrieval feedback, cross correlation DBs and others. Such techniques are not at the moment implemented and will be object of further works.

8. References

[1] (2004), MPEG-7 Website [Online], Available: http://www.chiariglione.org/mpeg/standards/mpeg-7/mpeg-7.htm.

[2] J. Garofolo, C. Auzanne, and E. Voorhees, The TREC spoken document retrieval track: A success story, *Proc. RIAO*, vol. 1, p. 1-20, 2000.

[3] N. Moreau, H.G. Kim, T. Sikora, Phone-based Spoken Document Retrieval in conformance with the MPEG-7 Standard, *AES 25th International Conference*, London, U.K., 2004.

[4] J. Löffer, K. Biatov, C. Eckes, J. Köhler, IFINDER: An MPEG-7-Based Retrieval System For Distributed Multimedia Content, *Proceedings of the ACM Multimedia Conference, 2002.*

[5] K. Ng, V.W. Zue, *Subword-based Approaches for Spoken Document Retrieval*, Elsevier, 1999.

[6] M.E. Maron, J.L. Kuhns, *On Relevance, Probabilistic Indexing and Information Retrieval*, Ramo-Wooldridge, New York, 1959.

[7] K. Spark Jones, S. Walker, S.E. Robertson, A Probabilistic Model of Information Retrieval: Development and Comparative Experiments, *ICASSP*, 2000.

[8] D. Abberley, D. Kirby, S. Renals, and T. Robinson, The THISL broadcast news retrieval sys-tem, *Proc. of ESCA ETRW Workshop on Accessing Information in Spoken Audio, Cam-bridge,* UK, April 1999.

[9] J. Choi, D. Hindle, J. Hirschberg, I. Magrin-Chagnolleau, C. Nakatani, F. Pereira, A. Singhal, S. Whittaker, *SCAN – Speech Content Based Audio Navigator: A System Overview*, AT&T Labs-Research report, 1999.

[10] B. Zhou, J.H.L. Hansen, SPEECHFIND: an experimental on-line Spoken Document Retrieval System for Historical Audio Archives, *ICSLP*, 2002.

[11] W. Walker, P. Lamere, P. Kwok, B. Raj, R. Singh, E. Gouvea, P. Wolf, and J. Woelfel, Sphinx-4: A flexible open source framework for speech recognition, *Technical Report SMLITR2004-0811*, Sun Microsystems, Inc., 2004.

[12] (2006), MPEG-7 AudioDB Website, [Online], Available: http://www.sourceforge.net/projects/mpeg7audiodb.

[13] J.R. Ullmann, An Algorithm for Subgraph Isomorphism, *Journal of the Association for Computer Machinery*, 1976.

[14] A. Sanfeliu, K.S. Fu, A Distance Measure Between Attributed Relational Graphs for Pattern Matching, *IEEE Transaction Systems, Man and Cybernetics*, 1983.

[15] R. Myers, R.C. Wilson, E.R. Hancock, Bayesan Graph Edit Distance, *IEEE Transaction of Pattern Analysis and Machine Intelligence*, 2000.

[16] E.S. Ristad and P.N. Yianilos, Learning string edit distance, *IEEE Transactions on Pattern Analysis and Machine Intelligence*, 20(2):522—532.

[17] J.S. Garofolo, L.F. Lamel, W.M. Fisher, J.G. Fiscus, D.S. Pallett, and N.L. Dahlgren, *DARPA TIMIT Acoustic-Phonetic Continuous Speech Corpus CD-ROM,* 1990, National Institute of Standards and Technology, NTIS Order No. PB91-505065.

[18] A. Hyvärinen, J. Karhunen, and E. Oja, *Independent Component Analysis*. Wiley & Sons, New York, 2001.

[19] A. Cichocki and S.I. Amari, *Adaptive blind Signal and Image Processing*, John Wiley & Sons, LTD, England, 2002.

Fundamentals of Verbal and Nonverbal Communication and the Biometric Issue
A. Esposito et al. (Eds.)
IOS Press, 2007

Overcomplete Blind Separation of Speech Sources in the Post Nonlinear Case Through Extended Gaussianization

Stefano SQUARTINI[1], Stefania CECCHI, Emanuele MORETTI, Francesco PIAZZA
Università Politecnica delle Marche, Ancona, Italy

Abstract. Blind Signal Processing techniques have been receiving an increasing attention by the scientific community also for their benefic impact to speaker recognition based biometric systems. This work deals with the speech separation problem in blind conditions and in presence of more sources than sensors and Post-Nonlinear (PNL) mixing, which likely represent a close-to-reality situation. The addressed method is made of three separate steps: compensation of nonlinearity, mixing matrix recovery and final unknown source estimation. It has been recently proposed and successfully evaluated in the case of synthetic mixtures of real world data (like speech signals). Here, the Extended Gaussianization is employed as first step instead of the common Gaussianization one in order to reduce the approximation error on the linearized mixture pdfs. Computer simulations allowed achieving a significant improvement of separation performances over the previous approach.

Keywords. Blind Source Separation (BSS), Speaker Recognition, Post Non Linear mixing, Extended Gaussianization, Under-determined BSS, Geometric ICA.

I. Introduction

Several biometric systems have been proposed in the literature so far; they can be basically classified in terms of the type of measurement involved, i.e. fingerprint, face, and voice. As pointed out in [1], many factors are in support to the voice as a compelling biometric. This explains why so much attention has been given by the scientific community on speaker recognition systems, both for the identification and the verification tasks.

The performances of the recognition procedures adopted strongly depends on the conditions of the environment where the speech signals are captured for processing. For instance, the presence of noise sources, reverberation and other speakers likely has a negative impact on the capabilities of the algorithm used to identify or verify a certain speaker. This justifies the employment of pre-processing speech enhancement techniques in order to enhance the quality of the speech signals to be processed [1], [2]. Furthermore, the beneficial contribution given by such an approach to the speaker

[1] Corresponding Author: A3Lab-DEIT Università Politecnica delle Marche, Via Brecce Bianche 31, 60131, Ancona, Italy, e-mail: sts@deit.univpm.it.

recognition objectives can also be extended over the several multi-modal techniques proposed in the literature so far in the same addressed field.

Recently, some efforts have been done along this direction involving the Blind Signal Processing (BSP) techniques [3]-[5]. The typical referred scenario is the occurrence of multi-talker speeches in a certain number of available signals, captured by a multi-microphone system. The basic idea for speech quality enhancement is separating the mixed speech sources in order to allow the speaker recognition algorithm to work properly with more accommodating signals. The separation task typically occurs in blind conditions, since we do not have the knowledge of the original signals and the system performing the mixing. The scientific community names this as the Blind Source Separation problem.

In the last decade, many efforts have been made in literature on this subject, proposing a large set of operating conditions, and related model assumptions and solving procedures. Indeed, different formulations of the BSS problem do exist in accordance with the nature of mixing. With reference to the number of sources and sensors over-complete means sources outnumbering the sensors [6]. In this case study, the separation procedure is typically based on sources reconstruction (Blind Source Recovery) beginning from the estimation of the mixing matrix (Blind Mixing Model Recovery). Among the large variety of solutions for finding out a demixing process, we can cite the Geometric ICA approach that yields good performances [6].

Furthermore we can distinguish between linear and non linear mixtures. In the first case, on the assumption that the independence implies the separation of the sources, it is said that the mixtures are separable. However, in general, non linear mixtures are not separable, so structural constraints or regularization techniques are needed. Among the class of nonlinearities, we have considered Post Non Linear (PNL) mixing that can be applied to model a system with linear transmission channels and with sensors introducing memory-less non linear distortions. Several approaches for PNL-BSS have been proposed but, up to our knowledge, all of them are focused on the complete model [7][8]. In [6] the under-determined BSS (uBSS) has been addressed with PNL mixing model employing Gaussianization method. The goal of this contribution is extending the previous work applying the Extended Gaussianization method [8] in the PNL-uBSS problem. As well as in [6], the real interest of this method is its applicability to speech mixtures (and so to voice measurement based biometric tasks, as aforementioned), in contrast to the other related few contributions appeared in the literature so far, up to the authors' knowledge [9].

II. Proposed Algorithm

The proposed algorithm is based on a three step approach for Blind Source Separation in the case of instantaneous, memory-less, over-complete, post non-linear mixtures [6]. In first step nonlinearity are recovered and compensated using an Extended Gaussianization two-step approach; in the second step the mixing matrix is estimated through known geometrical ICA method, and finally source estimation is performed using classical maximum likelihood (ML) technique.

A. Algorithm Overview

In this subsection we shall illustrate the general concept of proposed algorithm for solving instantaneous PNL-uBSS problem, with reference to three sources and two mixtures case, as shown in Fig 1 .

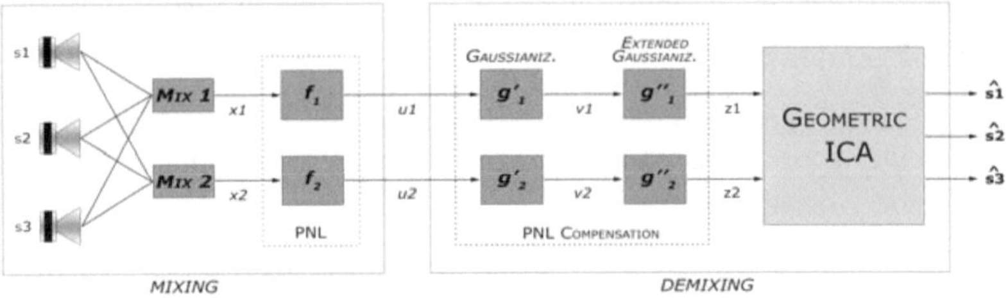

Figure 1: Schematic model for algorithm concept

Let s_i be the source vectors, with i=1,2,3 and let u_i be the mixture vectors, with i=1,2; in uBSS-PNL case, the relationship between s_i and u_i is

$$u_i = f_i\left(\sum_{j=1}^{3} a_{ij} s_j(t)\right) \quad i = 1,2. \tag{1}$$

Either of sources (s_i), linear mixing matrix (a_{ij}) and non linear distortion functions (f_i), are unknown parameters. Beginning from known mixtures vectors (u_i), and with assumption of sparsity and independence of s_i , we must recover all sources.

The first fundamental consideration is that the amplitude distribution of x_i (due to *"Central limit theorem"*) follow Gaussian distribution, but nonlinearity function tend to distort it. A statistical nonlinearity compensation approach could be defined through nonlinear functions g'_i that transform the mixture signals u_i in new signals v_i, with Gaussian amplitude distribution (Fig. 1). This transformation is called *"Gaussianizzation"* and was proposed in [10].

The second fundamental consideration is that (due to the practical limitation of *"Central limit theorem"*) the amplitude distribution of x_i is not exactly a Gaussian distribution, but we assume it as close to Gaussian distribution. In order to obtain better nonlinearity compensation we can define g''_i using a parametric model based on the Cornish-Fisher Expansion [8], so that the output signals z_i is as similar as possible to statistical amplitude distribution of x_i.

The global Post Non-Linear compensation functions are obtained by composition of either g'_i and g''_i ($g_i = g'_i \circ g''_i$). After nonlinearity compensation the signals z_i could be considered as linear mixtures, so that we could apply all known approach to solve Linear-uBSS problem.

B. Gaussianization

Gaussianization is an invertible transformation $T(\xi)$ that transforms a random variable into a standard Gaussian random variable with zero-mean and unit of variance. The Gaussianization transformation is [10],[11]

$$T(\xi) = \Phi^{-1}(F(\xi)) \tag{2}$$

whereas Φ(·) is the Standard Normal Distribution and F(·) is the cumulative distribution function (CDF) of input signal. F(·) is an unknown function, and it can be estimated in different ways; for the sake of easiness, we can use empirical CDF (ecdf) methods,

$$\hat{F}(\xi) = \frac{1}{N+1}\sum_{k=1}^{N} I\left(\xi^{(k)} \le \xi\right) \tag{3}$$

whereas N is the number of samples, $\xi^{(k)}$ is the k-th sample, and $I(\xi^{(k)} \le \xi)$ is a function with value 1 if $\xi^{(k)} \le \xi$, and 0 otherwise. A low-computational implementation method for Gaussian transformation, limiting the influence of outliers is the follow

$$v^{(k)} = \Phi^{-1}\left(c\frac{rank\left(u^{(k)}\right)}{N+1} + \frac{1}{2}(1-c)\right) \tag{4}$$

where $rank(u^{(k)})$ indicates the position of $u^{(k)}$ in sorted sample vector, ordered from smallest to largest, and c value is application dependent, but not critical.

C. Improved Nonlinearity compensation based on the Corner-Fisher Expansion

It is known that an ideal Gaussian distribution have cumulants equal to zero, while a distribution close to Gaussian, such as x_i amplitude distribution, can be modeled using an estimation of them. With Extended Gaussianization method, we want to create a parametric function to model remaining nonlinearity after Gaussian transformation, using v_i; as input and skewness and kurtosis as parameter.
Cornish and Fisher propounded a form of expansion in which the terms are polynomial functions of the standard Gaussian quantile, with functions of known cumulants as coefficients [8]. The four-term Cornish-Fisher expansion for α-quantile is

$$\xi(\eta_\alpha) \simeq \eta_\alpha + \frac{1}{6}\left(\eta_\alpha^2 - 1\right)k_3 + \frac{1}{24}\left(\eta_\alpha^3 - 3\eta_\alpha\right)k_4 - \frac{1}{36}\left(2\eta_\alpha^3 - 5\eta_\alpha\right)k_3^2 \tag{5}$$

where k_3 and k_4 are skewness and kurtosis of x respectively, η_α is the α-quantile of the standard Gaussian distribution, and $\xi(\eta_\alpha)$ is zero-mean and of unit variance. As the distribution of z_i is assumed to be close to Gaussian and v_i is result of Gaussian transformation, we can use the C-F expansion to approximate z_i in terms of v_i

$$z_i \simeq v_i + \frac{1}{6}\left(v_i^2 - 1\right)k_{3,i} + \frac{1}{24}\left(v_i^3 - 3v_i\right)k_{4,i} - \frac{1}{36}\left(2v_i^3 - 5v_i\right)k_{3,i}^2 \tag{6}$$

In [8] the parameters k_3 and k_4 come out from the adaptation process responsible for finding the independent components contained in the PNL mixtures, assuming to be in the complete case. In our over-complete case, we give a practical rough estimation of them by using the ones related to the v_i variables, exploiting the approximation limits of the Gaussianization process. This is not analytically exact but has the advantage of being very effective in terms of computational load and improvement of performances.

III. Computer Simulations

In order to investigate the performance of the previously described method, a MatLAB application has been implemented and tested with synthetic mixtures of speech signals. For this particular example, three 35328 sample long normalized speech sources have been used, containing a female voice and two male ones, sampled at 11025. The two mixtures have been yielded using the following mixing parameters:

$$A = \begin{bmatrix} 0,6216 & 0.9801 & 0.4536 \\ -0.7833 & 0.1987 & 0.8912 \end{bmatrix} \qquad \begin{matrix} f_1(x_1) = tanh(x_1) \\ f_2(x_2) = x_2^3 \end{matrix} \qquad c{=}0.97 \tag{7}$$

Fig. 2a shows the mixtures scatterplots respectively, without nonlinear distortion (linear), after the application of nonlinearities (nonlinear), after the recover and compensation of nonlinearities using the Gaussianization and the C-F Expansion methods. Fig 2b shows the potential functions obtained from the scatterplots, and we can note that using Extended Gaussianization approach, the detection of mixing matrix direction is better than Gaussanization results. Table 1 reports the SNRs and Itakura-Saito measure of the recovered sources. We have used the following definition of SNR

$$SNR = 10log\left(|s_k|^2 / |s_k - \hat{s}_k|^2\right) \tag{8}$$

where s_k is the k-th original source signal and $\hat{s}_k$ is the corresponding recovered source. The signal quality is evaluated by the Itakura-Saito distance, a widely used objective quality measure that supplies the distortion of the recovered sources. We must note that, after the recovery of the nonlinearities, higher SNR and lower IS can be obtained than the ones obtained from the "non linear" approach.

IV. Conclusions

The biometrical issue has been extensively gaining the attention of worldwide scientists and companies, for the many possible real-life applications are related to it. Nevertheless, the importance of voice as biometric measurement has been proven during the last decades by the existing flourishing literature on the speaker identification and verification tasks. However, in order to achieve good performances also in real conditions, the input speech signals need to be suitably pre-processed and the Blind Signal Processing techniques undoubtedly represent an effective choice when noise sources, reverberation and multiple speakers occur.

In particular the Blind Source Separation paradigm can be applied on purpose. In this work, the problem of separating speech sources in presence of Post Nonlinear under-determined mixing is tackled, and a proper approach proposed. It consists on applying the Extended Gaussianization concept to mixture signals in order to compensate the unknown nonlinearities. Experimental results confirm the effectiveness of the algorithm and a satisfactory recovery of the sources. Current efforts are on the development of a better estimation of the unknown C-F parameters for further improvement of separation performances, whereas future works will consider the presence of convolutive mixing with non linear distortion in order to achieve good performances also in real world data situations.

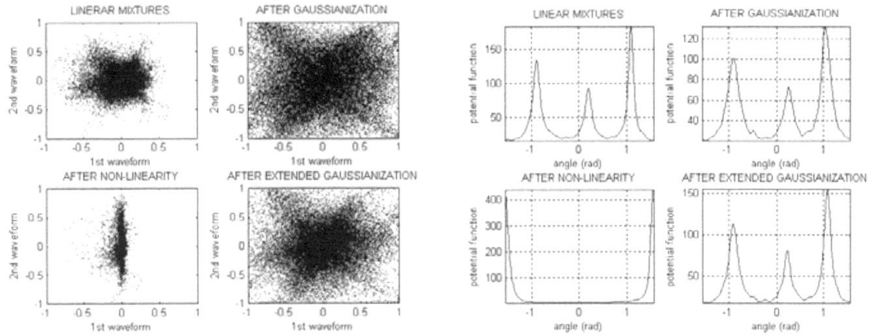

Figure 2: a) Scatterplots relative to the mixtures, b) Potential Function recovered by geometric ICA

Table 1 : a) SNR (dB) of recovered sources, b) IS indexes of recovered sources

	SNR(dB)			**ITAKURA-SAITO INDEXES**		
	Source 1	***Source 2***	***Source 3***	***Source 1***	***Source 2***	***Source 3***
Linear	12.457	11.094	11.269	0.38642	0.4987	0.71789
Non linear	5.3241	1.3497	5.7903	2.1143	3.6142	2.3026
Gaussianzation	9.9024	7.9426	9.4898	0.71756	0.76834	1.1159
Ext. Gaussianiz.	11.569	9.2781	10.824	0.50062	0.66472	0.85878

References

[1] F. Bimbot, J.F. Bonastre, C. Fredouille, G. Gravier, I. Magrin-Chagnolleau, S. Meignier, T. Merlin, J. Ortega-Garcia, D. Petrovsk-Delacrtaz, and D. Reynolds, "A tutorial on text-independent speaker verification," *EURASIP J. Appl. Signal Process.*, vol. 4, pp. 430-451, 2004.

[2] V. Mildner, S. Goetze, and K.D. Kammeyer, "Multi-Channel Noise-Reduction-Systems for Speaker Identification in an Automotive Acoustic Environment," *Audio Engineering Society, 120th Convention*, Paris, France, May 2006.

[3] G.J. Jang, S.J. Yun, and Y.H. Oh, "Feature vector transformation using independent component analysis and its application to speaker identification," *Proc. of the European Conference on Speech Communication and Technology (EUROSPEECH)*, Budapest, Hungary, pp. II.767-II.770, 1999.

[4] A. Şentürk and F.S. Gürgen, "The Usage of Independent Component Analysis for Robust Speaker Verification", *Artificial Intelligence and Applications*, Innsbruck, Austria, February 2006.

[5] E. Visser, and T.W. Lee., "Application of Blind Source Separation in Speech Processing for Combined Interference Removal and Robust Speaker Detection using a Two-Microphone Setup", *Proc. of ICA 2003*, pp.325-329, Apr. 2003.

[6] S.Squartini, A. Bastari, F.Piazza, "A practical approach based on Gaussianization for Post-Nonlinear Underdetermined BSS ",*Proc. ICCCAS,* Guilin, China 2006.

[7] A. Ziehe, M. Kawanabe, S. Harmeling, and K.R. Mueller, "Blind separation of post-nonlinear mixtures using Gaussianizing transformations and temporal decorrelation" *Proc. ICA2003*, pp.269–274, 2003.
[8] K. Zhang, L.W. Chan, "Extended Gaussianization Method for Blind Separation of Post-Nonlinear mixtures", *Neural Computation*, vol. 17, pp. 425-452,2005.
[9] F.J.Theis and S.Amari, "Postononlinear overcomplete blind source separation using sparse sources", *Proc. ICA 2004*, vol 3195 LNCS, Springer, pp.718-725, Granada, Spain 2004.
[10] S.S. Chen, R. A. Gonipath, " Gaussianization", *Proc. NIPS*, Denver,USA, 2000.
[11] A. Papoulis, S.U. Pillai, *"Probability, Random Variables and Stochastic Process. 4th Edition"*, McGrawHill, 2002

Section 3

Facial Expressions

Fundamentals of Verbal and Nonverbal Communication and the Biometric Issue
A. Esposito et al. (Eds.)
IOS Press, 2007

Visual Pragmemes and Pragmaphrasems in English, Croatian, and Polish Languages

Neda PINTARIĆ[1]
Filozofski fakultet, 10 000 Zagreb, Ivana Lucica 3, Croatia

Abstract. The author describes verbal and nonverbal possibilities of expressing eye-communication signs. Those signs are called pragmemes (pragmatic lexical units) and pragmaphrasemes (pragmatic idiomatical units). Eye-communication can consist of innate as well as cultural signs called culturemes. Each nonverbal sign has several corresponding verbal signs which means that language signs are more developed than nonverbal ones.

Keywords. Pragmatics, pragmemes, pragmaphrasemes, culturemes, nonverbal communication

Theoretical Introduction

The eye is the most important organ by the means of which man **watches** (or rather, **learns about**) the world around him. A Polish linguist [7] has defined four meanings of the lexeme *eye*: 1) the organ of sight; 2) a form similar to the eye (*a ring stone, fatty substance on a liquid surface, a loop in a knit or a net, a knot in a rope*); 3) a round water surface (lake – *Morskie* ***Oko***); 4) a new shoot on a plant. A Croatian etymologist [13] claims that the lexeme *eye* has originated from the stem *xuek- and in various languages and phases of their development it bears the meaning of both a noun and a verb *(eye; see, know, notice, realize)* and also points in the direction of the lexeme *learn* [13]. This shows that the eye enables us to see and gain experience and knowledge; in other words, the eye helps us to learn as well. The Modern English Language Dictionary ([9], p.296) outlines even more meanings. *Eye* is translated as *oko* and *vid* (sight) but this English word is also used to describe a great number of things that are round: a spot on a potato, an eyespot on a peacock's tail, the eye of a needle, the eye on a shoe, the eye on a tool, a loop on a rope, a hole in cheese or bread, a round opening, a small round window, a center, a goal and finally – gleam of a diamond. The eye needs light to be able to function and gleam is an echo of light.

The entire human culture is based primarily on the visual perception, starting with Altamira (cave paintings of the Paleolithic man) all the way to the computer icons. Why is that so? The answer to this question has been provided by the measurements of the information capacity of single receptors, which have shown that the optic channel has the greatest capacity (10^7 bit/sec!). It is followed by the auditory channel (10^6 bit/sec) and then by the tactile tract (10^5 bit/sec) [12].

[1] Corresponding Author: Filozofski fakultet, 10 000 Zagreb, Ivana Lucica 3, Croatia, e-mail: npintari@ffzg.hr .

Sight, hearing and touch are thus the main receptors through which we learn about the world. Now it is clear why the nonverbal code is so important in the history of human communication. It was precisely this receptor-based code that has allowed man to develop science, art and technology.

There are two fundamental proofs to support the thesis that the nonverbal code is older than the verbal (linguistic) one. These measurements deliver the first proof that the visual code has a larger capacity than the acoustic channel, which is secondary or additional in the nonverbal communication. The second proof was provided by biologists [1] who tried to teach animals how to speak, but not even the most developed monkeys managed to do so because their speech organs were not as developed as those of humans. These man-like apes were nevertheless able to easily learn the symbolic sign language, which is an obvious proof that the nonverbal communication preceded its verbal counterpart and that it is inherent in all living beings, including man.

The nonverbal code is based on three dominant channels: visual, tactile and the additional acoustic tract. That is why the elements of the nonverbal code – gestures – are primarily kinetic signals and the science that studies them is called **kinesics.** This name was coined by Birdwhistell who gave the same title to his 1952 book [6]. That same year a Croatian linguist Petar Guberina [14] wrote his own book called *Sound and Gesture in Language* in which he presented a thesis on the prevalence of gestures in communication. In the early beginnings of the human communication gestures were accompanied by onomatopoeic sounds (not by a voice!) and this has remained a constant in the nonverbal communication. Beside gestures, in kinesics it is necessary to stress *deixis* as a way of pointing (both verbally and nonverbally) to a spatial entity. This act of pointing is usually binary since there is a left and right, a front and back, an up and down to the speaker's perspective. This also represents a basic grammar for finding one's way in space [18]. In relation to space there are also the so-called proxemic signals [15], which indicate the models of human organization of space and time in each culture.

The second channel that is vital to communication is the tactile channel. Kauffman [16] called the science of tactile signs **tacesics.** Touch has an important role in communication (for example handshaking, clapping, stroking, shoulder patting and so forth). There are cultures in which participants in a communication tend to touch more often and more readily (the so called warm cultures) and those in which touching is allowed only in intimate situations (cold cultures) and where it is mostly sexually motivated. Today the clash between those cultures is especially transparent (for example, between the cold Anglo-Saxon and the warm Slavic ones, in which touching implies friendship and not only intimacy). But when strangers happen to meet in a public place (such as a bus, streetcar or shop) in both types of cultures they apologize to each other, which means that in both cultural contexts touching is desirable only among people who know each other well. Touching is an expression of positive emotions (if we put aside the emotionally negative touching of a fight, slap or sexual violence) and that is why it is used as a signal of love and friendship only by the people who know each other and wish each other well. The *culturem* of touching in family relations differs from its role in other types of social relations and varies from one culture to another [10].

The third path used in the nonverbal communication is the sound or acoustic channel. The sounds we make as we gesture carry their own meaning. Linguistics studies this channel in the context of verbal signs since one cannot be separated from the other. That is why they are called *paralinguistic signs* [17, 25] and are studied within the framework of **prosody**. This includes various forms of *intonation* in words and sentences, *accents* within words (especially those of the melodic variety, such as Croatian word accents) or emphasis of individual words in a sentence, color of voice or *timbre* as well as different unarticulated sounds (for example, laughter, crying, sobbing, hiccoughing and so forth). These are used to express positive or negative emotions, biological needs and imitations of the sounds originating in nature and society (the so called *unarticulated exclamations*). For the most part these paralinguistic signs are subconscious and cannot be controlled in a communication. Pszczołowska [24] though describes how color of voice can indicate a person's profession, their social standing and degree of education etc. and proceeds to illustrate this with the example of coachmen or newsboys whose voice is raspy because they cry out or shout in an open space, which causes their vocal cords to be enlarged and irritated. That is how a paralinguistic phenomenon can contain not only "biological" and "individual" but also "cultural information". Owing to the paralinguistic idiosyncrasies of the speaker, we are usually able to tell their age and gender. When we make use of paralinguistic signs consciously, we are using imitation or simulation [24]. Our use of paralinguistic signs is then conscious. That is for example what an actor does in a theatre or what we employ in a conversation when we quote a person we don't like to relate their words to a third party.

Argyle [2]distinguishes between the *speech-related* paralinguistic phenomena (speed variations in pronouncing parts of words, word stressing, pitch differences in pronouncing phonemes, different intensity in pronouncing parts of words and sentences in speech) from the *speech-unrelated* paralinguistic phenomena (noises used to express emotional states, speed and pitch differences, pronunciation intensity with regards to emotional states, interpersonal attitudes and individual characteristics relating to voice and accentuation [2]. This makes the first set of paralinguistic phenomena symbolic since they are used to express information about the speaker intentionally, whereas the other set is inherent and is used unintentionally, automatically. It provides information for the collocutor and the speaker is not aware of it. Owing to all of this, we can deduce that communication is a **multidimensional and multicoded event** in which all human senses participate, but sight, hearing and touch prevail [3, 20, 22]. Pavelin Lešić [21] calls communication a **global expression** that is realized nonverbally through *posturomimogestural manifestations*, whereas Arndt and Janney [3] have dubbed it an **intergrammar** that describes a **multimodal communication**.

With our eyes we watch other people and decipher their emotional states but others can decode us as well by reading the way we look at them. In a look we can find various subconscious messages we send (for example, when we like something, our pupils dilate automatically and we open our eyes wide, which Oriental merchants use when asking a certain price for their product from a tourist [15]. We often also consciously try to fake some positive feelings in spite of ourselves if we care about the collocutor and their present. But it is hard or even impossible to influence the signals our body emits automatically because we cannot, for example, automatically make our pupils dilate or blush with joy if we are not joyful or excited and so forth. By looking someone straight in the eye you can express different and even contradictory attitudes in different cultures. For instance, in the Croatian culture looking straight in the collocutor's eyes implies sincerity on the part of the one who is looking or it suggests that a person is listening to their collocutor intently. If they lower their gaze, it signals that the speaker or collocutor is ashamed or that they are lying. In the Croatian society a direct look of the speaker into the eyes of the collocutor thus represents a sincerity

culturem (the term culturem was coined by Alicja Nagórko to define a unit of cultural behavior in a communication). This sincerity culturem can be deduced from the original culturem, inherent among men in primitive societies, who have to look into each other's eyes when striking a deal verbally in order to confirm the truthfulness of their word. Truthfulness is also confirmed by the handshake signal. Morris [19] says that this was initially employed to indicate that participants in a communication were not armed, or rather, that they come in peace. In the Polish culture the culturem of looking straight in the eye means just the opposite: if you look straight into the eyes of the collocutor of a different gender, it may mean a sexual challenge. This was later translated onto communicators of the same sex. This Polish culturem is in keeping with the inherent behavior of girls and women in the communication with men, which Eibesfeldt [10] has identified in primitive peoples. By comparing the act of looking, one can conclude that the Croatian behavior culture is patriarchal and male-oriented, whereas its Polish counterpart is matriarchal because it accepts a prototypical female behavior. Evans [11] has developed "a cultural theory of emotions" in which he sets apart *natural inherent emotions* that include joy, sadness, anger, fear, surprise and disgust from the so called *higher emotions* (love, guilt, shame, concern, pride, envy, jealousy). Both types of emotions have a universal character and are expressed in the same nonverbal way but the latter ones are distinguished by the fact they are culturally conditioned (you can feel disgust not only for biological phenomena, which is the so called basic emotional type, but also for example, for some immoral human actions, and that is when such an emotional expression becomes a cultural signal, [11], p. 25, 46). The visual channel is dominant in expressing all these types of emotions. Through the eye it carries a piece of information to the cortical emotional centers, which Joseph LeDoux calls *the emotional brain* ([11], p. 55). In other words, the eyes reflect every individual type of emotion that can be expressed in the verbal code, i.e. translated into language.

Apart from its cultural meaning, the act of looking can also include emotional meanings that are expressed verbally. In that case these expressions become units of pragmatics and are called **pragmemes**. Here pragmatics will be used in the sense of the multicoded science that studies the meanings of human actions and experiences with the help of their emotional, modal and illocutive ways of expression [4]. Pragmems are the smallest pragmatic units that can be expressed verbally by the means of lexemes but can also be connected with larger linguistic units – phrasems, and are then called **pragmaphrasems** [23].

Since every nonverbal eye signal can be expressed in the verbal code as well, this article will also deal with the means of expressing visual pragmems and pragmaphrasems that are used to describe a look. These include nominal, adjectival and verbal visual lexemes (such as *eye, eyes, sight, look, perception; visual, seen, big-eyed; look, observe* and so forth) as well as syntagms and phrasems with a visual component. We shall be comparing their grammatical and pragmatical background in English, Polish and Croatian.

1. Grammatical and pragmatical analysis of visual pragmemes and pragmaphrasems

With regards to their grammatical constitution, we have divided visual pragmems into three subgroups. First we will be looking into the nominal, then the adjectival and finally into the verbal pragmems in the three languages we are comparing. Since English is an analytic language, as opposed to Croatian and Polish that are synthetic languages, word formation differences will be substantial. Croatian and Polish have several basic root words to which they add suffixes from both left and right, thus creating words of different meanings and different classes. The English language must employ a larger number of different lexemes but their form always remains the same, regardless of its position in the sentence and its word class. Polish and Croatian lexemes on the other hand undergo one more change: on top of all the suffixes, inflection endings are added onto them on the right side. That distinguishes nouns from adjectives and verbs; the only thing they have in common is the root word. Latin and Greek loans for visual forms are equally functional in all three languages. All three languages have compounds with two different roots. In all three languages several lexemes can center around one meaning and thus function as syntagms, phrasems and proverbs respectively (the latter ones are complete sentences unlike the previous two groups). In all three languages these are predominantly prepositional expressions.

1.1 Nominal lexemes and pragmeme

We shall first contrast simple nouns that consist of only one root. Then we shall proceed to analyze compounds and finally, nominal syntagms that encompass several lexemes.

1.1.1 Simple nouns

From the parallel translations of English nouns (that have been alphabetized) into their Croatian and Polish equivalents, we can see that in the English language there are forms that can also be used as verbal infinitives, whereas verbal nouns have the so called –ing form. In Croatian and Polish the nouns derived from the root verb often have suffixes *–enje/-anje* in Croatian and *–enie/-anie* in Polish. Croatian verbal nouns frequently employ the suffix *–ba* as well. Here it is visible only in the nouns *primjedba* and *zamjedba*. The nouns of Latin origin have kept their original form in English, whereas in Croatian and Polish they have been given a suffix, depending on their pronunciation (Cro. *-cija*, Pol. *-cja*). Some Croatian and Polish nouns have the zero suffix (*pogled, izgled; wygląd*). The suffix *-aj* (*opažaj*) appears only in Croatian. Since we have quoted mainly verbal nouns in these examples, and in Croatian and Polish there are perfective and imperfective verbs, these nouns also indicate from which verbal aspect they have been derived. In most cases these verbal nouns come from imperfective verbs (*buljenje, zurenje, gledanje* od *buljiti, zuriti, gledati*; *mruganie, gapienie* od *mrugać, gapić się*) but they can be derived from perfective verbs as well (*pogled, viđenje* from *pogledati, vidjeti*; *mrugnięcie* from *mrugnąć*).

In English nouns the perfective aspect is connected with a quick interrupted look (*glance, glimpse*) that can be related to perfective verbs in Polish and English. This is indicated by the translation of this noun into Croatian as *brz pogled – quick glance* ([9] p. 1083) or *letimičan pogled – cursory glance* ([8], p. 363) and by the -ing form, which is used to express the durative aspect of a verb (*staring, winking* as opposed to *stare, wink*). The so called -ing form is a constitutive element of the "continuous present tense", which denotes an action that is taking place as we speak and can thus be compared to Croatian and Polish durative imperfective verbs.

The nouns that are derived from adjectives form a special category in Croatian and Polish. Most of these nouns indicate a human characteristic that refers to a person's sight or the shape of their eyes but there are some general nouns that imply the act of watching in their root word.

Table 1: *Simple nouns*

English	Croatian	Polish
blink	Žmirkanje	mruganie, mrugnięcie
gaze	gledanje, piljenje, blejanje	wpatrywanie, oglądanie, gapienie
glance	pogled, treptaj	mgnienie
glimpse	pogled, treptaj	mgnienie
goggl/goggling	buljenje, zurenje	gapienie
gape/gaping	buljenje	gapienie
horizon	hrizont	horyzont
interview	intervju	wywiad
look/looking	pogled, izgled	wygląd
monitoring	promatranje, motrenje	obserwacja
observe/observation	opservacija, primjedba	obserwacja, uwaga
perception	percepcija, opažaj, uočavanje, zamjedba	spostrzeżenie, percepcja
perspective	perspektiva	perspektywa
overview	pregled	przegląd
review	osvrt	spojrzenie wstecz, rzut oka wstecz
sight	vidik, prizor	widok, scena
squint/squinting	žmirenje	mrużenie
stare/staring	zurenje	gapienie
view	pogled, viđenje	widzenie, pogląd, spojrzenie, widok
vision	vizija, viđenje	wizja
wink/winking	namigivanje	mruganie

In the Croatian language *okatost, škiljavost, razrokost, žmirkavost* point to negative characteristics in the shape of someone's eyes or their way of looking. The nouns *viđenost, uviđavnost, gledanost, obzirnost, očitost, očiglednost, uzornost, uzoritost* contain a visual component in the root word (viđ-, gled-, -zir/zor-) but unlike the former group evoke a positive emotional connotation. If we strike the suffix -ost, the nouns become adjectives with root words that have the same meaning as the nominal root words (*okat, škiljav, razrok, žmirkav; viđen, gledan, obziran, uzoran, uzorit, očit, očigledan*). Emotionally neutral are the nouns with the following type of visual root words: *uvid, očevid; uviđaj, opažaj.* In Croatian they are used as legal and psychological terms. They either have a zero suffix or have been given the suffix *–aj.*

The noun *interview (intervju)* has been adopted from the English language and it literally means *insight* but is used only as a term that defines a conversation between a questioner and an important person, which is employed to cast light on the life and work of the interviewed person.

The noun *osvrt* is terminologically especially pregnant. Its root word does not refer to a visual element but the noun itself still implies turning your head backwards in order to see something (engl. *re-view*). Together with a visual dimension in its deep structure, this noun also indicates a temporal category on the surface because we are turning back to something that had happened before it drew our attention. By moving our head backwards we are connecting it with some past event or with an event to which we are reacting with a delay by turning back. The English translation refers to a repeated look at something, which also implies that something has already been seen in the past, returning to this past event, reconstructing. The noun *overview* signifies the perfective aspect of looking or ability to see an entire area. For this purpose Croatian employs the suffix *pre-* in the word *pregled* and the same goes for the Polish equivalent: *przegląd*. This abstract noun, connected with the action of looking, is formed in the same way in all three languages (*-view, -gled, -gląd*).

In the body of nouns that have been quoted here we have set pragmems apart because they have an additional emotional meaning outside the context or because they are used to describe some nonverbal signal. There are three types of such pragmemic nouns. The first group encompasses the nouns describing a gesture (blink/ing – žmirkanje – mruganie; wink/ing – namigivanje – mruganie). In the second group we have included the nouns that describe the way of looking. Here we distinguish those that are emotionally negative (goggl/ing, gape/ing, stare/ing – buljenje, zurenje, piljenje, blejanje – gapienie) from those that, apart from their negative overtones (gaze – piljenje, blejanje, zurenje, buljenje – wpatrywanie się), can also be emotionally neutral and are thus not considered to be pragemems (gaze – gledanje – patrzenie). The third type are the nouns that are emotionally neutral but contain a gestural description of turning and a temporal dimension of an event from the past (review – osvrt – rozpatrywanie) or a spatial dimension of taking the entire space in with one look (overview – pregled – przegląd).

1.1.2 Compound nouns

Compound nouns are derived from two different roots or two words that are considered to be one, written either together or with a hyphen in English. These roots are mostly nominal but they can also be amalgams of other word classes with nouns (for example, *retro-spection* as an adverb-noun amalgam, *eye-wash* as a noun-verb amalgam, *eye-ful* as a noun-adjective amalgam etc). These examples show that the compounds in English can be translated into Polish and Croatian by using different forms: as individual nouns, compounds, syntagms or phrases (comp. *sljeparija, vidokrug, izbuljene oči, šaka u oko*; *wygląd, światopogląd, przenikliwe oczy*).

English compounds can be formed from two nouns (written together or hyphenated) as well as by combining prepositions and nouns (*outlook, insight*). The interesting thing about the shape of the eyes is that it often invites negative emotions and irony. An image of bulging eyes, for example, would translate into English literally as *saucer-eyes,* which implies not only a round shape but also a considerable size and those characteristics could be attributed to large round eyes as well. But the eyes that look like that do not invoke negative overtones, whereas saucer-eyes point to an element of ugliness more clearly. In Croatian there is a terminological metaphor *očna jabučica (eyeball)* that is used to indicate the protruding part of the eye. The Croatian and Polish nouns that consist of two separate words we shall consider syntagms (Engl. *modifiers*, [5]) because they have only one meaning. In English those syntagms are expressed as compounds and are translated into Croatian and Polish with two or more words. But the reverse may also be the case: an English compound (for example *eyebrow*) can be translated into Croatian and Polish as one lexeme (*obrva – brew*). In the third instance in English and Croatian we have compounds where there is only one lexeme in Polish (for example, Engl. *eyeful,* Cro. *vidokrug*).

It is interesting to note that in different English dictionaries certain compounds are hyphenated, whereas in other dictionaries the same words are spelled without a hyphen. For some English words some Polish dictionaries often have no translation and in the Croatian dictionary we have found the author's neologism under quotation marks (comp. *eye-opener*, Cro. «razbuđivač») as well as a phraseological translation (*ono što otvara komu oči*). It is just as frequent to have an ingenious English compound that can be translated into Croatian and Polish only descriptively (for example, *eye-popper*, Cro. *začuđujuća, zapanjujuća stvar*, Pol. *rzecz zadziwiająca*).

Table 2: *Compound nouns*

English	Croatian	Polish
eyeball	očna jabučica	gałka oczna
eyebrow	obrva	brew
eyeful	vidokrug	widnokrąg
eyeglass	monokl, okular	monokl, okular
eyehole	očna duplja	jama oczna, oczodół
eyelash	trepavica	rzęs
eyelid	vjeđa	powieka
eye-opener	'razbuđivač', ono što otvara oči	co otwiera oczy
eye-popper	začuđujuća stvar	rzecz zadziwiająca
eyeshot	vidokrug	widnokrąg
eyesight	vid	wzrok
eyesore	šaka u oko	rażący widok
eyetooth	očnjak	kieł
eyewash	sljeparija, opsjena,obmana	piasek w oczy
eyewitness	svjedok,očevidac	świadek
insight	uvid	wgląd
image	stil, imidž	styl, image
outlook	izgled,svjetonazor, pogled, vidik, horizont	Wygląd,,światopogląd horyzont, widok
saucer-eyes	iskolačene oči	wybałuszone oczy
sight-seeing	panorama, razgledavanje	panorama,widok, zwiedzanie
sight-seer	turist	turysta
viewership	gledateljstvo	widzowie
viewpoint	motrište, gledište, stajalište	punkt widzenia, stanowisko

There are cases where an English compound is translated into Croatian as a simple lexeme and into Polish as a phrasem (*eyewash*, Cro. *sljeparija, opsjena*, Pol. *piasek w oczy*). Some English compounds are translated into Croatian with a phrasem and into Polish with a descriptive syntagm (*eyesore*, Cro. *šaka u oko*, Pol. *rażący widok*). This last combination also shows that an English compound is translated into Croatian and Polish as a monolexeme (*eyesight*, Cro. *vid*, Pol. *wzrok*).

The Latin word *retrospection (retrospektiva, retrospekcija)* represents a visual noun in English, Croatian and Polish and in every language it is given a different suffix (in English the Latin suffix *–tion* is preserved, Croatian takes two suffixes *–iva* and *–cija,* whereas Polish employs the phonetic Latin form *–cja.* This noun could be treated as a compound since it consists of the adverb *retro* (in reverse) and *spection* (looking). It functions as a loan from Latin in all three languages. A second English compound that consist of an adverb and a noun is *out-look*, which in Croatian means *izgled, pogled* as well as *svjetonazor* and *vidokrug*. The first two translations are monolexemes, whereas the other two words have been derived from two nouns and each of them is built around one root word with a visual meaning (*-zor i vid-*).

The aforementioned nouns that are connected with the act of looking cannot all be considered pragmems. In each of the three languages pragmems are only those nouns that include an element of emotion, modality or illocutivity. We have introduced nouns with a visual element into this analysis in order to show which forms constitute pragmems outside the context. Placed in a certain context, each noun can be a pragmem if it is used in one of the three pragmatic meanings, i.e. if it is figurative or metaphoric. We have seen that English compounds are translated into Croatian and Polish mostly as phrasems (Engl. *eyesore* - Pol. *piasek w oczy,* Cro. *šaka u oko*) or by the means of a descriptive syntagm (Cro. *začuđujuća stvar,* Pol. *rzecz zadziwiająca*).

1.2 Adjectival visual lexemes and pragmemes

English visual adjectives are often indistinguishable from nouns (for example, *popping, blinking, monitoring*) or from verbs (for example, *considerate, seen, view, gape*) so that the class to which these forms belong can be defined only within a context and a sentence. In Polish and Croatian an adjective is apparent in its form even outside the grammatical context. This small body of adjectives has been ordered so that the English language comes first, followed by Croatian and then Polish. A hyphen separates each of the three languages.

After analyzing this body of adjectives, we can see that a large number of examples refer to a description of a wide-eyed person (agape, big-eyed, goggle-eyed, pop-eyed, lobster-eyed, popping, popping-eyes). That is also the most pronounced gesture that appears in different types of emotion: in astonishment, admiration, fear, anger, joy. A smaller group of emotions is associated with narrowing eyes (disgust, irony, sadness).

Table 3: *Adjectival visual lexemes and pragmemes*

English	Croatian	Polish
agape	zablenut	zagapiony
big-eyed, saucer-eyed	okat	który ma wielkie oczy
blinking	žmirkav	mrugający
cross-eyed	razrok, škiljav	zezowaty
foreseen, expected, envisaged	predviđen	spodziewany
goggle-eyed, pop-eyed	izbuljen, izbečen	wybałuszony
obvious	očit, očigledan	oczywisty
overlooked(not noticed)	previđen	przegapiony
overlooked	pregledan,	przejrzysty
popping	izbečen, zablenut	zagapiony
popping eyes	izbuljenih očiju	wyłupiasty
pop-eyed, goggle-eyed, lobster-eyed	buljook, okat	wyłupiasty
seen	viđen	widziany
thoughtful, considerate	uviđavan,obziran, pažljiv	oględny, rozważny, uważny, baczny
wall-eyed	razrok	zezowaty

Two pragmems (wall-eyed and cross-eyed) describe squinting as an eye defect. Compound adjectives consist of a preposition and an adjective (overlooked). The two forms of Croatian compound adjectives are connected with a visual root word: u-viđavan and ob-ziran. It is a similar situation with the Polish form *o-ględny*.

1.3 *Verbal visual lexemes and pragmemes*

For this thesis we have chosen the so-called *perception verbs* and *verbs of attitude and emotion* [5]. These groups include those reported in Table 4. We have included the verb *respect* because of its Latin root *spectare*, which also refers to looking, but in this case it is looking upwards, since we put those that we appreciate above all others. The rest of the verbs in English have the same form as nouns (for example, *glance, glimpse, look*). *To glimpse* is translated as *zabljesnuti, načas se ukazati, baciti pogled*, and in Polish it is: *przelotnie rzucić okiem*. The temporal dimension of a look that takes place instantaneously, just like a flash of light, is central here in all three languages.

Pragmemic visual verbs point to different ways of looking: *glance, glimpse*. These quick looks cannot be translated into Croatian and Polish with only one word; they require a phrasem, as we will show in the third chapter. Unlike them, durative verbs indicate that a look lingers on an object for a longer period of time. Those verbs are strongly pragmatically marked (*stare – buljiti – gapić się*).

Table 4: *Verbal visual lexemes and pragmemes*

English	Croatian	Polish
condone	promatrati	patrzeć
goggle	buljiti	gapić się
observe	promatrati	obserwować
respect	cijeniti	szanować
see	vidjeti, pogledati, gledati	widzieć, spojrzeć, zobaczyć; patrzeć, oglądać
stare	buljiti	gapić się
watch	gledati	patrzeć, oglądać

The verb *see* has a different form for the noun (*sight*) and for the adjective (*seen*). It is unique in Croatian and in Polish since it has several synonymous perfective forms and a completely different root for its imperfective aspect (comp. *gledati-vidjeti, patrzeć-zobaczyć* as well as all quoted examples).

The verb *watch* in English has the same form as the noun *watch* (sat, ura), whereas Croatian and Polish employ loans for this noun (Cro. *sat* from Turkish, Pol. *zegarek* from German). The verb *look* also has the same form as its nominal counterpart and Croatian and Polish add a verbal infinitive suffix to the prefixed noun root (Cro. *iz**gled**-ati,* Pol. *wy**gląd**-ać*), thus distinguishing verbs from nouns.

Reflexive verbs in Croatian denote a simultaneous act of looking, or rather, a mutual looking action: *gledati se, ogledati se, vidjeti se/viđati se.* The reflexive verb *osvrnuti se, obazrijeti se/obazirati se* does not denote a mutual action but a literal reflexive quality – the turning of the speaker's head. Croatian and Polish verbs have the perfective as well as the imperfective aspect, which is sometimes expressed by the means of suppletive forms: *gledati-vidjeti (patrzeć, oglądać-zobaczyć).*

There are also certain verbs in English that express an instantaneous action, whereas in Croatian and Polish this instantaneous quality cannot be translated with a verb (*glimpse* – pogledati na brzinu, u tren oka, baciti pogled). Some verbs cannot signify an instantaneous action or are rarely used in that sense (for example, *žmiriti/ zažmiriti* (imperfective and perfective forms), *žmirkati* – an iterative form,* *žmirnuti* is usually not used and it stands for an instantaneous action that has happened only once). Sometimes a visual verb can be given an entirely different metaphorical meaning (for example, *obazirati se* – may mean *osvrtati se* – *look around*, and its metaphorical meaning is *biti pažljiv – to be considerate* and *odgovarati na izrečeno – reply to what has been said, komentirati – comment*).

It seems that the aspect of English visual verbs is contained in their meaning. The verbs *squint, wink, observe (žmiriti, namigivati, promatrati),* for example, can thus be translated into Croatian with the imperfective verb aspect, whereas *glimpse, glance* imply the perfective and even instantaneous aspect in their meaning, so it is necessary to add an adjective that would describe this speed in Croatian and Polish (trenutačan pogled).

The verb *blink* stands for an iterative action of rhythmical flashes and it is often translated into Croatian with a Croatian infinitive of that English form: *blinkati*. Bujas [8, 9] translates this by using a perfective instantaneous verb *trepnuti*, which in Croatian implies a blink of an eye, whereas the imperfective or iterative form *treperiti* refers to a rhythmical repetition of interrupted light waves and not to lowering and raising of the eye lids. The same form can have two meanings in English, whereas in Croatian there is a distinction between the movements of the eyelids (that essentially also implies an interruption of the stream of light) and the interruption of light waves.

1.4 Visual sytagmas (collocations, modifiers)

A syntagm consists of two freestanding words that describe one concept but are not inseparable (for example, the noun *view* can collocate with *private view, fine view, general view, broad/wide view*). Those syntagms can be treated as compounds but their constitutive parts are written separately in order to emphasize the freedom they have in forming pairs (for example, *eyes* can be paired with the adjectives *popping, goggle* and others to express one meaning but the constitutive parts still retain their individual meaning). In syntagms, just like in all other cases, adjectives perform a descriptive function, whereas the noun has an autonomous lexical meaning. Biber et al. [5] call these syntagmas modifiers and distinguish *premodifiers* from *postmodifiers* (for example, ***a special*** *project* and *a block* ***behind*** ([5], p. 574-575). From our point of view though, a syntagm stands for a stronger bond between two words that constitute one concept, as the following English pragmemic syntagms illustrate:

Table 5: *Visual sytagmas (collocations, modifiers)*

English	Croatian	Polish
full view	potpun pregled	przegląd
fine view	lijep pogled	piękny widok
goggle eyes	izbuljene oči	wytrzeszczone (wybałuszone) oczy
double vision	dvostruka slika	podwójne widzenie
popping eyes	izbeljene oči	wybałuszone oczy
private view	osobni pogled, stanovište	osobiste spojrzenie, własny pogląd

1.5 Visual pragmaphrasems

Unlike syntagmas, where words can be loosely connected, words within phrasems cannot be separated and constitute only one meaning or one concept. But what functions as a phrasem in one language does not have to be a phrasem in the other. This following English sentence, for example, does not contain a phrasem (an idiom): *windows with a fine view of the sea*, whereas in Croatian there is one: *prozori s pogledom na more*. And the other way around, what constitutes an idiom in English (*to take a long view*), does not have to be expressed by one in Croatian: *biti dalekovidan*, *misliti unaprijed na posljedice*; *to take a short view – misliti kratkoročno, ne biti dalekovidan*. There is also a possibility to have a phrasem in both languages: *to keep in view – imati pred očima, imati u vidu, držati na umu*.

In Bujas' dictionary [8, 9] we have found certain phrasems that have been missing from the Polish-English Idioms by Wolfram-Romanowska et al. [26]. These include:

- at a **glance** – na prvi **pogled** – od pierwszego **wejrzenia,** na pierwszy rzut **oka**
- at **eye-blinking** speed – munjevitom brzinom, dok **okom trepneš**, brzinom munje; w **mgnieniu oka**
- at first **sight –** prima **vista** – prima **vista**
- catch a **glimpse** – uloviti **pogled** – uchwycić spojrzenie
- disapear **form sight** – nestati iz **vida** - zniknąć **z oczu**
- forty **winks** – sklopiti **oči**, zadrijemati - polj. zdrzemnąć się, uciąć drzemkę
- hang on by the **eyelids** – visiti o niti, očajnički se boriti, hvatati se za slamku – chwytać się jak tonący brzytwy
- have a **glance** at – baciti oko, baciti pogled – rzucić okiem
- in a **wink** - u **tren oka - w** tej **chwili,** w tym momencie, nagle
- **look** into each other's **eyes** – gledati se **u oči** – **patrzeć** sobie **w oczy**
- **look** right **into** somebody's **eyes** – **gledati** komu ravno **u oči – patrzeć** komu prosto **w oczy**
- lost from **view** – izgubiti se iz **vida** - stracić **z widoku,** zniknąć **z widoku**
- maintain **eye-contact** – zadržati pogled – stać **patrząc,** trzymać **wzrok**
- not sleep **a wink** – (ni) **oka** ne sklopiti – nie zmrużyć **oka**
- open one's **eyes** wide – širom otvoriti (izbuljiti) **oči** – **zagapić się**
- pick somebody's **eyes** out - iskopati komu oči -
- **wink** at me – **namigni** mi – **mrugnij okiem** do mnie
- up to one's **eyeballs** – do **ušiju**, do vrata, do grla, totalno - po **uszy**

According to the English-Polish Idiom Dictionary [26], we have singled out only phrasems that include *eye* and its compounds. We have compared English and Polish with the Croatian phrasemes. The Polish authors have included some proverbs among phrasems and we have set those apart because, unlike phrasemes, proverbs constitute entire sentences.

In all three languages there are some **identical phrases** with the constitutive element *eye*. In Croatian and Polish this element can be in nominative or some other case (instrumental, genitive) and in both singular and plural.

Phrasems containing *eye* in singular and in various cases:

- have a (good) **eye** for something – imati dobro **oko** za što (imati dobro zapažanje) - mieć (dobre) **oko** do czegoś
- in the twinkling of an **eye** – dok **okom** trepneš – w **oka** mgnieniu
- notice/ see something out of the corner of one's **eye** – vidjeti/ gledati što krajičkom **oka** – zauważyć/ zobaczyć coś kątem **oka**
- with naked **eye** – golim **okom** – gołym **okiem**

Phrasemes can include the lexeme *eye,* but in English it is in singular and in Polish and Croatian in plural:

- look somebody straight in the **eye** – gledati ravno u **oči** – patrzeć /spojrzeć komuś prosto w **oczy**

Many phrasems **differ** only in their visual lexemes. While in English we have a double visual lexeme *bird's* eye and *view,* other visual lexemes are represented in Croatian:

perspektiva, pogled. The visual lexemes *widok* and *wzrok* appear in the Polish translation.

- a bird's-**eye view** – iz ptičje **perspektive** – **widok** z lotu ptaka
- as far as the **eye** can reach – dokle **pogled** seže – jak (daleko) **okiem** sięgnąć
- measure somebody/ something with one's **eye**; size somebody/ something up – mjeriti koga **pogledom** – zmierzyć kogoś/ coś **wzrokiem**

English idioms with the plural component *eyes* that correspond to Croatian and Polish idioms with the component *oči; oczy* in nominative:

- close/shut one's **eyes** (to something) – zatvarati **oči** pred čime (na što) - przymykać / zamykać na coś **oczy**
- cry one's **eyes** out – **oči** isplakati za kim – wypłakiwać sobie **oczy**
- open somebody's **eyes** (to something) – otvoriti komu **oči** (na što) – otworzyć komuś **oczy** (na coś)
- throw dust in somebody's **eyes** – sipati komu prašinu u **oči** – sypać komuś piaskiem w **oczy**

Phrasems with components in different cases (instrumental and genitive) and in plural:

- bags under one's **eyes** – vrećice pod očima, natečene **oči** – worki/ wory pod **oczyma**
- (do something) with one's **eyes** closed – raditi što zatvorenih **očiju** – robić soś z zamkniętymi **oczami**
- in somebody's **eyes** – u čijim **očima**, po čijem mišljenju – w czyichś **oczach**
- keep one's **eyes**/ears (eyes and ears) – otvorenih **očiju** i ušiju, otvoriti **oči** i uši – mieć **oczy**/uszy (oczy i uszy) otwarte
- not believe/credit one's **eyes**/ ears – ne vjerovati svojim **očima**/ušima – nie wierzyć własnym **oczom**/uszom
- not take one's **eyes** off somebody/something – ne odvajati **oči** od koga/od čega – nie odrywać od kogoś/od czegoś **oczu**/wzroku
- scratch somebody's **eyes** out – skočiti komu u **oči** – wydrapać komuś **oczy**
- the scales fell (dropped) from somebody's **eyes** – nekomu su se otvorile **oči**, netko je progledao – łuski spadły komuś **z oczu**
- with fear/ desperation/ sadness etc. in one's **eyes** – sa strahom/ s očajem/ s tugom itd. **u očima** – ze strachem/ z rozpaczą/ ze smutkiem itp. **w oczach**

Where Croatian and Polish have different visual components, English employs *eyes*:

- drop (lower) one's **eyes**– spustiti **pogled** – opuścić/ spuścić **oczy/ wzrok**
- feast one's **eyes** (up)on something – **ne moći se nagledati** koga/čega, uživati gledajući koga/što – sycić **wzrok** /oczy czymś
- have shadows under the (one's) **eyes** – imati **podočnjake** – mieć cienie pod **oczyma**
- strain one's **eyes** – treba dobro otvoriti **oči**, dobro gledati da se što vidi, napinjati **oči** – natężać/ wytężać **wzrok**

From the dictionary we have singled out the same proverbs in which the component *eye* varies with regards to the case and number in Croatian and Polish, whereas the only variation in English is the number:

- an **eye** for an eye (a tooth for a tooth) – **oko** za oko, zub za zub – **oko** za oko, ząb za ząb

- crows will not pick out crows' **eyes** – vrana vrani **oči** ne iskopa – kruk krukowi **oka** nie wykole

If the sentence is shuffled around somewhat, the proverbs containing the *eye* component can vary in Croatian. The following two are identical in English and Polish but different from the one in Croatian:

- what the **eye** doesn't see, the heart doesn't grieve for (over) – daleko od **očiju**, daleko od srca – czego **oko** nie widzi, tego sercu nie żal
- the master's **eye** fattens the horse (the master's **eye** makes the horse fat) – gospodarevo **oko** konja goji - pańskie **oko** konia tuczy

EYEBALL – očna jabučica – gałka oczna

- eyeball to **eyeball** – **u** četiri **oka** - **oko w oko**

EYESHOT – pogled - wzrok

- within **eyeshot** (within sight) – biti na **oku** – w zasięgu **wzroku**

EYELID – vjeđa, očni kapak - powieka

- **wink** at (**connive** at, **condone**) – zatvoriti **oči** na što, gledati komu kroz prste - patrzeć z przymrużonymi **oczami**
- without batting an **eyelid** – da ni okom ne trepne – bez zmrużenia powieki (powiek, oka), bez mrugnięcia powieką

2. Nonverbal eye-communication (Neverbalna očna komunikacija)

Many examples we have shown here imply facial gestures that are expressed through visual verbal signals. Now we shall state the type of relationship between the emotion and its nonverbal expression.

What can we do with our eyes? We widen them, open them, narrow them, dilate or constrict the pupils, we roll our eyes so that their white becomes obvious, we raise and lower our eyes (gaze), we close our eyes and lower our eyelids. Our eyes can be bright or glazed. They show joy, sadness, anger, disgust, disapproval. We hold our eyes steady, lower them, look aside or turn our head and look behind. We raise our eyebrows or we lower them. We scowl, which means that we wrinkle our forehead and draw our eyebrows close. All of these things we can put in words and demonstrate by moving our eyes.

These movements are often accompanied by nonverbal sound expressions that substantiate visual eye movements. These sound expressions cannot be transcribed phonetically so we call them unarticulated. This is why we have used italics to emphasize the description and approximation of these sounds, after which we have described the eye signals and finally pinned a meaning to it.

- *loud sighs* (yhhh) go with the wide-open eyes and raised eyebrows, which can have a homonymous meaning: positive and negative surprise, fear, astonishment;
- *mumbling* is accompanied with a lowered gaze, scowling, drawing the eyebrows together, which is used to express disagreement, disapproval or denial;
- *crying, sobbing* goes with lowered eyelids and dimmed tearful eyes, the hands are used to cover the eyes or to wipe the tears off, which is an expression of sadness or dejection;

- *hooray* is a cry associated with laughter that is also reflected in the eyes, the eyes are glowing, the eyebrows are raised straight up, the eyes and pupils dilate, tears of joy begin to flow – all in expression of merriment and joy;
- *muttering in a low voice* is connected with scowls, the eyebrows are slightly raised, the eyelids drop halfway down, the head is sinking and the eyes are sad, which signals disappointment;
- *a-ha* usually accompanies the collocutor's gaze that is fixed on the speaker, the gaze holds if the agreement is complete and shortens if the agreement is pretended, which means that agreement is often connected with sincerity;
- *phooey, phew* is a shout that indicates an unpleasant odor, accompanied by lowering the gaze, averting the eyes from the source of the foul smell, by a scowl and the eyebrows drawn close together.
- *aaa(h)* is a shout that expresses delight in a pleasant smell or general satisfaction and is accompanied by opened eyes and dilated pupils, the gaze is directed upwards and the head raises; and
- *silence or a curt goodbye,* looking away i.e. turning the head from the collocutor is a way of saying that the speaker does not wish to communicate with them any more because they are angry with them;
- *silence* that is accompanied by lowered eyes and head as well as lowered eyelids expresses obedience, submissiveness and contrition, humility;
- *a short hem* with a decisive and short action of raising and lowering of the head, accompanied with a lowered gaze in the direction of the person that the speaker is underestimating, indicates a feeling of superiority on the part of the speaker.

This review of nonverbal sound and gestural signals shows that they can be used not only to express certain basic emotions but also complex and culturally founded feelings such as overestimating, underestimating, putting an end to communication, agreeing or disagreeing with the stated views.

It also shows the possibility of nonverbal expression of basic emotions such as joy, happiness or sadness and sorrow. There are degrees in expressing those emotions just as there are verbal degrees of making those emotions known. The degrees of verbal and nonverbal signals indicate the intensity of the emotion in question. For example, sorrow or moroseness can be reflected in dimmed eyes, a dejected (depressed) person rejects communication, i.e. withholds the gaze, lowers the eyes or simply stares. These signs can also indicate sadness, along with sobbing, crying, weeping, hiding the eyes with one's hands, wiping the tears off with a palm of the hand or a handkerchief and other nonverbal signals that we have not mentioned here since this description is limited to visual eye signals only. Some nonverbal signals can be homonymic, i.e. the same signal can be used to express different emotions, as is the case of the wide-open eyes, for example. Nonverbal signals though can also be synonymic since different means of expression can be used to indicate the same meaning (for example, facial gestures set all parts of the face in motion in order to express one single emotion: laughter is reflected on the lips as well as in the eyes, eyebrows, cheek muscles etc.).

What is the purpose of looking? "*From a psychobiological perspective, gaze serves essentially a monitory function /.../.*"[3]. But there are communicative reasons of gaze. According to Allen and Guy [27] the main purpose of gazing while speaking is to obtain feedback. Gaze can vary according to cognitive load, such as the topic of conversation; it is «higher when speaking about familiar topics than» during the greater reflection. «Higher eye contact has been correlated with attentiveness, interpersonal involvement, the desire to obtain or communicate information, the attempt to act friendly and the attempt to be persuasive.» It is also higher when listening than when speaking. The distance can be the reason of duration of gaze: increase of distance tends to increase the eye contact.» ([27],p.300-301). There are affective interpretative alternatives (positive and negative) for high and low eye contact. «High eye contact tends to be interpreted more positively in connection with positive verbal messages; speakers combining high eye contact with negative messages are liked less.» Positive high eye contact is used when we are interested, concerned, surprised, attentive, attention-seeking, sincere etc., negative gaze we use while being upset, worried, shocked, aggressive, dominating, authoritarian etc. Low eye contact can also be positive (when we are disinterested, preoccupied, reflective, respectful, deferential, insecure) and negative (when we are confused, overtaxed, perplexed, inattentive, aloof, insincere etc.) ([27], p.302-303)

Nonverbal signals are a foolproof indicator of an emotion that is expressed in various movements and sounds. The same emotion can be expressed verbally through various synonyms that nevertheless display different shades of meaning. For example, *iznenađenje* can be translated into Polish as *niespodzianka* (surprise) and *zaskoczenie* (astonishment), which is associated with the Croatian form *začuđenost, čuđenje*. Both words can have a negative and a positive meaning, depending on the context. The Polish word *zdumienie* can be translated as *začuđenost, iznenađenost* (amazement). We can see that meanings overlap, differ in shades and can be expressed in synonyms in all three languages. The same goes for the words *negodovanje, prosvjedovanje, revolt, indignacija, nezadovoljstvo* and other similar ones. In Polish we have *niezadowolenie, oburzenie, bulwersowanie (zbulwersowany)* and in English for example, *discontent, dissatisfaction, indignation, revolt. Divljenje* (Pol. *podziw*) is listed in a dictionary as *admiration* but Croatian has a synonym with a stronger shade of meaning: *obožavanje*, which implies admiration of a person as if they were a deity (Pol. *adoracja, adorowanie*). All these examples demonstrate that a nonverbally expressed emotion cannot be easily translated into verbal signals. Each language has a different set of means for outlining the qualities of the emotion in question, or rather, sees different shades of meaning in one and the same emotion.

3. Conclusions

All of this leads us to conclude that basic emotions are common to all people and are connected with inherent nonverbal signals, such as laughter, crying and other sensations we pick up through our various senses. But many emotions have developed into culturems and vary from one culture to another. As far as the eye communication is concerned, we have seen that there are cultures where looking straight into someone's eyes is considered polite, whereas in other cultures it is just the opposite. Those culturems are used and learned consciously, whereas those inherent to us we must often restrain and suppress because society sees them as impolite and unwanted (for example, a long gaze in someone's direction is described with negatively connoted verbs in all three languages we have included here: piljiti, buljiti, blejati and so forth). We have seen that nonverbal pragmems are far fewer in number than their verbal counterparts because one and the same emotion can find a number of different linguistic expressions (a simple lexeme, a compound, several lexemes, collocations, syntagms, modifiers and finally phrasems that can have a direct and a metaphorical gestural meaning). Nonverbal signals can be translated into the verbal variety but the eye communication is most frequently expressed subconsciously so that it can rarely be excluded from the simultaneous verbal expression. Nonverbal eye signals can be used deliberately, as, for example, actors do when they cry or laugh because their role requires them to. But in everyday communication our body language is mostly spontaneous so that eye signals come naturally as well. If we want to pretend, it is precisely the body language that gives away our insincerity. Nonverbal signals are used to intensify verbal expression (if they are compatible with the meaning expressed), and if they point in the direction opposite to the meaning expressed, they indicate the insincerity on the part of the speaker or their illness (there is medical research into the correlation between gesticulation and speech since it has been proven that aphasia affected patients do not coordinate their gestures with speech). Nonverbal signals are important in indicating a positive or a negative affective state of the speaker and collocutor, which is useful in deciphering the meaning they are trying to express.

References

[1] J. Aitchison: *The Articulate Mammal. An Introduction to Psycholinguistics,* London, New: Routledge, 4th edition, 1998.
[2] M.Argyle: *Bodily Communication*, International Universities Press, New York.1975.
[3] H. Arndt, R.W. Janney: *InterGrammar*, Mouton de Gruyter, New York, 1987.
[4] A. Awdiejew: *Klasyfikacja funkcji pragmatycznych*, Polonica IX, Kraków, 53-88, 1983.
[5] D. Biber, J.S., Leech, G. Conrad, S. Finegan, E. Finegan: *Grammar of Spoken and Written English*, Longman, London. 1999.
[6] R. L. Birdwhistell: *Introduction to Kinesics*, University Press, Louisville. 1952
[7] W. Boryś: *Słownik etymologiczny języka polskiego*, Wydawnictwo Literackie, Kraków, 2005.
[8] Ž. Bujas: *Veliki englesko-hrvatski rječnik*, Nakladni zavod Globus, Zagreb, 2001.
[9] Ž. Bujas: *Veliki hrvatsko-engleski rječnik*, Nakladni zavod Globus, Zagreb, 2001.
[10] I. Eibl-Eibesfeldt: *Miłość i nienawiść*, PWN, Warszawa, 1987.
[11] D. Evans: *Emotion*, Oxford University Press, 2001.
[12] H. Frank: *Kybernetische Grundlagen der Pädagogik,* vol. II, Agis Verlag, Baden-Baden, 1969.
[13] A. Gluhak: *Hrvatski etimološki rječnik*, «August Cesarec», Zagreb, 1993.
[14] P. Guberina: *Zvuk i pokret u jeziku*, Matica hrvatska, Zagreb, 1952.
[15] E. T. Hall: *The Silent Language*, ABE, 1973
[16] L.E. Kauffman: *Tacesics: The Study of Touch: A Model for Proxemic Analysis,* Semiotica, 4, Mouton the Hague. 1971.
[17] [G.V. Kolšanskij: *Paralingvistika*, Nauka, Moskva, 1973.
[18] [18] A. Krawczyk-Tyrpa: *Frazeologi somatyczna w gwarach polskich,* PAN, Ossolineum, Wrocław Warszawa-Kraków-Gdańsk-Łódź, 1987.
[19] D. Morris: *Manwatching*, Triad Granada, 1978.
[20] M. Nowakowska: *Komunikacja werbalna i niewerbalna jako język wielowymiarowy*, Studia Semiotyczne IX, 1979.
[21] B.Pavelin Lešić: *Strukturacija interakcije licem u lice*, Govor, 2, 155-164, 2005.
[22] N. Pintarić: *Pragmemi i pragmafrazemi prehrane u poljskom i hrvatskom jeziku*, Strani jezici, 1, 2006.
[23] N. Pintarić; *Pragmemi potvrđivanja u hrvatskom i poljskom jeziku,* Riječki filološki dani, 6, 131-140, 2006.
[24] L. Pszczołowska: *O parajezičnim pojavama u književnom djelu*, Umjetnost riječi, Zagreb, 2-4, 279-287,1974.
[25] N. Rot: *Znakovi i značenja*, Nolit, Beograd, 1976.
[26] D. Wolfram-Romanowska, K. Przemysław, M. Parker: *Idiomy polsko-angielskie*, PWN, Warszawa, 2005.
[27] D. E.Allen, R. F. Guy: *Conversation* Analysis: *The Sociology of Talk.* The Hague: Mouton, 1974.

A. Esposito et al. (Eds.)
IOS Press, 2007

Eye localization: a survey

Paola CAMPADELLI, Raffaella LANZAROTTI, Giuseppe LIPORI [1]
Dipartimento di Scienze dell'Informazione.
Università degli Studi di Milano.
Via Comelico, 39/41 - 20135 Milano (Italy)

Abstract. In this survey we first discuss the importance of eye localization as the initialization step for many face processing techniques, then we further show the influence of the localization precision on the rate achieved by some baseline face recognition techniques. To do so, we give the definition of the objective error measures already known in literature.

A brief review of some state-of-the-art eye localization methods is subsequently given, focusing on those articles that tackle the problem of precise localization and that use some objective criterion for performance evaluation. Finally their localization results are compared on a number of different datasets.

Keywords. Face and Eye Localization, Face Recognition, Localization Error Metrics

Introduction

The task of facial feature extraction requires to localize the facial components of interest (eyes, nose, mouth, etc.) and to estimate their scale. Facial feature extraction is essential for the initialization of many face processing techniques like face tracking, facial expression recognition or face recognition. In particular face recognition is a research area where it has been made a great effort in the last years to design and compare different systems [1,2]. The most general scenario for face recognition is the identification case: given a set of faces (stored in a *gallery* under an appropriate representation) whose ID is already known, use them to attribute an ID to the subjects depicted in a new set of images (the *probe* set). This task does not require cooperation of the subjects under investigation and tries to answer the question "who is this person?". The face recognition rate is the score at rank 1, that is the percentage of probe images correctly identified.

Not all the face recognition techniques (FRTs) presented in literature clearly state the assumptions they make on their initialization; many of them skip the feature extraction step, and assume perfect localization by relying upon manual annotations of the facial feature positions [3,4,5,6]. This attitude is not feasible as it will not necessarily lead to a robust FRT: the error made at the localization stage influences the recognition rate in a way that is not always predictable. This is the reason why it is fundamental to achieve an automatic, robust and precise extraction of the desired features prior to the application of a FR technique.

[1]Corresponding Author: e-mail: campadelli@dsi.unimi.it

In literature there is a general agreement that the eyes are the most important facial features, so it is not surprising that the greatest research effort has been devoted to eye detection and localization. The prominency of eyes with respect to the other facial features is due to several reasons:

- Eyes are a crucial source of information about the state of human beings. When looking at a picture of a person, people tend to devote the greatest attention to the eyes. The shape, size, and color of the eyes provide cues in recognizing individuals; it has been shown that the upper part of the face plays a more important role for face recognition than its lower part. The eye movement may reveal "interest", or "attention" of a person, and it may contribute to recognize his/her emotional state.
- Faces contain two eyes (if not occluded) that have in general a very similar appearance, so they can be searched simultaneously with the same technique. This allows to roughly identify the scale of the face (the inter-ocular distance is relatively constant from subject to subject) and its in-plane rotation. Sometimes this information is already sufficient to achieve the image normalization required for the initialization of certain face processing techniques.
- The eye appearance is less variant to certain typical face changes. For instance they are unaffected by the presence of facial hair (like beard or mustaches) and are almost unaltered by small in-depth rotations of the face (though they look closer to each other due to the perspective change), while the nose appearance greatly changes. The same can be said about facial expressions: even if the eyes are considered important for the recognition of human expressions, they are evidently less articulate than the mouth.
- The accurate eye localization permits to identify all the other facial features of interest [7].
- Eyes (the retina) show particular reflection properties when exposed to *near infrared* (NIR) light. This characteristic is not sufficient for a robust eye localization due to the presence of many false positives in uncontrolled environments, therefore it is often used in combination to more traditional (visible light) image analysis techniques. NIR systems (also called *active* systems) require a dedicated hardware which is not universally available.

Despite of the great effort made in recent years, the problem of (eye) feature extraction is still open, as shown by the number of contributions to the area.

In Sec. 1 we introduce the localization error metrics and discuss the relation between the eye localization error and the face recognition rate. In Sec. 2 we briefly describe some recent and interesting eye localization methods [8,9,10,11,12,13,14,15] that clearly affirm the importance of precise eye localization and that adopt some sort of objective evaluation measure. In Sec. 3 we discuss their localization performance. The last section is destined to draw some conclusions about the eye localization problem.

1. The importance of eye localization

Given the true positions of the eye centers (by manual annotation), the eye localization accuracy is expressed as a statistics of the error distribution made over each eye (usually the mean or the maximum), measured as the Euclidean pixel distance (see Fig. 1). In order to make these statistics meaningful, so that they can be used to compare the results obtained on any dataset, it is necessary to standardize the error by normalizing it over the face scale.

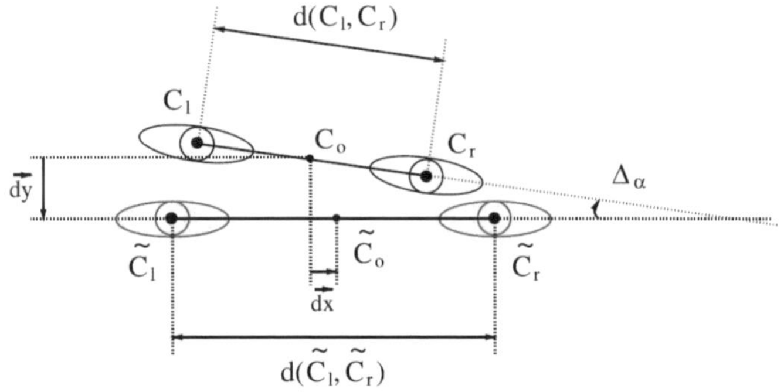

Figure 1. Localization error: (C_l, C_r) are the ground truth positions, $(\tilde{C}_l, \tilde{C}_r)$ are the results of automatic localization.

One popular error measure has been introduced by Jesorsky et al. [8] in 2001, and it has been already adopted by some research works on feature localization. The measure, which can be considered a worst case analysis, is defined as

$$d_{eye} = \frac{\max(||C_l - \tilde{C}_l||, ||C_r - \tilde{C}_r||)}{||C_l - C_r||}$$

There is a general agreement [8,9,16] that $d_{eye} \leq 0.25$ is a good criterion to flag the eye presence (to claim eye detection). This precision roughly corresponds to a distance smaller than the eye width. However, this accuracy level may not be sufficient when the localized positions are used for the initialization of subsequent techniques.

Following the idea presented in [9], we studied the relationship between d_{eye} and the face recognition rate of some baseline methods available in the CSU package [17] and our FR algorithm presented in [18]. To mimic the behavior of eye localization techniques that achieve different levels of precision, we carried out four recognition experiments by increasingly perturbing the ground truth quality; both C_r and C_l have been randomly displaced inside circles of radii 0.05, 0.1 and 0.15 with uniform distribution. In Fig. 2 we report the results of this study on the XM2VTS database [19], which contains 1180 pictures grouped into 4 sessions of 295 subjects. The experiment is defined as follows: session 1 is used for the gallery, session 2 for the probe, sessions 3 and 4 constitute the training set. Differently from [9] where only the probe set is affected by synthetic error, all three sets (gallery, probe and training) have been perturbed as it would happen in a completely automatic system.

The graphs of Fig. 2 clearly show that the precision of eye localization is critical for the alignment of faces, even if it does not affect all the methods in the same way.

The article [20] is entirely devoted to the so called *curse of misalignment*. There it is reported the high dependence of the Fisherface method [21] performance on the alignment precision, especially with respect to rotation or scale errors. The authors also pro-

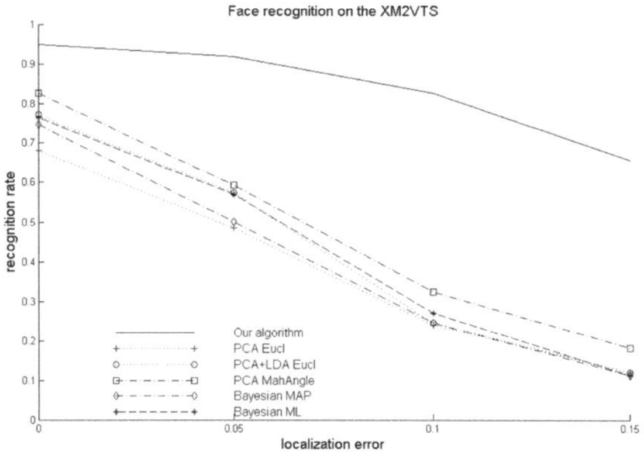

Figure 2. Face recognition vs. eye localization precision (d_{eye}).

pose to evaluate the *overall face recognition rate* with a measure, $rate^*$, that integrates the FR rate over all possible misaligned initializations, weighted by their probability:

$$rate^* = \int_{e \in \text{errors}} rate(e)P(e)\text{d}e$$

They measure the robustness of a FRT to errors as the overall FR rate normalized with respect to the ideal case of absence of error, i.e. $rate^*/rate(0)$.

The very recent work [22] further developed the issue suggesting a new error measure which is more discriminative than d_{eye} as it permits a quantitative evaluation of the face recognition degradation with respect to different error types. Instead of considering only the Euclidean distance of the detections from the ground truth points, it considers four kinds of error: the horizontal and the vertical error (both measured between the midpoints $C_o, \tilde{C}_o$ of the segments $\overline{C_r C_l}$, $\overline{\tilde{C}_r \tilde{C}_l}$, see Fig. 1), the scale and the rotation error.

$$\Delta_x = \frac{dx}{||C_l - C_r||} \quad (horizontal) \qquad \Delta_s = \frac{||\tilde{C}_l - \tilde{C}_r||}{||C_l - C_r||} \quad (scale)$$

$$\Delta_y = \frac{dy}{||C_l - C_r||} \quad (vertical) \qquad \Delta_\alpha = \widehat{\overrightarrow{C_l C_r} \overrightarrow{\tilde{C}_l \tilde{C}_r}} \quad (rotation)$$

In fact it happens that some FR systems are more sensitive to certain types of error. In particular, the baseline PCA method is extremely sensitive to all types, while the FR system described in the article (referred to as DCT/GMM) seems to be almost indifferent to translational errors (Δ_x, Δ_y), while its performance notably degrades when the error is due principally to scale or rotation inaccuracy (Δ_s, Δ_α). The authors conclude that it is not possible to define an absolute concept of precise localization: each FR will have a different tolerance to errors and it should clearly state the level and type of precision required for its initialization.

A totally different approach is that of [23] where, instead of imposing the maximum level of acceptable localization error, it is proposed to deal with it by learning its distribution directly into the statistical model of each subject. The method requires a quantitative estimate of the localization error distribution to be used to perturb each image accordingly, generating a certain number of new images characterized by the possible displacements. These enriched samples constitute the classes to be modelled (one for each subject) either with a Gaussian or with a mixture of Gaussians in case of more complex distributions. Such models are then used for face recognition, being robust to localization error by construction. A similar approach has also been proposed by [24].

2. Methods

In this section we briefly describe some methods that, to our knowledge, are state-of-the-art solutions for the problem of eye localization and, what is more, adopt similar criteria of objective performance evaluation. The next section reports the localization accuracy declared by each of them.

2.1. Robust face detection using the Hausdorff distance

In 2001 Jesorsky et al. [8] were among the first to highlight the strong dependency of the FR performance and the accuracy of the face detection. In their article they proposed both an eye localizer and a measure to evaluate quantitatively the performance of this class of methods.

The method is a model-based, coarse-to-fine approach adopting the modified Hausdorff distance. Both the coarse and the refinement level consist of a segmentation and a localization phase, while referring to different models (the whole face and the eye region respectively) and working on portion of images progressively reduced. The segmentation steps extract and binarize the edges; the localization stages compare the models previously built with the edge images, searching for the best match according to the Hausdorff distance, and allowing translation and scaling of the models. Finally the found positions are refined applying a Multi-layer Perception (MLP) trained with pupil centered images.

2.2. Robust precise eye location under probabilistic framework

The article [9] proposes a three stage method for robust and precise eye localization. All stages are based on cascaded AdaBoost classifiers [25], the output of which are interpreted under a probabilistic framework. The first two stages follow a top-down scheme: first a standard Viola-Jones face detector gives the face bounding box, then the face region is vertically divided in two (by looking at the vertical projection); each half is separately scanned with an eye detector with threshold set to a low value in order to collect many detections. The eye detections are subsampled according to proximity (by exploiting the face size) and to their probability. This step is necessary to reduce the number of detections to consider. The last stage proceeds in a bottom-up fashion: the sampled detections are paired in all possible ways and a specifically trained eye-pair classifier identifies the three most probable eye pairs. Finally, the average positions of the three pairs gives the eye localization.

The method is declared to work on upright frontal faces with in and out of the plane rotations up to $25°$.

2.3. Robust precise eye location by AdaBoost and SVM techniques

The research work [10] performs eye localization by combining cascaded AdaBoost and Support Vector Machine (SVM) classifiers. The system is very similar to the one described in Sec. 2.2 and it also has three levels: an AdaBoost face detector, an AdaBoost eye detector and a SVM post classifier.

First the original AdaBoost is applied to find the face region, then its upper part is scanned with another AdaBoost classifier trained on eyes. The discard of the lower part is motivated by a priori knowledge of the face appearance and it permits to reduce the computational complexity. The authors note that the eye detector is not discriminative enough to reliably separate the true eyes from false detections, so the classifier threshold is set to a low value to accept many detections. Finally the last word is left to a SVM trained on geometrical features regarding the eye positions within the face region. All pairs of detections are evaluated and those classified as admissible configurations are averaged together.

Due to the geometrical constraints, the method is declared to work on images with head rotation up to 10° both in and out of the plane, regardless of background clutter and light conditions.

2.4. Feature-based affine invariant localization of faces

Hamouz at al. [11] propose a bottom-up, feature-based method oriented to face authentication scenarios. This assumption allows them to skip the face detection step (one large face is assumed to be present in the image), and to apply the feature search step to the whole image.

The method consists of four subsequent steps. At first a feature detection is applied: 10 points on the face (features) are searched for, applying a bank of Gabor filters and comparing the response matrix in each single spacial location with the feature models previously trained for each feature. Such step generates a large number of false positives, thus requiring further refinement: face hypothesis generation, registration, and appearance verification.

In particular, the face hypothesis generation considers each triplet of features and discards the ones which lead to improbable geometrical configurations. The registration step consists in applying an affine transformation to the remaining candidate configurations, in order to normalize them with respect to the scale and in-plane rotations. Finally the method applies a verification test to the obtained candidate configurations, constituted of two SVMs acting in a coarse-to-fine manner.

2.5. Automatic eye detection and its validation

The contribution proposed in [12] uses the AdaBoost algorithm, while trying to overcome the limited discriminant capability of the Haar wavelet.

AdaBoost has been chosen by the authors for its effectiveness in selecting the critical features and training weak classifiers as well as updating the sample weights. Indeed, it has been proved that as long as the weak classifiers are slightly better than random guessing, the final classifier will have a much better accuracy after combining all the weak classifiers together.

The authors propose to adopt the Recursive Nonparametric Discriminant Analysis (RNDA) to extract more effective features, besides providing sample weights useful for the AdaBoost algorithm.

The final classifier is constituted of two layers: the first has only two features, thus being very fast, while the second has about 100 features to refine the search.

2.6. A generative framework for real time object detection and classification

Fasel et al. [13] propose a generative framework for real time object detection and discuss in particular detail the case of eye localization. The algorithm makes inferences about the presence of certain objects within certain patches of the image that have been previously segmented. The approach requires the development of likelihood-ratio models to distinguish the patches that contain the objects of interest from the background ones. The models are learned using GentleBoost [26] trained on a vocabulary of Haar-like features.

As in many other methods, eye localization is achieved decomposing the problem in a top-down manner that involves three models: a face detector, a rough and a fine eye localizer. Regarding the rough localizer, very interestingly the authors study the relation between the detection rate and the size of the eye patches; the optimal choice is to take square patches centered in the eye center and having side equal to twice the inter-ocular distance. The fine eye localizer is not applied to detect the eyes separately, instead its model is based on patches that covers both eyes.

The technique has been developed in the context of real time video applications, however it can be used also in case of still images.

2.7. 2D Cascaded AdaBoost for eye localization

The very recent method presented by Niu et al. [14] is based on AdaBoost classifiers, while bootstrapping on both positive and negative examples. The training procedure starts building a first classifier on the basis of a small set of positive examples randomly chosen, and progressively bootstrapping the negative ones, thus reducing the false alarm rate. Afterwards, in order to augment the detection rate, the method bootstraps also the positive examples. The whole process is iterated n times, determining n classifiers characterized by an increasing discriminatory capability, but also by an increasing time consumption (the first uses few dozens of Haar features, while the last needs thousands of them).

The authors propose two different localization procedures, one obtained weighting all classifiers, the other cascading them. The first performs effectively, while being time consuming. The second applies the obtained classifiers in cascade, stopping in correspondence to the first one which detects at least one region, thus drastically reducing the time consuming. In case of more than one detection, both procedures compute the corresponding centroid.

2.8. Regression and classification approaches to eye localization in face images

In [15] three approaches to eye localization are proposed and compared: a regression method which aims to directly minimizing the prediction error; a Bayesian approach in which two distinct probabilistic models of the eye and non-eye appearance are built independently; a single strong classifier trained for eye detection using AdaBoost. All the methods are trained and tested on the same images and applied in cascade to a face detector module (Viola-Jones). Given the face bounding box, the eye search methods are applied in a very narrow portion of it, strongly relying on a correct and precise face detection. Results show that the simple Bayesian model outperforms the others, pointing out the difficulty of using classifiers for the task of localization.

2.9. *Precise eye localization through a general-to-specific model definition*

In this section we briefly outline the technique we employ to perform precise eye localization, presented in [27]. The method can be applied to the output of any face detector that returns a rough estimation of the face position and scale (in our experiments we have used both [25,28]).

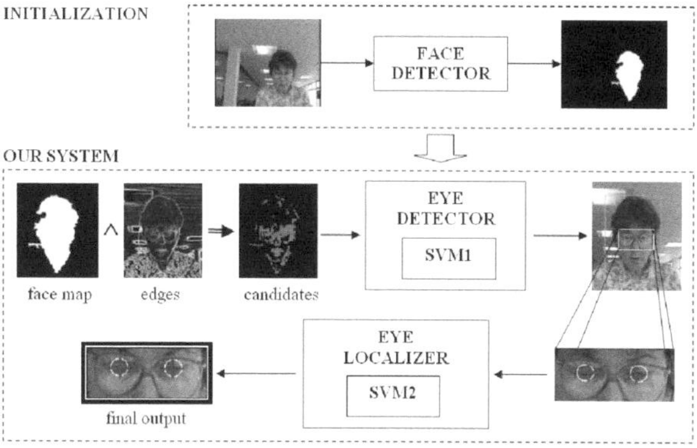

Figure 3. General outline of the eye localization system.

The system is top-down and consists of two modules: the *eye detector* and the *eye localizer*. The eye detector is initialized on the output of a face detector and serves two distinct objectives: it not only produces a rough localization of the general eye features, it also validates the output of the face detector (all face detectors manifest a certain false detection rate that must be dealt with). The eye localizer can be applied in cascade to the eye detector in order to refine the localization precision by using the specific eye pattern definition.

Each module is built around a SVM trained on optimally selected Haar wavelet coefficients. As sketched in Fig. 3, the scan strategy proceeds by progressively reducing the search domain from the face region to a subsampling of the edge map (by exploiting the face size), to the regions where eyes are located with high probability (using the SVM output as a measure of confidence). Possible false detections are discarded by reinforcing verticality and rewarding candidates with high SVM output.

The system proves to be robust to background clutter, to moderate illumination changes and to head rotations up to 20° both in and out of the plane.

2.10. *Face shape and appearance models*

Strictly speaking, the face shape and appearance models (ASM, AAM [29,30]) are not pure eye localization techniques, although they can be improperly used for this aim.

These methods share the idea of modelling the shape of the whole face together with its appearance (or the appearance of its components), trying to match the model (by deformation) over the input image. A shape is a collection of labelled points, that is points associated with a semantics, plus some constraints among them.

Both the shape and the appearance are learned from a training set via PCA: the mean shape and appearance, and a number of modes of variation, can be calculated from this set of examples; then the matching is sought by letting the mean shape deform according to these modes of variation, adapting to the input data (minimizing a proper energy function). The adaptation is driven by the image edges and by the facial feature appearance. The algorithm is iterative and needs a good initialization in order to converge to the correct solution; this can be obtained by placing the model in correspondence to the output of a face or eye localizer.

The Constrained Local Models (CLM) [31] are a very recent evolution of face shape and appearance models. The shape used by CLM aims to track the position of 17 points (eye centers, eye corners, eyebrows extremes, moth corners, nostrils, etc.). On one hand it can be argued that this is a much stronger objective than just localizing the two eye centers. On the other hand it has to be pointed out that the request of precision is less demanding than in the specific case of eye localization: the authors of CLM adopt the relative error metrics $m_e = \sum_{i=1}^{17} \frac{\|\tilde{p}_i - p_i\|}{\|C_l - C_r\|}$ that averages the localization error over all features without particular stress on the points semantics.

3. Localization results

The graphs in Fig. 4 display the performance of our eye detector (SVM1), our eye localizer (SVM2) and, when available, we report the performance achieved by the methods presented in Sec. 2.4 (denoted as '1 face on the output' in the original article) and 2.10.

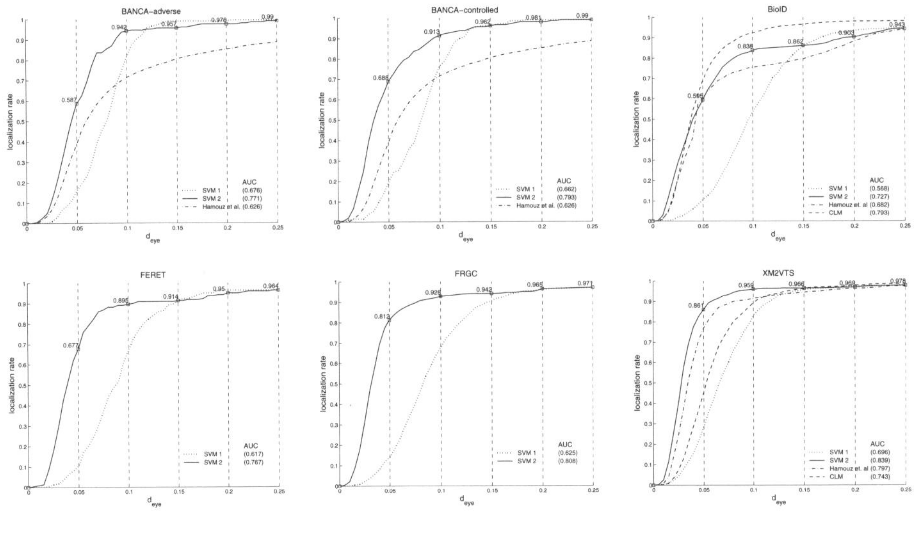

Figure 4. The cumulative distributions of eye detection and localization over different databases.

Regarding CLM, the curves plotted on the BioID [32] and XM2VTS [19] graphs have been extrapolated from results kindly provided by the authors of the method.

The numbers reported in parenthesis on the graphs represent the Area Under the Curve (AUC), therefore they give a global estimation of the performance of each localization method over that particular dataset. Regarding eye detection ($d_{eye} \leq 0.25$) the SVM1 alone permits to achieve rates of 99.0%, 96.1%, 96.4%, 97.1% and 97.8% over the datasets BANCA [33], BioID, FERET [34], FRGC v. 1.0 [35] and XM2VTS respectively. As expected, the addition of the second classifier greatly improves the precision of the detection and the curves are systematically above the rates declared by Hamouz et al. (Sec. 2.4). Regarding CLM, we note that it is very effective in localizing the eyes over the BioID database, while on the XM2VTS it achieves a lower rate[2].

Also the works in Sec. 2.1, 2.2, 2.3 and 2.7 use the error measure d_{eye} in order to assess the quality of eye localization. The fist work exhibits a localization performance that is lower than that reported by Hamouz and et al. The second one presents a cumulative curve that looks similar to the performance of our SVM1 but it is obtained referring to a mix of databases with no intersection with the ones we considered, making impossible a direct comparison. The third paper reports results on the BioID, tabulating only the values corresponding to $d_{eye} \leq 0.1$ and $d_{eye} \leq 0.25$ (91.8% and 98.1% respectively), while omitting the curve behavior under this value. Finally, the last work presents results on XM2VTS and BioID. We do not report the curves in figure since the values are not clearly tabulated, however we note that the performance on XM2VTS is comparable to ours, while on the BioID their results are significantly better.

Other works face the same problem, while adopting a different metrics. Wang et al. (Sec. 2.5) adopt a normalized mean error (not the maximum, which is a worst case analysis) and give an error of 2.67% on the entire FRGC. By adopting this measure on the FRGC subsets defined before, we observe an error of 2.65% and 3.88% on the controlled and uncontrolled images (respectively). If we consider that the entire FRGC contains twice as many controlled images as uncontrolled ones, we see that these results are very similar. Fasel et al. (Sec. 2.6) provide the localization results on the BioID in terms of the mean relative error, expressed in irises. Noting that the iris diameter is slightly shorter than the 20% of the inter-ocular distance, their measurement corresponds to a mean relative error of about 0.04 with standard deviation of 0.01, which is quite a good result. On the BioID we report a mean relative error of exactly 0.04 with a standard deviation of 0.03. The method described in Sec. 2.8 carries out the experiments on the FERET database: in the 90% of images the mean relative error is reported to be smaller or equal to 0.047, which is remarkable (for the same level of precision, on the FERET we count about the 78% of images).

Conclusions

In the last years many research works have pointed out the importance of a correct, precise and automatic eye localization as the fundamental step for the initialization of other methods, mostly face recognition techniques. Indeed the localization error influences the baseline techniques in a non linear way, with their rate rapidly decaying as the localization quality decreases. Moreover, not all types of error affect the subsequent process-

[2]The authors attribute this behavior to the major similarity of BioID images to the images used to train CLM.

ing in the same way: for instance scale errors usually affect a FR technique more than translational misalignment.

The survey of recent eye localization techniques clearly shows that we are far from perfect localization and there is still room for improvement. We note that many different methods tend to solve the problem in an analogous way, by decomposing it in a top-down fashion and thus incrementally reducing its toughness: first the face is roughly localized, then it is searched for eyes. Some methods proceed in this way by pointing directly to the eye locations, some other prefer to keep a low threshold to generate many (false) detections and then select the correct positions by averaging them out.

We strongly support the adoption of objective error measures like d_{eye} or others, in order to make a meaningful comparison of different eye localization methods on the many publicly available databases. Since every FR technique exhibits a different robustness to certain types and amount of error, there exists no absolute threshold for precise localization. We encourage the authors of face recognition techniques to investigate the robustness of their methods with respect to misalignments, and to clearly state the tolerance to error that they assume when communicating the face recognition rate.

References

[1] W. Zhao, R. Chellappa, P.J. Phillips, and A. Rosenfeld. Face recognition: A literature survey. *ACM, Computing Surveys*, 35(4):399–458, 2003.

[2] P.J. Phillips, P.J. Flynn, T. Scruggs, and K.W. Bowyer. Overview of the face recognition grand challenge. *Proc. IEEE Conf. Computer Vision and Pattern Recognition*, 2005.

[3] C. Liu. Gabor-based kernel PCA with fractional power polynomial models for face recognition. *IEEE Trans. Pattern Analysis and Machine Intelligence*, 26(5):572–581, 2004.

[4] G. Givens, J.R. Beveridge, B.A. Draper, P. Grother, and P.J. Phillips. How features of the human face affect recognition: a statistical comparison of three face recognition algorithms. *Proc. IEEE Conf. Computer Vision and Pattern Recognition*, 2004.

[5] K.I. Chang, K.W. Bowyer, and P.J. Flynn. An evaluation of multimodal 2D+3D face biometrics. *IEEE Trans. Pattern Analysis and Machine Intelligence*, 27(4):619–624, 2005.

[6] S. Romdhani and T. Vetter. Estimating 3D shape and texture using pixel intensity, edges, specular highlights, texture constraints and a prior. *Proc. IEEE Conf. Computer Vision and Pattern Recognition*, 2005.

[7] R. Lanzarotti. *Facial feature detection and description*. PhD thesis, Università degli Studi di Milano, 2003. Address: `http://lanzarotti.dsi.unimi.it/`.

[8] O. Jesorsky, K.J. Kirchberg, and R.W. Frischholz. Robust face detection using the hausdorff distance. *Lecture Notes in Computer Science*, 2091:212 – 227, 2001.

[9] Y. Ma, X. Ding, Z. Wang, and N. Wang. Robust precise eye location under probabilistic framework. *Proc. IEEE Int'l Conf. Automatic Face and Gesture Recognition*, 2004.

[10] X. Tang, Z. Ou, T. Su, H. Sun, , and P. Zhao. Robust Precise Eye Location by AdaBoost and SVM Techniques. *Proc. Int'l Symposium on Neural Networks*, pages 93–98, 2005.

[11] M. Hamouz, J. Kittler, J.K. Kamarainen, P. Paalanen, H. Kälviäinen, and J. Matas. Feature-based affine invariant localization of faces. *IEEE Trans. Pattern Analysis and Machine Intelligence*, 27(9):1490–1495, 2005.

[12] P. Wang, M. Green, Q. Ji, and J. Wayman. Automatic eye detection and its validation. *Proc. IEEE Conf. Computer Vision and Pattern Recognition*, 2005.

[13] I. Fasel, B. Fortenberry, and J. Movellan. A generative framework for real time object detection and classification. *Computer Vision and Image Understanding*, 98:182–210, 2005.

[14] Z. Niu, S. Shan, S. Yan, X. Chen, and W. Gao. 2D Cascaded AdaBoost for Eye Localization. *Proc. of the 18th Internationl Conference on Pattern Recognition*, 2006.

[15] M.R. Everingham and A. Zisserman. Regression and classification approaches to eye localization in face images. *Proceedings of the 7th International Conference on Automatic Face and Gesture Recognition (FG2006)*, 2006.

[16] Z.H. Zhou and X. Geng. Projection functions for eye detection. *Pattern Recognition Journal*, 37:1049–1056, 2004.

[17] J.R. Beveridge, D. Bolme, B.A. Draper, and M. Teixeira. The CSU face identification evaluation system. its purpose, features, and structure. *Machine vision and applications*, 16:128–138, 2005.

[18] S. Arca, P. Campadelli, and R. Lanzarotti. A face recognition system based on automatically determined facial fiducial points. *Pattern Recognition*, 39(3):432–443, 2006.

[19] XM2VTS DB. Address: `http://www.ee.surrey.ac.uk/Research/VSSP/xm2vtsdb/`.

[20] S. Shan, Y. Chang, W. Gao, and B. Cao. Curse of mis-alignment in face recognition: problem and a novel mis-alignment learning solution. *Int'l Conf. Automatic Face and Gesture Recognition*, pages 314–320, 2004.

[21] P.N. Belhumeur, J.P. Hespanha, and D.J. Kriegman. Eigenfaces vs. fisherfaces: Recognition using class specific linear projection. *IEEE Trans. Pattern Analysis and Machine Intelligence*, 19(7):711–720, 1997.

[22] Y. Rodriguez, F. Cardinaux, S. Bengio, and J. Mariéthoz. Measuring the performance of face localization systems. *Image and Vision Computing*, 2006.

[23] A.M. Martinez. Recognizing imprecisely localized, partially occluded, and expression variant faces from a single sample per class. *IEEE Trans. Pattern Analysis and Machine Intelligence*, 24(6):748–763, 2002.

[24] Jaesik Min, Kevin W. Bowyer, and Patrick J. Flynn. Eye perturbation approach for robust recognition of inaccurately aligned faces. *Proc. of the Internationl Conf. Audio and Video based Biometric Person Authentication (AVBPA)*, LCNS 3546, 2005.

[25] P. Viola and M. Jones. Rapid object detection using a boosted cascade of simple features. *Proc. IEEE Conf. Computer Vision and Pattern Recognition*, 2001.

[26] J. Friedman, T. Hastie, and R. Tibshirani. Additive logistic regression: a statistical view of boosting. Technical report, Stanford University, 1998.

[27] P. Campadelli, R. Lanzarotti, and G. Lipori. Precise eye localization through a general-to-specific model definition. *Proceedings of the British Machine Vision Conference*, 2006.

[28] P. Campadelli, R. Lanzarotti, and G. Lipori. Face localization in color images with complex background. *Proc. IEEE Int'l Workshop on Computer Architecture for Machine Perception*, pages 243–248, 2005.

[29] T.F. Cootes and C.J. Taylor. Active shape models - their training and application. *Computer Vision and Image Understanding*, 61(1):38–59, 1995.

[30] T.F. Cootes, G. Edwards, and C.J. Taylor. Active appearence models. *Proc Int'l Conf. Computer Vision*, 2:484–498, 1998.

[31] D. Cristinacce and T.F. Cootes. Feature detection and tracking with constrained local models. *Proc. the British Machine Vision Conf.*, 2006.

[32] BioID DB. Address: `http://www.humanscan.de/support/downloads/facedb.php`.

[33] BANCA DB. Address: `http://www.ee.surrey.ac.uk/Research/VSSP/banca/`.

[34] FERET DB. Address: `http://www.itl.nist.gov/iad/humanid/feret/`.

[35] FRGC DB. Address: `http://www.frvt.org/FRGC/`.

Face Recognition: An Introductory Overview

Marcos FAUNDEZ-ZANUY[1]
Escola Universitaria Politècnica de Mataró, Spain

Abstract. Face recognition is one of the most successful ways for biometric recognition. One of its advantages is the high social acceptance by users, because it does not have the criminal stigma associated to other biometric traits, such as fingerprints. On the other hand, face is also suitable for facial expression analysis and emotion analysis, which can be useful for user convenience systems, where the computer can interact in a more intelligent way with the user. In addition, the cost of a webcam can be below 20 EUR, which is certainly lower than other biometric acquisition devices. This paper provides an overview of face recognition with special emphasis on biometric recognition (identification/ verification).

Keywords. Face recognition, biometrics, face acquisition.

1. INTRODUCTION

Face recognition is probably the most natural way to perform biometric authentication between human beings. Additionally, it is the second most popular biometric trait, after fingerprints. This paper is organized as follows: section 1 summarizes some important facts about acquisition, detection, recognition and applications. Section two is focused on the face recognition problem. Section three provides some experimental results obtained with public domain databases for biometric recognition. Section four summarizes the main conclusions.

1.1 Acquisition

Face recognition can rely on single still images, multiple still images, or video sequence. Although traditionally most efforts have been devoted to the former one, the last ones are quickly emerging [30], probably due to price reduction in image and video acquisition devices. For instance, a sequence of images can provide a unimodal data fusion scheme [10], where the verification relies on a set of images, rather than on a single one. These set of images can be acquired with several cameras in parallel, which can provide snapshots from different angles. In this case, a three dimensional model can be constructed using the different views. Obviously this approach requires several cameras. Another possibility is to acquire successive acquisitions of a same face using a single camera. Although this case is simpler than the several camera acquisition approach, the resulting photographs will be more correlated among them. On the other

[1]Corresponding Author: Avda. Puig i Cadafalch 101-111, 08303 MATARO (BARCELONA) SPAIN
E-mail: faundez@eupmt.es, http:www.eupmt.es/veu. This work has been supported by FEDER and the Spanish grant MEC TIC2003-08382-C05-02.

hand, this approach is quite close to the video sequence acquisition, although much less frames will be provided by the multiple still images system. Three dimensional face model reconstruction and recognition is beyond the scope of this paper. Interested readers are referred to [1].

1.2 Face detection and alignment

Next step after image acquisition is face detection. It consists of segmenting the face areas from the background [16]. In the case of video, the detected faces may need to be tracked using a face tracking component. In this case, the temporal evolution of consecutives images will alleviate the problem of face detection, which is more complicated when using a single image (for image sequences the background use to remain stationary, while some movement can be appreciated for the face).

The detector should provide information about face location, size and pose. This information will be useful to normalize the image to a predefined size, compensating translations, rotations, zooms and pans. Figure 1 shows some variation sources in two-dimension image acquisition.

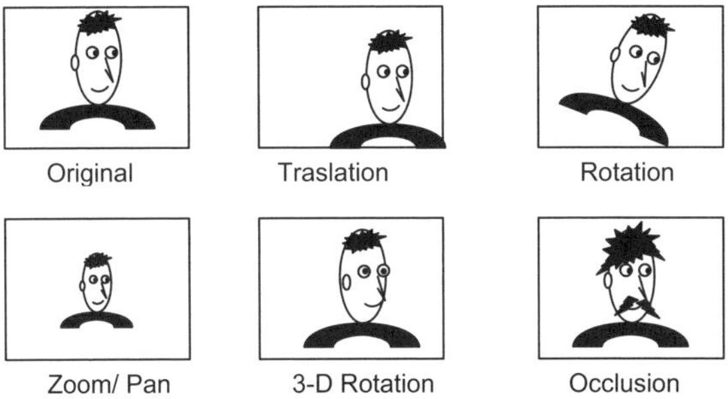

Figure 1. Some variation sources in two-dimension image acquisition.

After the normalization procedure, the face image should be fitted to a standard box with a predefined size.

Thus, face recognition problem using a fast and low-complexity recognition algorithm is: how can we reduce the variability between different image acquisitions? Fortunately, for cooperative user scenarios, we can take advantage of a similar strategy applied for iris acquisition in some sensors, as described in [11]. It consists of a special glass, which performs as transparent glass in one direction and mirror in the other one. Figure 2 shows a schematic representation of this proposed device and figure 3 the first implementation of this system using an ARM core processor [9].

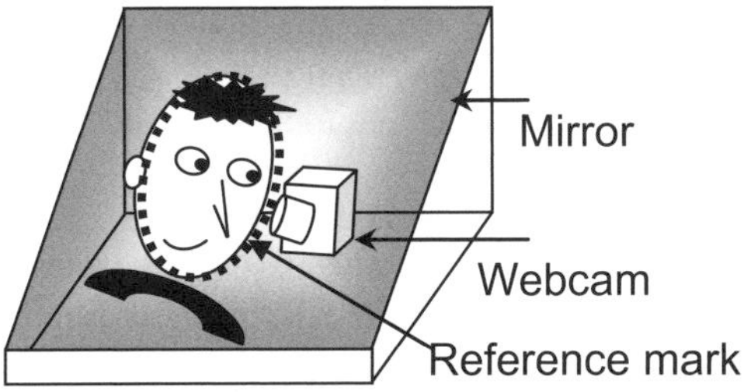

Figure 2. Face image acquisition device

Basically, image acquisition is performed with a low-cost webcam hidden behind the glass. We have plotted an ellipse on the mirror surface, and the user just needs to fit the reflex of his face inside the contour. This solution provides same orientation and distance for all the acquisitions, making the posterior feature extraction and matching much more easily. Our proposed system [9] can manage all the variation sources described in figure 1 except for occlusions.

Figure 3. Face image acquisition and recognition device presented in [9]

1.3 Face recognition and applications

The face recognition problem applied to biometric recognition will be described in more detail in the next sections. However, there are other applications of face recognition not related to security. Some of them are:

Multimedia content indexing: human faces frequently appear in films, news, etc. Together with speech recognition, face recognition is a powerful tool for automatic indexing, retrieval and access to digital multimedia content, in a similar way than text browsing using google, yahoo, etc.

Human computer interaction: to achieve user-friendly human-computer interaction. The face can be considered as a natural input "device" [16]. It is argued that for the computer to be able to interact with humans it must have the communication skills of humans, and one of these skills is the ability to understand the emotional state of the person. In fact, the most expressive way humans display emotions is through facial expressions [16]. However, for emotion analysis, higher level knowledge is required, and it goes one step further from facial expression analysis. Although facial expressions can convey emotion, they can also express intention, cognitive process, physical effort, or other intra or interpersonal meanings [16]. Facial analysis analyzes the facial actions regardless of context, culture, gender, etc.

Face synthesis and animation:

In the film industry we would like to create virtual human characters that are indistinguishable form the real ones. In games, people have been trying to create human characters that are interactive and realistic. In addition, some human-machine dialog systems use realistic-looking human faces as visual representation of the computer.

1.4 Face databases

Competitive algorithm testing on a database shared by dozens of research laboratories is a milestone for getting significant technological advances.

A major advantage of database availability is to set up the evaluation conditions that can avoid some common mistakes done in system designs [3]:

1. "Testing on the training set": the test scores are obtained using the training data, which is an optimal and unrealistic situation.
2. "Overtraining": The whole database is used too extensively in order to optimize the performance. This can be identified when a given algorithm gives exceptionally good performance on just one particular data set.

Databases include different material for training and testing in order to avoid the first problem. In addition, the availability of several databases helps to test the algorithms over new stuff and thus to check if the algorithms developed by a given laboratory can generalize their results (solving the second problem).

Nagy presents with irony in "Candide's practical principles of experimental pattern recognition" [21] some rules, which certainly should not be done by an honest researcher:

1. Theorem: there exists a set of data for which a candidate algorithm is superior to any given rival algorithm. This set may be constructed by omitting from the test set any pattern which is misclassified by the candidate algorithm.
2. Casey caution: do not ever make your experimental data available to others; someone may find an obvious solution that you missed.

Thus, database availability is important in order to validate a given algorithm, to make feasible comparison between different algorithms, and also to develop an algorithm. This kind of evaluations is known technology evaluation [19]. The goal of a technology evaluation is to compare competing algorithms from a single technology. Testing of all algorithms is carried out on a standardized database collected by a "universal" sensor. Nonetheless, performance with this database will depend upon both the environment and the population in which it is collected. Testing is carried out using offline processing of the data. Because the database is fixed, the results of technology tests are repeatable.

Some important aspects of a given database are:

- Number of users (a large number of users permits to study the discriminability capability of a given biometric trait).
- Number of recording sessions (several sessions performed in different days permit to study the inter-session variability).
- Number of different samples per session (several acquisitions per session permit to study the intra-session variability).

Figure 4 shows a sample of the JAFFE (Japanese Female Facial Expression Database) [17].

For an excellent up-to-date overview of face databases, see chapter 13 of [16]. In this paper, we have used the ORL [26] and FERET [14] databases.

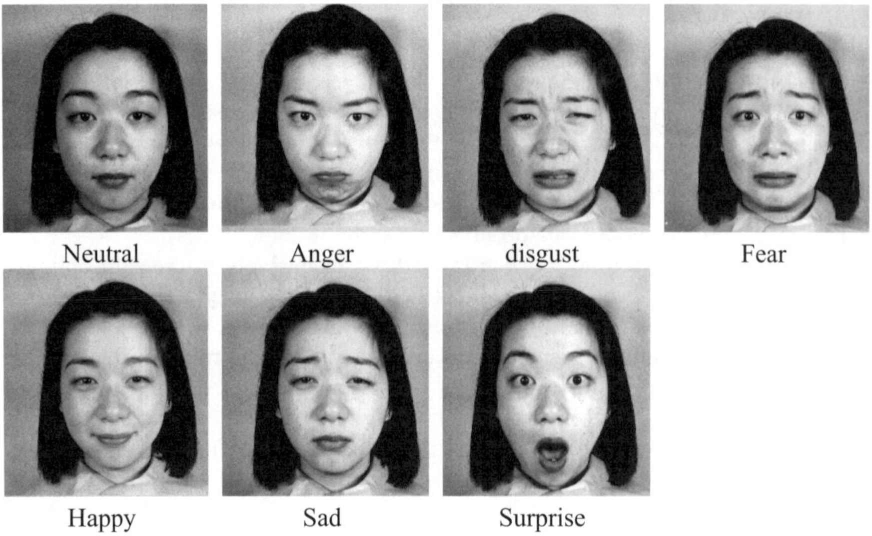

Figure 4. Example images from the JAFFE database.

2. THE FACE RECOGNITION PROBLEM

Comparative studies reveal that the problem of face recognition cannot perform as accurate as other biometric traits such as fingerprint or iris [18]. However, we think that part of this problem is due to the high degree of variability that a three-dimensional object exhibits when represented in a two-dimensional way (such as a photograph). Fingerprint acquisition is simpler, because we are always getting the ridge and valley pattern over the sensor surface. Thus, there is not zoom/ panning nor three dimensional rotations, nor shadows nor objects blocking (hair, glasses, etc.). We perform thousands of face recognition, and we do not realize how complicated it is for a human brain to do this. For instance, the following experiences described by neurologists about people with sight restoration after long-term blindness [25] are very interesting:

"...an English surgeon removed the cataracts from the eyes of a thirteen-year-old boy born blind. Despite his high intelligence and youth, the boy encountered profound difficulties with the simplest visual perceptions. He had no idea of distance. He had no idea of space or size. And he was bizarrely confused by drawings and paintings, by the idea of a two-dimensional representation of reality. As he had anticipated, he was able to make sense of what he saw only gradually and insofar as he was able to connect visual experiences with tactile ones. It has been similar with many other patients in the two hundred and fifty years since his operation: nearly all had experienced the most profound, Lockean confusion (difficulty of not being able to make sense of what they saw) and bewilderment... He had no idea what he was seeing. There was light, there was movement, there was color, all mixed up, all meaningless, a blur. Then out of the blur came a voice that said, well? Then, and only then, he said, did he finally realize that his chaos of light and shadow was a face...sometimes he would get confused by his own shadow (the whole concept of shadows, of objects blocking light, was puzzling to him) and come to stop, or trip, or try to step over it... Moving objects presented a special problem, for their appearance changed constantly. Even his dog, he told me,

looked so different at different times that he wondered if it was the same dog... it was in the two-dimensional representation of reality, pictorial or televisual, that he was still completely at sea".

The rest of us, born sighted, can scarcely imagine such confusion. For we, born with full complement of senses, and correlating these, one with the other, create a sight world from the start, a world of visual objects and concepts and meanings. When we open our eyes each morning, it is upon a world we have spent a lifetime learning to see. We are not given the world: we make our world through incessant experience, categorization, memory, and reconnection. We achieve perceptual constancy –the correlation of all the different appearances, the transforms of objects– very early, in the first months of life. It constitutes a huge learning task, but it is achieved so smoothly, so unconsciously, that its enormous complexity is scarcely realized (though it is an achievement that even the largest supercomputers cannot begin to match). In [13], we reported similar situations, and even in the Bible, in Mark's description of the miracle at Bethsaida, at first, the blind man saw "men as trees, walking" and only subsequently was his eyesight fully restored (Mark 8:22-26).

However, even for those born sighted, face recognition is an extremely complicated task.

Before having an automatic system able to solve our identification problems with the accuracy and scale that we desire, we can examine the situation when the operator is a human being. Probably if a person can perform this task, and we can summarize the procedure done by him/ her, the answer is yes. But what does really mean that "a human can do the task easily"? In the next section we describe some facts that can help understand the complexity of biometric face recognition.

2.1 Face Recognition: an easy task for human beings born sighted?

Although common sense points out that face recognition is an easy-task for human beings, it is more complex than we think. It is not an innate ability, at least when using two-dimensional representation of faces (photos, drawings, etc.), and it must be acquired when we are young. A British anthropologist, named Nigel Barley, went to study the Dowayo people (see figure 5), a strangely neglected group in North Cameroon, and he reports in his book [2]:

"*...In the end I managed to lay my hands on some postcards depicting African fauna. I had at least a lion and a leopard and showed them to people to see if they could spot the difference. Alas, they could not. The reason lay not in their classification of animals but rather in the fact that they could not identify photographs. It is a fact that we tend to forget in the West that people have to learn to be able to see photographs. We are exposed to them from birth so that, for us, there is no difficulty in identifying faces or objects from all sorts of angles, in differing light and even with distorting lenses. Dowayos have no such tradition of visual art; theirs is limited to bands of geometric designs. Nowadays, of course, Dowayo children experience images through schoolbooks or identity cards; by law, all Dowayos must carry an identity card with their photograph on it. This was always a mystery for me since many who have identity cards have never been to the city, and there is no photographer in Poli. Inspection of the cards shows that often pictures of one Dowayo served for several different people. Presumably the officials are not much better at recognizing photographs than Dowayos...*

...The point was that men could not tell the difference between the male and the female outlines. I put this down simply to my bad drawing, until I tried using photographs of lions and leopards. Old men would stare at the cards, which were perfectly clear, turn them in all manner of directions, and then they say something like "I don't know this man". Children could identify the animals...."

Figure 5. Dowayos looking at drawings

Thus, the face recognition ability of human beings is not as trivial as we could think. It is the result of a long learning process, which is not really well understood, varies along our life, and even relies on different parameters [28]. Probably it is similar to learn to pronounce some phonemes. It is not a problem when we are child, but after several years, it is almost impossible (for instance, a French person cannot pronounce the /rr/ phoneme of the Spanish word "perro", which means dog). Taking into account the experience of this anthropologist, we can see that really the enrollment phase, where the computer tries to learn a model for each person using some (limited amount of) training data, is fundamental.

Even when we are used to seeing photos and recognizing faces, there are difficult situations. The first coming to our mind is that related to changes in hairstyle, makeup, facial hair, addition or removal of eyeglasses, hats, scarves, etc. Figure 6 shows six different photographs of the actor Val Kilmer on the film "The saint". Although it is an extreme case, where the user is trying to avoid his identification, most faces databases contemplate changes on facial expression, illumination, etc. because it is an actual problem in real scenarios.

Figure 6. Six different snapshots of Val Kilmer.

Another difficult situation is when trying to recognize a human face different from the race that we are used to deal with in our normal life. For instance, [24] reports that "*The last half-century's empirical studies of cross racial IDs has shown that eyewitnesses have difficulty identifying members of another race. . .*"

On the other hand, human beings are also subject to errors on our face identification decisions, even more than we think! This is well-known on justice courts. One of the most famous cases [29] of mistaken identity occurred in the 1896 English trial of Adolf Beck. Beck was convicted based on the identifications of ten women. He was convicted and spent 7 years in prison, all the while maintaining his innocence. He claimed he was mistaken for a man named John Smith. While Beck was still in prison, more of these offenses occurred. Smith was eventually arrested and Beck was released. Afterwards it was discovered that Beck had spent seven years in prison for a crime he did not commit, a committee was formed. The committee found that "*evidence as to identity based on personal impressions, however bona fide, is perhaps of all classes of evidence the least to be relied upon, and therefore, unless supported by other facts, an unsafe basis for the verdict of a jury.*"

Thus, [29] concludes that "*No person should be deprived of his liberty solely on the basis of eyewitness testimony unless the jury is fully aware of the ways in which such testimony may be flawed*".

The conclusion of this section should be that things are not as easy as they seem, and for some recognition systems, we cannot expect 100% identification rates neither 0% false acceptance and rejection rates.

2.2 Face recognition: when common sense can fail.

Common sense says that color information is certainly useful for face (pattern) recognition. However, common sense is based on our own experience, where we have been "trained" to classify patterns using color information, and our skill is slightly degraded without color information. In fact, Oliver Sacks, a professor of neurology at the Albert Einstein College of Medicine, reports in one of his books [25] some interesting experiences about human recognition and color in special situations:

"I recently heard of an achromatopic botanic in England said to be even better than color normals at swiftly identifying ferns and other plants in woods, hedgerows, and other almost monochromatic environments. Similarly, in World War II, people with severe red-green colorblindness were pressed into service as bombardiers, because of their ability to "see through" colored camouflage and not be distracted by what would be, to the normal sighted, a confusing and deceiving configuration of colors. One veteran of the Pacific theater reports that colorblind soldiers were indispensable in spotting the movement of camouflaged troops in the jungle. (All of these things may also be clearer to color normals at twilight)."

In [23] we presented some experiments with the FERET database. FERET database is in RGB format, and the relevance of color can be established. In our case, we studied the identification rates and minimum detection Cost Function (verification application) using a single primary color R (Red), G (Green) and B (Blue), and several data fusion [10] strategies:

a) Feature level data fusion: R, G and B are combined in order to get the luminance of each pixel: $Y=0.3R+0.59G+0.11B$. Thus, recognition is performed using luminance. This result is comparable to ORL images.
b) Score level data fusion: We work out the recognition rates for each color separately, and we also provide results for score fusion.

Table 1 summarizes the obtained results for FERET database. Our experiments show that color information provides minor improvements when compared with luminance images. At least, when using our parameterization and recognition algorithm.

On the other hand, skin color can offer a strong cue for efficient localization and tracking of human body parts.

Table 1. Recognition rates (Identification rate and verification error) using FERET database.

Input signal	Identif.	Min(DCF)
R	73.08%	5.60%
G	68.55%	5.99%
B	65.63%	6.09%
Y	69.46%	5.76%
Score fusion: R+G+B	71.57%	5.31%
Score fusion: 0.3R+0.59G+0.11B	70.97%	5.45%

2.3 Are human beings better than machines for biometric identification?

We should not be surprised if automatic face recognition systems exhibit some amount of errors. This is not exclusive of face recognition technology. In [7-8] we found that human beings outperform computers for the task of locating the characteristic points of a fingerprint. Thus, there is still room for improvement on this technology, and even better results can be expected. However, in fingerprints it is easy to define what we are looking for. Mainly it is terminations and bifurcations on the ridges of the fingerprint. For other biometric patterns, a human being can experiment troubles in identifying people, because they are not used as identifiers in our normal life. For instance, figure 7 shows four snapshots of four hands. How many different people can we differentiate? If, instead of four hand scanned images, there were four photos of their face, the answer would be trivial.

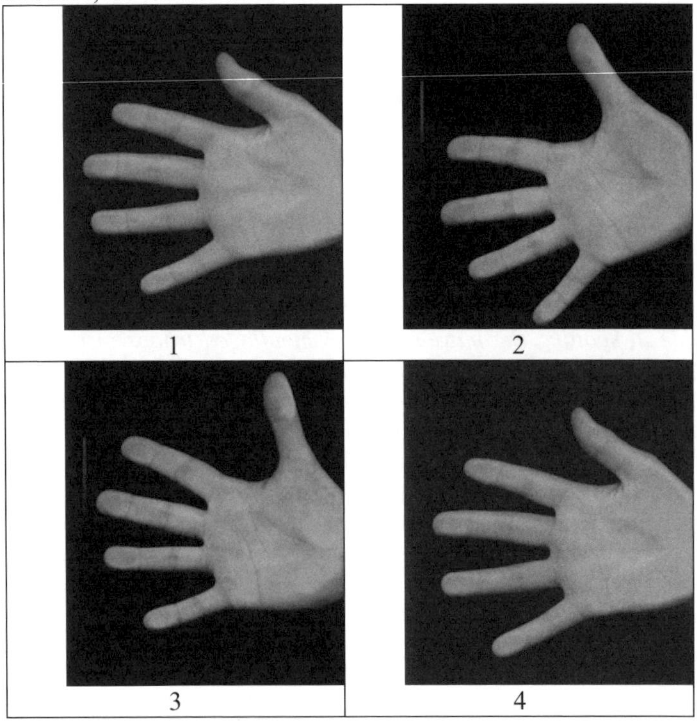

Figure 7. Four different snapshots of a hand. How many different persons are?

For this quiz, plots number 1 and 4 really belong to the same female person, and 2 and 3 to another different one. In this case, it is easier for a computer to work out some parameters like finger length, width, etc. and assign identities, than for a human being. Thus, there certainly are situations where a computer can outperform a person. In any case, the adoption of a biometric solution can dissuade criminals, and it is nowadays good enough for a huge set of civilian applications. Although it is not a trivial problem, let us hope much better solutions in a near future.

3. The Face Recognition Approaches

3.1 Identification vs. verification

Figure 8 shows the general scheme of a biometric system. These systems can be operated in two ways:

a) Identification: In this approach no identity is claimed from the person. The automatic system must determine who is trying to access. One-to-many comparisons (1:N) are required.
b) Verification: In this approach the goal of the system is to determine whether the person is who he/she claims to be. This implies that the user must provide an identity and the system just accepts or rejects the users according to a successful or unsuccessful verification. Sometimes this operation mode is named authentication or detection. One-to-one comparisons (1:1) are required.

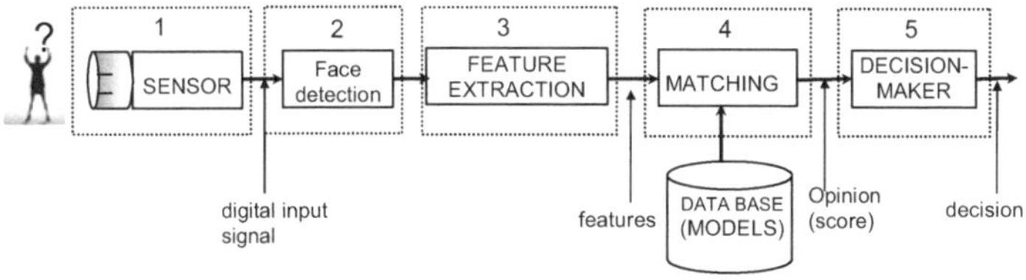

Figure 8. General scheme of a biometric recognition system.

Identification applications with a large set of users cannot provide good identification rates. In these cases, verification applications should be chosen. We have performed [23] identification and verification using the same algorithm and two different databases: ORL (40 subjects) [26] and FERET (994 subjects) [14]. The use of a small-size database is interesting for developing algorithms, because experimental simulations are much less intensive. Additionally, its worth is remarkable for those scenarios where few users are expected, such as accessing a Personal computer or a PDA without providing any identity (PIN-less access or matching based on 1:N comparisons). Theoretically, for a large set of users, PIN-less mode is not recommended [3], because False Match Errors will be high. For verification applications, the database size should not present any relevance, because each matching is based on 1:1 comparison. However, in practice, the obtained results are much more statistical significant when using larger databases, especially for low error rates. In [15] the minimum size of the test data set, *N*, which guarantees statistical significance in a pattern recognition task, is derived. The goal in the abovementioned work is to estimate

N so that it is guaranteed, with a risk α of being wrong, that the error rate P does not exceed that estimated from the test set, $\hat{P}$, by an amount larger than $\varepsilon(N,\alpha)$, that is,

$$\Pr\left\{P > \hat{P} + \varepsilon(N,\alpha)\right\} < \alpha \tag{1}$$

Letting $\varepsilon(N,\alpha) = \beta P$ and supposing recognition errors as Bernoulli trials (i.i.d. errors), we can derive the following relation after some approximations:

$$N \approx \frac{-\ln\alpha}{\beta^2 P} \tag{2}$$

For typical values of α and β (α =0.05 and β =0.2), the following simplified criterion is obtained:

$$N \approx \frac{100}{P} \tag{3}$$

If the samples in the test data set are not independent (due to correlation factors that may include variations in recording conditions, in the type of sensors, etc.), then N must be further increased. The reader is referred to [15] for a detailed analysis of this case, where some guidelines for computing the correlation factors are also given.

Table 2 provides the experimental results for the ORL database (see [23] for additional details). If we compare table 1 and 2 we will realize that identification rates are not very high using the FERET database. However, verification errors are comparable in both cases.

Table 2. Recognition rates for ORL database.

Error criterion	Identif. rate	Min(DCF)
MSE	91%	5.84%
MAD	92.5%	6.28%

In both cases, illumination and time between acquisitions of each image can significantly affect face recognition performance. [22] describes and interesting experiment with different testing scenarios, described in table 3. It reveals, for a given algorithm, the dependency with illumination and temporal drifts. For this reason, robust feature extraction and/ or illumination compensation must be used. It is interesting to observe that mismatch just affects the probability to reject a genuine user, but it does not affect the probability to accept an impostor.

Table 3. Example of face verification performance for a given algorithm [22].

Scenario	FAR (%)	FRR (%)
Same day, same illumination	2	0.4
Same day, different illumination	2	9
Different days	2	11
Different days over 1.5 years apart	2	43

3.2 Identification

For identification, if we have a population of N different people, and a labeled test set, we fill up a matrix S, where the elements are interpreted in the following way:

$$s_{ijk} = O[j]\Big|_{\vec{x} \in person\#i} \quad ,k = 1, \cdots \# trials \tag{4}$$

Where trials is the number of different testing images per person (in our experiments, *#trials*=5 for ORL and 1 for FERET), and s_{ijk} is the similarity from the k realization of an input signal belonging to person i, to the model of person j.

This matrix can be drawn as a three dimensional data structure (see figure 9).

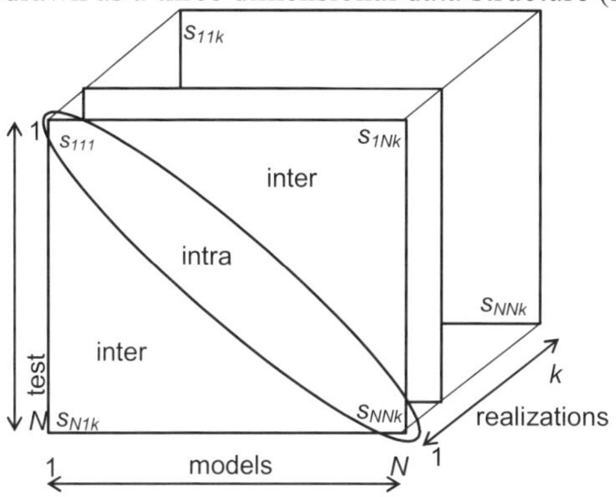

Figure 9. Proposed data structure.

Thus, the identification rate looks for each realization, in each raw, if the maximum similarity is inside the principal diagonal (success) or not (error), and works out the identification rate as the ratio between successes and number of trials (successes + errors):

```
for i=1:N,
    for k=1:#trials,
            if(s_iik>s_ijk) ∀j≠i, then
              success=success+1
            else error=error+1
            end
  end
end
```

3.3 Verification

Verification systems can be evaluated using the False Acceptance Rate (FAR, those situations where an impostor is accepted) and the False Rejection Rate (FRR, those situations where a user is incorrectly rejected), also known in detection theory as False Alarm and Miss, respectively. There is trade-off between both errors, which has to be usually established by adjusting a decision threshold. The performance can be plotted in a ROC (Receiver Operator Characteristic) or in a DET (Detection error trade-off) plot [20]. DET curve gives uniform treatment to both types of error, and uses a logarithm scale for both axes, which spreads out the plot and better distinguishes different well performing systems and usually produces plots that are close to linear. DET plot uses a logarithmic scale that expands the extreme parts of the curve, which are the parts that give the most information about the system performance. For this reason the speech community prefers DET instead of ROC plots. Figure 10 shows an example of DET plot.

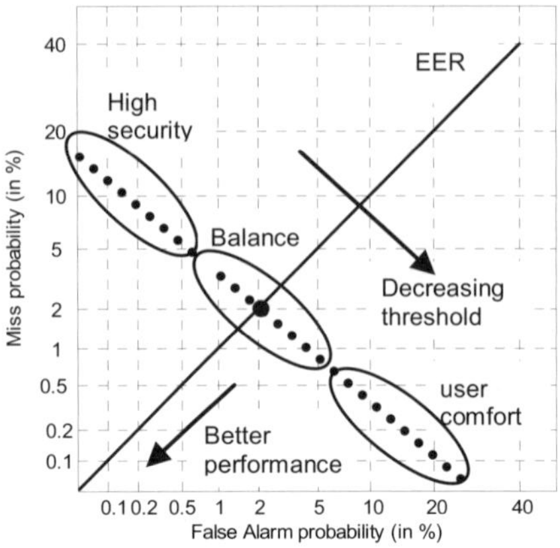

Figure 10. Example of a DET plot for a user verification system (dotted line). The Equal Error Rate (EER) line shows the situation where False Alarm equals Miss Probability (balanced performance). Of course one of both errors rates can be more important (high security application versus those where we do not want to annoy the user with a high rejection/ miss rate). If the system curve is moved towards the origin, smaller error rates are achieved (better performance). If the decision threshold is reduced, we get higher False Acceptance/Alarm rates.

We have used the minimum value of the Detection Cost Function (DCF) for comparison purposes. This parameter is defined as [20]:

$$DCF = C_{miss} \times P_{miss} \times P_{true} + C_{fa} \times P_{fa} \times P_{false} \tag{5}$$

Where C_{miss} is the cost of a miss (rejection), C_{fa} is the cost of a false alarm (acceptance), P_{true} is the a priori probability of the target, and $P_{false} = 1 - P_{true}$, $C_{miss}= C_{fa} =1$.

Using the data structure defined in figure 9, we can easily apply the DET curve analysis. We just need to split the distances into two sets: intra-distances (those inside the principal diagonal), and inter-distances (those outside the principal diagonal).

3.4 Face Recognition

It is quite straight forward to obtain identification and verification results using the scores in the format of matrix presented in figure 9. We will use the general name recognition when we do not want to specify identification or verification. Thus, the recognition problem is to find the scores between the models and the testing faces. Two main approaches exist for face recognition:

a) Holistic approaches consider the image as a high dimensional vector, where each pixel is mapped to a component of a vector. Due to the high dimensionality of the vectors some vector dimension reduction algorithm must be used. Typically the

Karhunen Loeve transform is applied. In fact, it is a simplified version named eigenfaces [27].

b) Geometry-feature-based methods try to identify the position and relationship between face parts, such as eyes, nose, mouth, etc., and the extracted parameters are measures of textures, shapes, sizes, etc. of these regions.

In this paper, experimental results are based on the first approach. The matching techniques can be split into the next categories:

1. Template matching methods: The input and model faces are expressed as feature vectors and compared using a distance measure between them.
2. Statistical methods: The features extracted from the training faces are used to work out a statistical model. During testing, the similarity of input and reference is established. The most popular methods belonging to this category are Hidden Markov Models (HMM), Gaussian Mixture Models (GMM), etc.
3. Neural Networks: For instance, a Multi-Layer Perceptron can perform as a classifier. This was the approach that we used in our previous work [5], [6]. Although it offers very good results, the computational burden is extremely high, when dealing with large databases.
4. Other methods: Bayesian networks, Support Vector Machines, etc.

In this paper, we will use a template matching method, where the classifier consists of a Nearest Neighbor (NN) classifier using the Mean Square Error (MSE) or the Mean Absolute Difference (MAD) defined as:

$$MSE(\vec{x},\vec{y})=\sum_{i=1}^{(N')^2}(x_i-y_i)^2 \qquad (6)$$

$$MAD(\vec{x},\vec{y})=\sum_{i=1}^{(N')^2}|x_i-y_i| \qquad (7)$$

Where *N'* is the dimensionality of the vectors that represent faces. According to our previous work [6], we have chosen *N'*=100.

Table 1 shows the obtained results using the ORL database. Each vector, which represents one face, has been obtained with the procedure described in [6].

The MAD criterion is faster and simpler to work out. Thus, we have chosen the MAD criterion in our simulations with FERET database.

References

[1] P. W. Hallinan, G. Gordon et al. "two and three dimensional patterns of the face". A. K. Peters limited 1999

[2] N. Barley "The innocent anthropologist" Waveland Press, Inc. Prospect Heights, Illinois 1983.

[3] Bolle R. M., Ratha N. K., Pankanti S., "Performance evaluation in 1:1 Biometric engines". Springer Verlag LNCS 3338, pp.27-46 S. Z. Li et al. (Eds.) Sinobiometrics 2004.

[4] Bolle, R., Connell, J., Pankanti, S., Ratha, N., Senior, A.: Guide to biometrics. Springer professional computing, New York 2004

[5] Espinosa V., Faundez-Zanuy M. "Face identification by means of a neural net classifier" 33° IEEE International Carnahan Conference on Security Technology. Pp.182-186. Madrid, 1999

[6] Faundez Zanuy M. "Face recognition in a transformed domain". 37° IEEE International Carnahan Conference on Security Technology. Pp.290-297. Taipei (Taiwan), October 2003.

[7] Faundez-Zanuy, M., "Are inkless fingerprint sensors suitable for mobile use?" IEEE Aerospace and Electronic Systems Magazine. Vol. 19 nº 4, pp. 17-21, April 2004.
[8] Faundez-Zanuy, M. "Technological evaluation of two AFIS systems". IEEE Aerospace and Electronic Systems Magazine. Vol. 20 nº 4, pp. 13-17, April 2005.
[9] Faundez-Zanuy, M. Espinosa-Duro, V., Ortega-Redondo J.A."A low-cost webcam & personal computer opens doors". IEEE Aerospace and Electronic Systems Magazine. Vol.20 nº 11, pp.23-26, November 2005.
[10] Faundez-Zanuy M., "Data fusion in biometrics" IEEE Aerospace and Electronic Systems Magazine. Vol. 20 nº 1, pp. 34-38, January 2005.
[11] Faundez-Zanuy, M., "Biometric security technology". IEEE Aerospace and Electronic Systems Magazine. Vol. 21 nº 1, June 2006
[12] Faundez-Zanuy, M., "On the vulnerability of biometric security systems". IEEE Aerospace and Electronic Systems Magazine. Vol. 19 nº 6, pp. 3-8, June 2004.
[13] Faundez-Zanuy, M., "Biometric recognition: why not massively adopted yet?" IEEE Aerospace and Electronic Systems Magazine. Vol.20 nº 8, pp.25-28, August 2005.
[14] "Color FERET. Facial Image Database.", Image Group, Information Access Division, ITL, National Institute of Standards and Technology. Oct. 2003.
[15] Guyon, I., Makhoul, J., Schwartz, R., and Vapnik, V.: 'What size test set gives good error rate estimates?', IEEE Trans. Pattern Anal. Mach. Intell., 1998, 20, (1), pp. 52–64.
[16] S. Z. Li and A. K. Jain Handbook of face recognition. Springer 2006
[17] Lyons M. J. , Akamatsu S., Kamachi, M., Gyoba J. "Coding facial expressions with Gabor wavelets . Proc. Third IEEE int. Conf. on automatic face and gesture recognition, pp.200-2005, Nara, Japan. April 1998
[18] T. Mansfield, G. Kelly, D. Chandler, and J. Kane (2001, Mar.) biometric product testing final report. Available online: http://www.cesg.gov.uk/site/ast/biometrics/media/BiometricTestReportpt1.pdf
[19] Mansfield A. J., Wayman J. L., "Best Practices in Testing and Reporting Performance of Biometric Devices". Version 2.01. National Physical Laboratory Report CMSC 14/02. August 2002.
[20] Martin A., Doddington G., Kamm T., Ordowski M., and Przybocki M., "The DET curve in assessment of detection performance", V. 4, pp.1895-1898, European speech Processing Conference Eurospeech 1997
[21] Nagy G. "Candide's practical principles of experimental pattern recognition". IEEE Trans. On Pattern Analysis and Machine Intelligence Vol. 5 No. 2, pp.199-200, March 1983.
[22] Phillips, P. J., Martin A., Przybocki M. "An introduction to evaluating biometric systems". IEEE Computer February 2000, pp.56-63.
[23] Roure J., Faundez-Zanuy. M. "Face recognition with small and large size databases". 39th International Carnahan Conference on Security Technology ICCST'2005 Las Palmas de Gran Canaria. Pp.153-156. October 2005.
[24] J. P. Rutledge "They All Look Alike: The Inaccuracy of Cross-Racial Identifications", 28 American Journal of Criminal Law. pp. 207-228. Spring 2001
[25] Sacks O., "An anthropologist on Mars". First Vintage books edition, USA, Feb. 1996
[26] Samaria F., Harter A. "Parameterization of a stochastic model for human face identification". 2nd IEEE Workshop on Applications of Computer Vision December 1994, Sarasota (Florida).
[27] Turk M., Pentland A., "Eigenfaces for Recognition" Journal Cognitive Neuroscience, Vol. 3, nº 1 pp 71-86, Massachusetts Institute of Thecnology.1991.
[28] J. J. Weng and D. L. Swets "Face Recognition" pp.65-86 in Biometrics personal identification in networked society. Ed. Kluwer 1999.
[29] S. Woller "Rethinking the role of expert testimony regarding the reliability of eyewitness identifications in New York" New York Law School, June 2004. Available at http://www.nyls.edu/docs/v48n1-2p323-352.pdf
[30] Zhou S.K. "Face recognition using more than one still image: what is more?" Lecture Notes In Computer Science LNCS 3338, pp. 212-223, A. Z. Li et al. Ed., Sinobiometrics. Springer Verlag 2004

Fundamentals of Verbal and Nonverbal Communication and the Biometric Issue
A. Esposito et al. (Eds.)
IOS Press, 2007

Detection of Faces and Recognition of Facial Expressions

Praveen KAKUMANU, Nikolaos BOURBAKIS[1]
ITRI / Department of Computer Science and Engineering
Wright State University, Dayton Ohio 45435 USA

Abstract: Face detection is the foremost task in building vision-based human-computer interaction systems and in particular in applications such as face recognition, face identification, face tracking, expression recognition and content based image retrieval. A robust face detection system must be able to detect faces irrespective of illuminations, shadows, cluttered backgrounds, facial pose, orientation and facial expressions. Many approaches for face detection have been proposed. However, as revealed by FRVT 2002 tests, face detection in outdoor images with uncontrolled illumination and in images with varied pose (non-frontal profile views) is still a serious problem. In this article, we describe a Local-Global Graph (LGG) based method for detecting faces and for recognizing facial expressions in real world image capturing conditions both indoor and outdoor, and with a variety of illuminations (shadows, high-lights, non-white lights) and in cluttered backgrounds. First, a neural color constancy based skin detection procedure to detect skin in complex real world images is presented. Second, the LGG method for detecting faces and facial expressions with a maximum confidence from skin segmented images is presented. The LG Graph embeds both the local information (the shape of facial feature is stored within the local graph at each node) and the global information (the topology of the face). The LGG approach presented here emulates the human visual perception for face detection. In general, humans first extract the most important facial features such as eyes, nose, mouth, etc. and then inter-relate them for face and facial expression representations. Facial expression recognition from the detected face images is obtained by comparing the LG Expression Graphs with the existing expression models present in the LGG database. The methodology is accurate for the expression models present in the database.

Keywords: Neural Color Constancy, Skin Detection, Face and Expression Detection, Local-Global Graphs.

Introduction

For face detection, a number of methods have been previously proposed [1]. To achieve a good performance, many of these methods assume that the face is either segmented or surrounded by a simple background and the images are well-illuminated with frontal-facial pose, which is not always true. The robustness of these approaches is challenged by many factors such as changes in illumination across the scene, shadows, cluttered backgrounds, image scale, facial pose, orientation and facial expressions. In this article, we present a Local-Global (LG) Graph approach for detecting faces and for

---

[1] Corresponding Author: ITRI/Department of Computer Science and Engineering, Wright State University, Dayton, Ohio, 45435. Email: nikolaos.bourbakis@wright.edu

recognizing facial expressions. The overall approach to face and facial expression detection presented here - the LG graph approach combined with the skin detection procedure is robust against cluttered backgrounds, uncontrolled illuminations, shadows, and to a variety of facial poses and orientations. Given an image, the neural color constancy based skin detection technique (section 2) provides skin-color similar regions in the image. These segmented skin regions form potential candidate face regions. However, all skin-color similar regions do not represent faces and non-face regions must be rejected. First, to detect facial feature regions, a fuzzy-like segmentation method is applied on the segmented skin regions. A curve fitting procedure is applied on the segmented regions that result in a set of line segments which are later used to define a local graph that represents the shape of the region (section 3). To reduce the number of regions and to highlight the key facial features, a skin region synthesis procedure based on color information is applied (section 4). From the remaining facial regions, a local-global graph which incorporates the spatial geometry of facial features is constructed. To detect a face, the candidate LG graph thus formed is finally compared with the face-model LG graph present in the LG database using a computationally effective graph matching technique (section 5). The LG graph approach in general can be applied to any object detection provided that there is a corresponding object model in the LG database. For recognizing facial expressions, new candidate Expression LG graphs are formed and then compared with the existing expression models present in the LG database (section 7).

1. Neural color constancy based skin detection

A survey of different methods for skin detection using color information is presented by Kakumanu et al. [2]. To detect skin from images, a two stage color constancy based approach is used [3]. The first stage - the color correction stage consists of estimating the image illuminant using a skin color adapted Neural Network (NN) and then color correcting the image based on the NN illuminant estimate. The second stage - skin detection stage classifies the skin and non-skin pixels using a threshold based on the achromatic value of the color corrected images.

The color correction step involves transforming an image under an unknown illuminant to the corresponding colors under a known canonical illuminant. This consists of estimating the illuminant color and then color correcting the image pixel-wise based on the estimate of the illuminant. Given a good estimate of the illuminant, the color correction step is applied using the vonKries diagonal model of chromatic adaptation. According to this model, the mapping of an image under an unknown illuminant to a known canonical illuminant is obtained by scaling each channel independently as given by the following equation:

$$\begin{bmatrix} L' \\ M' \\ S' \end{bmatrix} = \begin{bmatrix} K_L & 0 & 0 \\ 0 & K_M & 0 \\ 0 & 0 & K_S \end{bmatrix} \begin{bmatrix} L \\ M \\ S \end{bmatrix} \tag{1}$$

where (L, M, S) and (L', M', S') are the tristimulus values of the *Long, Medium* and *Short* cone responses of the unknown and adapted illuminants respectively and

(K_L, K_M, K_S) are the corresponding scaling coefficients. The scaling coefficients can be expressed as the ratio of the cone responses to a white patch under the reference canonical illuminant and to those of unknown illuminant. In practice, the LMS values are approximated by the CIE XYZ tristimulus values or the image RGB values. However, it has been shown that the vonKries diagonal model holds in an optimal sense for chromatic adaptation only if the sensor responses are expressed in LMS cone space. Hence, a series of transformations are applied to transform the image RGB values to LMS cone space and the color correction is applied in LMS cone space. The illuminant estimate for a given image is obtained as the response of the NN trained to adapt to the skin color [3]. The NN is trained on randomly selected images from a database consisting of images collected under various illumination conditions both indoor and outdoor, and containing skin colors of different ethnic groups. The color correction step assigns achromatic color (gray) to skin pixels. This property is used to detect skin pixels. A lower color variance indicates skin pixels. A suitable threshold value is used to classify skin and non-skin pixels. The color variance is defined as:

$$Var = \frac{1}{3} \cdot \sum_{k=1}^{3} \left(I(k) - \hat{I} \right)^2 \tag{2}$$

where $I(k)$ denotes the normalized color value in red, green and blue channels for k = 1, 2, 3 respectively and $\hat{I}$ denotes the average color value. To remove the noise and to fill the holes created by the skin detection procedure (for example, eye and eye-brow regions are not detected by the skin detection procedure), morphological operations such as erosion, dilation and hole-filling are applied on these images. Figure 1 shows the results of the skin detection procedure on example images collected from AR face database [4] (row 1) and other custom collected images.

The advantage of the NN method for color adaptation is that it does not have any inherent assumptions about the object surfaces in the image or the illumination sources as the input to the NN is only the color from the image. The overall approach for skin detection is computationally inexpensive and is feasible for real-time applications. However, if the goal is to detect faces, skin detection alone is not sufficient. The skin detection step provides face-like regions in the image and serves as a primary step for the LG graph procedure described below. On the skin segmented images, a fuzzy-like segmentation method [5] is applied. Figure 3(b) shows the corresponding FRG segmented image for the image in Figure 1(d). Clearly, the FRG method segments important facial feature regions such as eyes, eye-brows, nose and mouth.

(a) Original Images *(b) NN corrected Images* *(c) Skin detection* *(d) Skin-similar regions*

Figure 1. Neural color constancy based skin detection results

2. Local region graph (Facial feature region graph)

The application of FRG segmentation method generates a set of color regions. On these segmented regions, a curve fitting procedure is applied which results in a set of connected line segments that defines the shape of each region. The shape of the region is used in generation of a local region graph and later in the synthesis of neighboring regions (section 4). The representation of a region by a set of line segments is not sufficient as it does not hold all the connectivity relationships, which is very important for accurate description of the geometrical shape of the region. An accurate description of the shape of facial features is critical for robust face detection/recognition. A good solution to this problem is to build a local graph which encodes the spatial relationships of these line segments [6]. Thus the shape S of the region is represented as:

$$S = L_1 R_{12}^C L_2 R_{23}^C L_3 R_{34}^C \ldots L_{n-1} R_{n-1n}^C L_n \ldots \qquad (3)$$

where L_i represents the line segment and R_{ij} represents the connectivity relationship between line segments L_i and L_j of the region shape. R_{ij} is characterized by connectivity *(c)*, parallelism *(p)*, symmetry *(s)*, relative magnitude *(rm)*, and relative distance *(rd)*.

$$R_{ij} = \{c, p, s, rm, rd\} \qquad (4)$$

Each line segment (L_i) is characterized by its properties (P_i) such as starting point *(sp)*, length *(l)*, orientation *(d)*, and curvature *(cu)*.

$$L_i = \{sp, l, d, cu\} \qquad (5)$$

At this stage, we have defined the shape of the region with a local graph, with the line segments as the nodes and the connectivity between these line segments is described by the edges.

2.1. Facial feature matching using local graph

To detect a face, we first need to match the key facial regions. In the LG graph method, each facial region is represented by a local graph. Hence, to identify the segmented regions as facial regions, we need to match the local graph of the segmented regions to the stored model local graphs in the database. Assume that the model facial feature region and the candidate region are represented by the curves $f(t)$ and $g(t)$. To match a region, we need to find the geometric transformation that best maps the curve $g(t)$ onto $f(t)$. The basic geometric transformation is composed of translation *(T)*, rotation *(M)* and scaling *(S)* [7]. In the case of affine transformation which includes reflection and shearing, there are six free parameters. These model the two components of the translation of the origin on the image plane, the overall rotation of the coordinate system, the global scale, together with the parameters of shearing and reflection. The transformation between two closed curves, $f(t)$ and $g(t)$ is defined:

$$g(t) = S \cdot R \cdot f(t) + T \qquad (6)$$

Ignoring the parameters of shearing and reflection, the transformation in matrix notation is:

$$\begin{bmatrix} x'(t) \\ y'(t) \\ 1 \end{bmatrix} = \begin{bmatrix} 1 & 0 & s_x \\ 0 & 1 & s_y \\ 0 & 0 & 1 \end{bmatrix} \bullet \begin{bmatrix} \cos\theta & -\sin\theta & 0 \\ \sin\theta & \cos\theta & 0 \\ 0 & 0 & 1 \end{bmatrix} \bullet \begin{bmatrix} x(t) \\ y(t) \\ 1 \end{bmatrix} + \begin{bmatrix} 1 & 0 & t_x \\ 0 & 1 & t_y \\ 0 & 0 & 1 \end{bmatrix} \qquad (7)$$

Most of the imaging equipments have the same scale factor to both x and y directions. Hence, we assume that the scale factors in both x and y directions the same $(s \equiv s_x \equiv s_y)$. Since, in geometric transformation, rotation and scaling does not change the centroid, the translation is defined as:

$$T = F_{cen}(g(t)) - F_{cen}(f(t)) \tag{8}$$

where $F_{cen}(f(\cdot))$ is a function which computes the centroid of the curve $f(\cdot)$. We compute the centroid (x_0, y_0) of each region and move the centroid's to the origin. To find the scale parameter, we compute the momentum of the object. The momentum is a measure of the mass distribution of an object and is defined as:

$$Mom = \frac{1}{N}\sum_{i=1}^{N} m_i \cdot \left\| p_i - p_{centroid} \right\|^2 \tag{9}$$

where N is the number of curve points, m_i is the mass weight at point p_i, $p_{centroid}$ is the centroid point of the closed curve. ||•|| is a kind of norm, such as Euclidean norm. The translation and rotation do not change the shape of an object. So, *Mom* is identical before and after translation and rotation. Thus the scaling factor is defined as:

$$s = \left(\frac{Mom'}{Mom}\right)^{1/2} \tag{10}$$

To find the rotation parameter, we suppose that the curve, $f'(t)$ is generated by rotating the curve $f(t)$ by a certain angle θ. The rotation is obtained on the centroid of the curve $f(t)$. However, the problem associated with Eq. (6) is that we do not know the point-to-point correspondence to calculate the rotation angle θ. To by-pass the point-to-point correspondence, wavelet coefficient of the region borders are used [6, 8]. Figure 2 shows a complete example of single region matching.

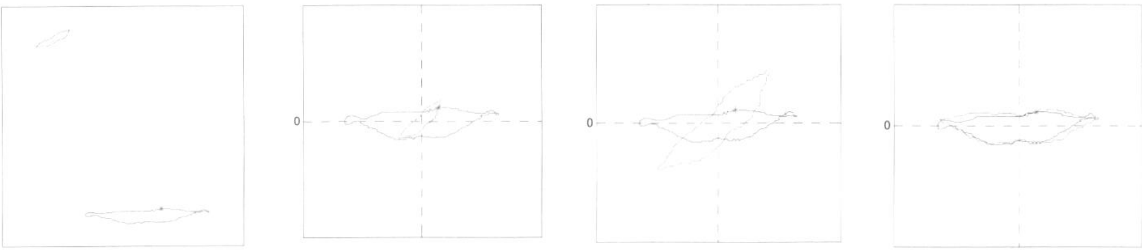

(a)Object(red) and model(blue) regions (b)Translation adjusted (c)Scaling adjusted (d)Rotation adjusted

Figure 2. Single region matching and the matching results at each stage (b, c, d)

3. Skin region Synthesis

Different regions of the face are illuminated differently and hence the segmentation procedure might produce a large number of regions on the face. Also, due to the parameter settings of the segmentation procedure (such as region-filling thresholds) and the relative facial image size, the segmentation procedure might produce a number of regions on the face (Figure 3(b)). Not all regions generated by the segmentation procedure are needed. It should be noted that the image region used for segmentation was identified by the skin detection procedure previously as skin. To identify these skin regions as a face, it is sufficient if we can identify the distinctive facial features such as eyes, eyebrows, nostrils and mouth. Hence, to reduce the number of regions (nodes) and to simplify the successive graph matching procedure, we first apply a skin region synthesis procedure to merge neighborhood regions based on relative *RGB color similarity* as described below.

3.1. Neighbor region searching and region synthesis

Two regions are defined as neighbors, if their borders have common shape and the common parts are very close in Euclidean space. If the correspondent common parts have distance greater than zero, then these two regions are pseudo-neighbors. For skin region synthesis we consider only those regions whose RGB color value is close to average skin color value. The skin region synthesis procedure is described as below:

1) Initialize the first region which is closer to average skin color as the active skin region. The degree of closeness is calculated as the color difference in RGB space.
2) Select the next region which is closer to average skin color. Find the common edge between this region and the active region.
3) If a common edge is found, synthesize the current region and the active region. Assign new region to active region.
4) If all regions have been processed, region synthesis completes; otherwise go to step 2.

If two regions are neighbors, they have at least one common border. Since, the local graph represents the shape in high-level, we use the local graph to find common edge between two regions. Once the common edge is detected, a region synthesis process is started. This process merges all the neighbor regions and retains distinctive facial features as the color of these features is far from average skin color value. Given two regions, R_1 and R_2, the rules to synthesize a new shape R_{12} are defined as

$$shape(R_{12}) = \begin{cases} shape(R_1)shape(R_2), & if\ REL(R_1, R_2) = contiguous \\ shape(R_1), & if\ REL(R_1, R_2) = contain \\ shape(R_2), & if\ REL(R_1, R_2) = contained \\ \phi, & if\ REL(R_1, R_2) = separate \end{cases} \tag{11}$$

In order to improve the performance of the region synthesis process, we build a table to accelerate the finding common-edge process. We do not need to compute the relationship between each node pair. Some of the relationships can be deduced from other relationships. Figure 3(c) shows the skin synthesized image. The application of this step retains important facial features such as eyes, eyebrows, nostrils and mouth since the color value of these facial features is relatively different from the average skin color. The facial features can now be effectively used to represent a face and are further used for generating image LG graph.

4. Matching multiple regions with Local Global (LG) Graph method

To perceive the group of facial features as a face, the spatial relationships between the corresponding facial features is an important constraint. It is just not enough to identify the facial features alone but their geometrical positions and placements are also very important. The location of spatial features serves as a natural choice (as landmarks) in relating multiple views of faces. Rather than using the shape constraints to establish similarity correspondence, we use the constraints provided by the spatial adjacency of the regions. These constraints are relaxed by separately triangulating the data and model regions. We use the neighborhood consistency of the correspondences in the triangulations to weight the contributions to the similarity function. In this section, we describe how the relational consistency is used in the matching process. In particular, we abstract the representation of correspondences using a bipartite graph. Because of its well-documented robustness to noise and change of viewpoint [9-11], we use the Voronoi Tessellation method, the Delaunay triangulation and the Local-Global graphs as our basic representation of the image structure.

The process of Delaunay triangulation generates relational graphs from the two sets of point-features. More formally, the point-sets are the nodes of data graph and model graph are:

$$G_D = \{D, E_D\},\ G_M = \{M, E_M\} \tag{12}$$

where D and M are the node sets of data(image) and model respectively. $E_D \subseteq D \times D$ and $E_M \subseteq M \times M$ are the edge-sets of the data and model graphs. The key to the matching process is that it uses the edge-structure of Delaunay-Graphs to constrain the correspondence matches between the two point-sets. This correspondence matching is denoted by the function $f : D \to M$ from the nodes of the data-graph to those of the model-graph. According to this notation, $f(i) = j$ indicates that there is a matching between the nodes $i \in D$ of the data-graph and the node $j \in M$ of the model-graph. The Delaunay graph is robust to noise and geometry transformation. In other words, if the node set undergoes any kinds of transformations, the new Delaunay graph is the transformed version of model Delaunay graph, with the same transform parameters.

4.1. Image representation with Local Global (LG) Graph

In the LG graph scheme, each node in the graph represents a facial region, not just a facial point and hence contains information about the region. Each node is defined with its spatial location, same as the center of gravity *(x, y),* color of the region *(color),* local graph associated with this node *(L)*, the number of pixels in the region *(size)* and the contour pixel set *(border).*

$$node = \{(x, y), color / texture, L, size, border\} \tag{13}$$

After introducing the local graph information into the global graph scheme, the new method can now handle both local (region) and global (object) information in the matching process. We combine the segmented skin regions as described in section 2 with the LG graph scheme. The skin region synthesis procedure retains only few facial regions. Every region is a characteristic of the face. As previously defined, the object and model LG graphs are represented as:

$$G_D = \{D, E_D\},\ G_M = \{M, E_M\} \tag{14}$$

Where D and M are node sets, E_D and E_M are edge sets. The node-set *(NS)* and edge-set (E) are defined as:

$$NS = \{node_i, i = 1,2,3...n\} \qquad n\ is\ the\ total\ number\ of\ nodes \tag{15}$$

$$E(i,j) = \begin{cases} 1 & if \ \ node \ i \ connects \ \ with \ \ node \ j \\ 0 & otherwise \end{cases} \tag{16}$$

Figure 3(d-e) shows the model LG graph. As shown in the figure, for representing faces, we consider only eyes, eyebrows, nostrils and mouth regions.

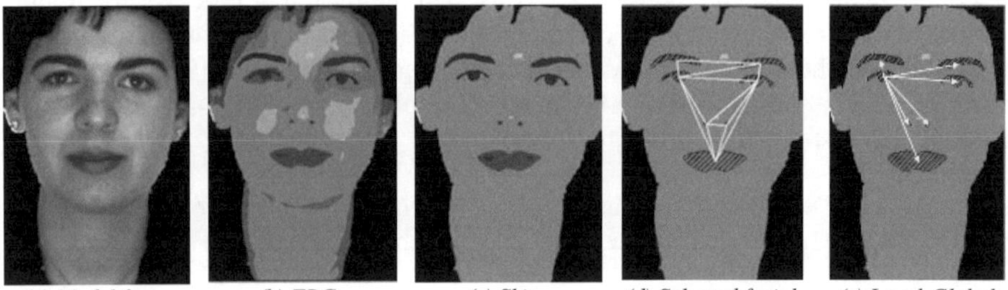

(a) Model face image *(b) FRG segmented image* *(c) Skin synthesized image* *(d) Selected facial regions and Delaunay graph* *(e) Local-Global Graph view*

Figure 3. Model face image and the corresponding Delaunay graph

4.2. LG Graph matching

Matching two graphs comprises of establishing point correspondences between two graph node sets that maximizes the likelihood between two graphs, given the spatial constraints [12]. Given only the geometry position of the nodes, the matching process in general requires a permutation or a recursive method. In the proposed LG scheme, the node correspondences is solved by using the local graph similarity constraint (color and shape similarity) bypassing the computationally expensive recursive or permutation methods. Thus the matching complexity can be reduced to a reasonable extent.

4.2.1. Establishing node point correspondences

In the LG graph scenario, each node corresponds to a region. Here, every region is a meaningful part or a characteristic of the face. We define that only regions with similar characteristics can be considered as Potential Correspondent Region Pair (*PCRP*). Random matching of one node in the graph with nodes in another graph is not reasonably acceptable. In the human perception system, color plays an important role in recognizing objects. It is reasonable to link node pairs that have similar color. The method proposed here does not allow one node to correspond to nodes with very different color. Suppose a model set *M* has *N* nodes. A threshold th_c is chosen to filter out dissimilar nodes in the data graph. This process is described below:

1) Initialize *PCRP* table, which has n entries, the region count in the model graph.
2) For each region in the model graph, do the following
   a) Compute color distances between all regions in the data graph.
   b) For any region with distance < $th_{c,,}$ add it to current *PCRP* table entry.
   c) If no *PCRP* is found for a current region, return an *ERROR*.

Figure 4(a-b) shows an example of the skin segmented data image (image taken from FERET Database [13]) and the regions retained after the skin synthesis procedure.

4.2.2. Graph Similarity

Once the node correspondences are established, the next step is to compute the similarity between two graphs. The graph similarity can be determined by the relative spatial connectivity between the nodes. The translation, rotation and scaling do not change the graph spatial structure. The spatial structure of the graph is mainly represented by the angle, i.e., if the correspondent angles between arcs are similar, the graphs are similar too. We define the angle similarity function $S_{ANGSIM}(\Delta\theta)$ by using two thresholds. One is $\boldsymbol{\theta_{th1}}$, lower bound of angle difference; another is $\boldsymbol{\theta_{th2}}$, the upper bound of angle difference. The angle similarity function, $S_{ANGSIM}(\Delta\theta)$ defined as follows:

$$S_{ANGSIM}(\Delta\theta) = \begin{cases} 1, & \Delta\theta < \theta_1 \\ \dfrac{\theta_2 - \Delta\theta}{\theta_2 - \theta_1}, & \theta_2 \geq \Delta\theta \geq \theta_1 \\ 0, & \Delta\theta > \theta_2 \end{cases} \tag{17}$$

where θ_1 is the lower bound of the angle difference and θ_2 is the upper bound of the angle difference. The total similarity between two graphs, SIM_{LG} is defined as

$$SIM_{LG} = \frac{1}{N}\sum_{i=1}^{N}\sum_{j=1}^{N} E(i,j).S_{ANGSIM}(\theta_{ij} - \theta_{i0}) \tag{18}$$

where N represents the total number of nodes, θ_{ij} is the angle of the *edge(i, j)* and θ_{i0} is the base angle for every node. The base angle can be selected as the angle of the first arc associated with node *i*. The PCRP graphs built as shown in Figure 4 are compared with the face model graph shown in Figure 3. Table 1 lists the similarity values computed between the PCRP graphs and the model graph.

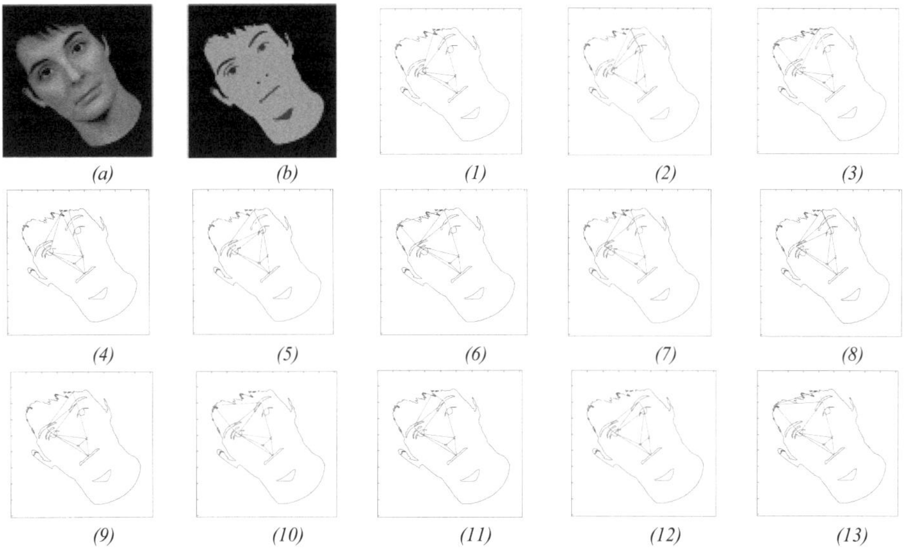

Figure 4. Data Image (b) and the PCRP graphs (1-13)

Table 1. Graph similarity comparison result.

Graph Index	*1*	*2*	*3*	*4*	*5*	*6*	*7*
Graph Similarity Value	*0.46*	*0.71*	*0.47*	*0.46*	*0.76*	*0.32*	*0.33*
Graph Index	***8***	*9*	***10***	***11***	***12***	***13***	
Graph Similarity Value	*0.47*	*0.61*	*0.60*	*0.63*	*1*	*0.4*	

4.2.3. The LG Graph Relation Checking

The PCRPs described in section 5.2.1 are selected only based on their color similarity. The combinations of all PCRPs have 13 possible graphs. Figure 4 shows all the *PCRP* graphs. However, only one graph among these can be a right match. In the extreme case, we may not have the right match in the selected graphs. This indicates that only graph spatial constraints cannot ensure finding the right match. Thus we need to examine the validity further. In Eq. (18), the graph similarity is measured based on the assumption that every node pair has the same weight. However, if the shapes of two nodes as defined by local graphs are different, it is very likely that it is a wrong PCRP. In this case, even though the PCRP has high node similarity with its counterpart in the model graph, we reduce its contribution to the graph similarity by introducing a shape similarity weight factor to the graph similarity measure. We weigh the graph similarity measure with the corresponding node pair shape similarity measure calculated as described in section 3.1. High weight is given to the node pair with high shape similarity while low weight is given to the node pair with low shape similarity. After introducing the shape similarity factor, the graph similarity is defined as

$$SIM_{LG} = \frac{1}{N}\sum_{i=1}^{N}\sum_{j=1}^{N} E(i,j).S_{ANGSIM}(\theta_{ij} - \theta_{i0}).Weight(i,j) \tag{19}$$

In other methods, graph nodes are only points in 2D or 3D space and have only geometry relationships among them [12]. In the LG Graph scenario, every node represents a region, i.e., a characteristic region of the face. These nodes have certain pre-determined constraints. We define four relationships between two regions – contiguous, contain, contained and separate. The relationship checking table is shown in Table 2. For any two regions *R1* and *R2*, if their relationship in the image is r_{12} and in the model graph is r_{12}', then the similarity relationship between these two regions is

$$sim_{REL} = T_{RE}(r_{12}, r_{12}') \tag{20}$$

The relationship between two graphs is determined by checking every relation between any two connected graph nodes.

$$SIM_{REL} = \prod_{i=1}^{N}\prod_{j=1}^{N} T_{RE}(r_{i,j}, r_{i,j}') \tag{21}$$

Table 2. Relationship checking table

Model Relationship / Image Relationship	Contiguous	Contain	Contained	Separate
Contiguous	*1*	*0*	*0*	*0*
Contain	*0*	*1*	*0*	*0*
Contained	*0*	*0*	*1*	*0*
Separate	*0*	*0*	*0*	*1*

4.2.4. LG Graph Similarity

The overall LG graph error is defined as:

$$ERR_{LG} = (1 - ERR_{rel}) \bullet (ERR_G + ERR_{shape}) \quad (22)$$

where ERR_{LG} is the total matching error between object and model, ERR_{rel}–the matching error of relationships among regions, ERR_G-the matching error between global graphs, and ERR_{shape}-the matching error between two objects' shapes. The overall matching scheme is defined as:

1) First compare the relationships between the object and the model. If all the node relationships comply, then ERR_{rel} returns *0*; otherwise return *1* and the process ends.
2) Build the global graphs for the object and model graphs. The global graph error ERR_G is obtained by comparing the object and model graphs. To cope up with differences in various individuals, a relative big threshold Th_{graph} is set for ERR_G. If the $ERR_G > Th_{graph}$, it means that the current object has a very different structure from the model. We will not consider them as similar and quit current step.
3) Finally, objects are merged using the region synthesis procedure described in section 4. This is in particular applied for recognizing expressions described in section 7. After synthesizing, we compute shape similarity between the object and the model. If the shape matching error is less than predetermined threshold Th_{shape}, i.e., $ERR_{shape} < Th_{shape}$, then the object is similar to the model.

If all the three terms are computed and fall in the acceptable range, the face is located. The total similarity measure is as computed by equation 22. The lower the ERR_{LG}, the more similar the object and model are.

5. Experimental results

Figure 5 shows the application of LG graph based face detection on sample images from AR face database [4]. The AR face database includes frontal view images with different facial expressions, illumination conditions and occlusion with by sunglasses and scarf. It should be noted that, at present the images with sunglasses are not used in evaluating the proposed method. Also, the results were shown on custom-collected images (Figure 5(e, f)).

5.1. Effect of feature shape and spatial deformations on LG graph matching

One of the major problems with detecting faces is that face is dynamic in nature and the shape of the facial features deform with the facial expressions. The facial features also appear deformed depending on the non-frontal views of the face. Added to this, there are differences between the faces from one person to another. If the target is to detect faces, all these deformations must be accounted for. In the above described LG graph method for face detection, the facial feature information is stored at the local-level (within the local graph at each feature node) and the spatial geometry of the facial features is stored at the global level (within the local-global graph with Delaunay Triangulation). The PCRP selection method described in section 5 allows us first to select different candidate facial features depending on the color, but not on the shape similarity. The validity of the selected facial feature's relationships is then constrained by the graph and shape similarity relationships. Selecting the nodes based on color similarity and then selecting a big threshold for graph similarity, Th_{graph} (section 5.2.4) allows us to cope up with the deformations in facial features and the differences between the individuals from the model graph. Selecting a big threshold for individual feature shape similarity allows us to cope up with the facial feature deformations. Figure 5(a-c) shows the example of faces detected from images containing various facial expressions.

5.2. Effect of the missing facial features (nodes) and partial LG graph matching

Another problem associated with face detection is that due to partial occlusions, all the facial features used to represent the model graph might not be present in the data image. Many of the existing appearance based methods fail to detect partially occluded faces. To detect partially occluded faces, we use partial LG graph matching. At the present stage, all the possible node combinations are considered by eliminating each node at a time. Since the node correspondences are solved using shape and color similarity, the overall computationally complexity is kept within an acceptable level. Node elimination and partial graph matching should be computationally expensive, if the PCRP regions selected in the data image (section 5.2.1) are relatively huge in number. However, this is rarely the case as the overall search space is first reduced by the skin detection procedure and then further reduced by skin region synthesis. Figure 5 (d, f) shows the example of detecting partially occluded faces.

(a) *(b)* *(c)* *(d)*

(e) *(f)*

Figure 5. Example of face detection on sample images. Shown in figure - the original images, the skin segmented images and the detected faces.

6. Extending LG Graph method for recognizing facial expressions

For recognizing facial expressions, the series of tasks to be performed are face detection, facial feature recognition and then facial expression recognition. In the proposed LGG method, after the face detection step, the various facial features from the image can be retrieved by the local-global graph node correspondences. The extra step needed for recognizing facial expressions is that we need to compare the image LG graph with the existing Expression LG graphs. To recognize facial expressions, the LG data-base also consists of the corresponding Expression LG graphs. It should be noted that each facial expression is characterized by a specific configuration of the facial features. This particular configuration of each facial feature is represented by the Expression LG graphs, similar to the Face LG graph. The advantage of the proposed method over other existing methods is that each individual facial feature configuration is represented by a more detailed graph representation which allows us to represent even more subtle expressions. For recognizing expressions, each node in the LG graph is modified as shown below:

$$node = \{(x,y), color/texture, L, size, border, LG_{EXPR1},..LG_{EXPRi}\} \quad (23)$$

where LG_{EXPRi} represents an Expression LG graph. At present, the features used are left and right eyebrows, left and right eyes and mouth for representing each expression. Figure 5 shows the corresponding Expression LG graphs for neutral, happy, angry and scream expressions present in the AR database. After the face detection step, an approximate position of the facial features in the data image can be found from the node correspondences established during LG graph matching step. A part of the image region around the facial feature in the data image is considered. This region is segmented using the FRG segmentation method and an image Expression LG graph is built as described before. To detect an expression, we compare each image facial feature to the existing LG Expression graphs corresponding to that facial feature (node). The combination of the LG Expression graphs for a particular expression for which we obtain the minimum average shape error is selected as the corresponding expression for that image. Table 3 shows an example.

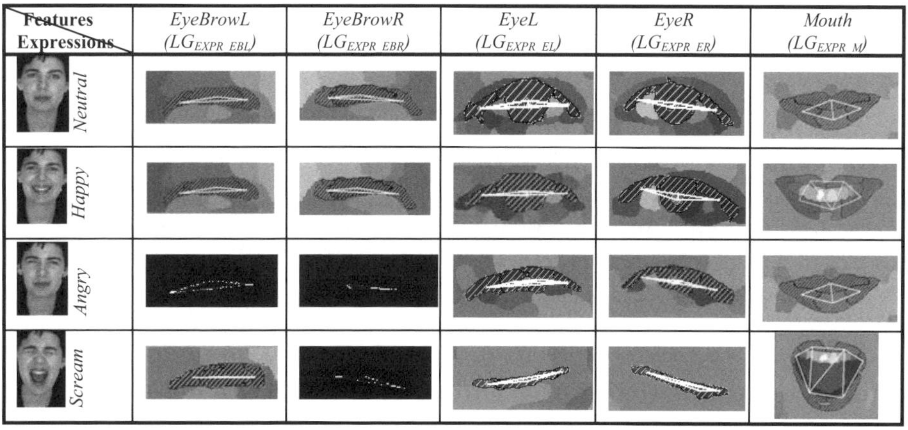

Figure 6. Figure showing the LG Expression Graphs for the features considered for neutral, happy, angry and scream expressions respectively.

7. Conclusions

In this article, we proposed a novel face detection and facial expression recognition method based on Local-Global Graphs. A face is described by the Face – Local Global Graph model. To detect faces, the model Face-LG graph and the corresponding test image Face-LG graph are constructed and then compared with an efficient LG graph matching technique. The corresponding matching error is evaluated as the similarity between the two graphs. For representing facial expressions, the specific configurations of facial features corresponding to a particular expression are represented by the Expression LG graphs. Since, the LG graph embeds both the structure of the local facial features (at the nodes of the graph) and the whole geometric structure of the face; it is a more accurate way of representing the face and is in accordance with the psychological way of perceiving human faces and facial expressions. The proposed LG graph method does not require any training as in other methods and is feasible in situations where we do not have many training samples. The graph method is invariant to scale, rotations, to a certain extent to pose and is shown to perform robustly under various illumination conditions. The methodology is also shown to perform accurately for recognizing expressions corresponding to the models present in the database.

Table 3. Table showing the Expression LGG errors for the expressions considered

Expression LG Graph Errors		 *Neutral*	 *Happy*	 *Angry*	 *Scream*
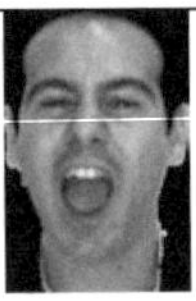	*EyeL*	*0.72*	*0.53*	*0.22*	***0.26***
	EyeR	*0.78*	*0.52*	*0.27*	***0.28***
	EyeBrowL	*0.33*	*0.34*	*0.40*	***0.42***
	EyeBrowR	*0.37*	*0.37*	*0.39*	***0.42***
	Mouth	*0.85*	*0.67*	*0.82*	***0.08***
	Avg. Error	*0.61*	*0.49*	*0.42*	***0.29***

8. References

[1] M.H. Yang, D.J. Kriegman and N. Ahuja, "Detecting faces in images: A survey," *PAMI*, 24(1), 2002.

[2] P. Kakumanu, S. Makrogiannis and N. Bourbakis, "A Comparative Survey of Skin Color Modeling and Detection Approaches," *Pattern Recognition, (to appear, 2007).*

[3] P. Kakumanu, A Face Detection and Facial Expression Recognition Method Applicable to Assistive Technologies and Biometrics, PhD Dissertation, Wright State University, Dayton, Ohio, March, 2006

[4] A.M. Martinez and R. Benavente, "The AR Face Database," *CVC Technical Report #24*, June 1998.

[5] X. Yuan, D. Goldman, A. Moghaddamzadeh and N. Bourbakis, "Segmentation of Colour Images with Highlights and Shadows using Fuzzy-like Reasoning," *Pattern Analysis and Applications*, 4(4), 2001.

[6] N. Bourbakis and P. Yuan et al., "Object Recognition using Local – Global Graphs," *Proc. ICTAI,* 2003

[7] D. Hearn and M.P. Baker, *Computer Graphics, C version*, NJ: Prentice Hall, 1997.

[8] R. Alferez & Y. Wang, "Geometric & Illumination Invariants for Object Recognition," *PAMI,* 21(6), 99

[9] N. Ahuja, "Dot Processing Using Voronoi Neighborhoods," *PAMI,* 4(3), 1982.

[10] N. Ahuja, B. An and B. Schachter, "Image Representation Using Vornoi Tesseletation," *Computer Vision, Graphics and Image Processing*, 29, 1985.

[11] K. Arbter, W. E. Snyder, H. Burkhardt and G. Hirzinger, "Application of Affine Invariant Fourier Descriptors to Recognition of 3D Object," *PAMI,* 12(7), 1990.

[12] A.D.J. Cross & E.R. Hancock, "Graph matching with a Dual Step EM Algorithm," *PAMI*, 20(11), 1998.

[13] P.J. Phillips, H. Moon, S.A. Rizvi, and P.J. Rauss, "The Feret Evaluation Methodology for Face-Recognition Algorithms", *PAMI,* 22(10), 2000.

[14] B. Fasel & J.Luettin, "Automatic Facial Expression Analysis:A Survey," *Pattern Recognition,* 36, 2003.

A. Esposito et al. (Eds.)
IOS Press, 2007

Face Recognition Experiments on the AR Database

Marco GRASSI[1a], Marcos FAUNDEZ-ZANUY[b]
[a]*Università Politecnica delle Marche, Ancona, Italy*
[b]*Escuela Universitaria Politecnica de Matarò, Adscrita a la UPC*

Abstract. Biometric recognition and authentication based on face recognitions can actually be used in many real-time applications such as surveillance, security systems, access control, and much more. For these purposes systems have to grant a fast computational speed but also robustness to illumination and facial expressions variations. In this paper we propose a very fast image pre-processing by the introduction of a linearly shaded elliptical mask centered over the faces. Simulation results, over the AR Database, show that the proposed method leads to a significant performance improvement in identification, using both eigenfaces and DCT with Nearest Neighbour and RBF Neural Network, without modifying the global computational weight.

1. Introduction

Face recognition represents one of the more interesting modalities of biometrics for the wide number of applications in which it can be used (surveillance, security systems, access control, etc…) as well as for its low intrusiveness and the easy and economic way in which it's possible to acquire images.

On the other hand it presents many problems resulting from face localization inside an image to pose, illumination, expression variations other than to the natural face variation with time and age. Much work has been done in recent years in this field and many different approaches have been proposed for face recognition [1]. In general they can be divided in two main categories: holistic approach which considers the image as an high dimensional vector, where each pixel is mapped to a component of a vector, and geometry-feature-based methods that try to identify the position and relationship between face parts, like eyes, nose and mouth and the extracted parameters that are measures of distance, shape, size of these regions.

---

[1] Corresponding Author: Marco Grassi, Facoltà di Ingegneria Elettronica, Università Politecnica delle Marche, Ancona. Italy, e-mail: margra75@hotmail.com .

The former approach results being faster than the latter, once that the high dimensionality of the vectors is reduced by some vector dimension reduction algorithm. The second is more robust against rotation, scale and illumination variations and against the change of the face with time and aging. Obviously the choice of the method used for the face recognition problem is strictly correlated with the kind of application to which it is intended.

In the case of a real-time access control system, in a cooperative user scenario, like the one we are working to implement using the device we dispose [2], the critical parameter is the computational time, so an holistic approach for characteristics extraction has to be preferred. In particular the DCT (Discrete Cosine Transform) method appears particularly efficient, allowing to concentrate the information, reducing the dimensionality of the problem and granting faster and better performances in comparison to alternative classic methods, like the eigenfaces one, especially if used in combination with an RBF neural network classifier [3].

In our system a 2D frontal, still, greyscale images acquisition is performed by a low cost webcam hidden behind a special glass. This glass performs as a transparent glass in one direction and as a mirror on the other and over it has been applied an elliptical mask, approximating the oval of a human face. So the user has to fit the reflex of his face inside the ellipse and in this way the problems of image localization, translation, rotation, zoom and pans can be, if not completely eliminated, significantly reduced.

This process is very important for the system performances because, clearly, if the user doesn't comply it correctly the process of recognition will be a failure. For this reason it is important that the user be guided by an administrator when he is entered in the database and his first pictures captured, allowing the user to make more than one proof to familiarize himself with the device and granting the introduction of correct pictures into the database. Afterwards, the device being completely unsupervised and automatic, on one hand will be the user himself that will learn with experience to place his face correctly and avoid assuming non-neutral facial expressions, in order to achieve a successful verification and open the door. On the other hand the system itself will have to understand when there is an user trying to access or simply someone joking in front of the webcam, so that user and system could perform a correct communication.

Furthermore in this real situation in which a user may undergo daily face recognition processes for acceding a door, the change of face constitutes a central problem, especially using an holistic approach for features extraction instead of a geometry feature based method. So, therefore a constant database update has to be realized but also it is fundamental to limit, as much as possible, the face analysis to the face part which is more insensible to daily variation, i.e. eyes, nose and mouth, excluding the parts which can include hair and beards. Moreover in order to provide robustness to illumination and expression variation, it is necessary to dispose of a significant image database which can consider these variations for every single subject.

The purpose of this work is to propose a fast image pre-processing method, consisting in the application of a linearly shaded elliptical mask centred over the faces. The elliptical form is the one that best approximates the shape of the human face and many algorithms for face localization deal with identification of ellipses in the images [4], so the method appears suitable to be applied also in combination with these approaches. Such a mask will let unchanged the part of the face containing the eyes, the nose and mouth and will blacken the outside part. The application of a rough mask, that simply turns to black the part of the image that we want to eliminate, determines non-linearity problems in frequency domain, for this reason the proposed mask presents a linear transition from the completely transparent part to the completely black one.

Before creating our own database, this pre-processing approach has been tested over the AR face database, using both the eigenfaces method and the DCT for features extraction with nearest-neighbours and radial basis function neural network as classifiers and the results compared with those coming from the application of a simple black mask.

1. Image pre-processing

The first step in face recognition is the face detection inside an image and the subsequent image normalization [5]. The detector should provide information about face position, size and pose. This information will be used to normalize the image to a predefined size, compensating translation, rotations, zoom and pans.

In our work the face detection inside the image has been done manually by eye identification, extracting the pixel position of the pupils, that are easy identifiable in all the pictures. Once that the eye position is known it results very easy to proceed to automatic image normalization. Eventual head rotations can be compensated by detecting the angle between the direction connecting the two eyes and the horizontal direction and operating a counter-rotation. The eye distance can be calculated and set to a fixed value by image scaling and finally the image can be cropped eliminating the background adapting the crop box to the shape of the face [Figure 1].

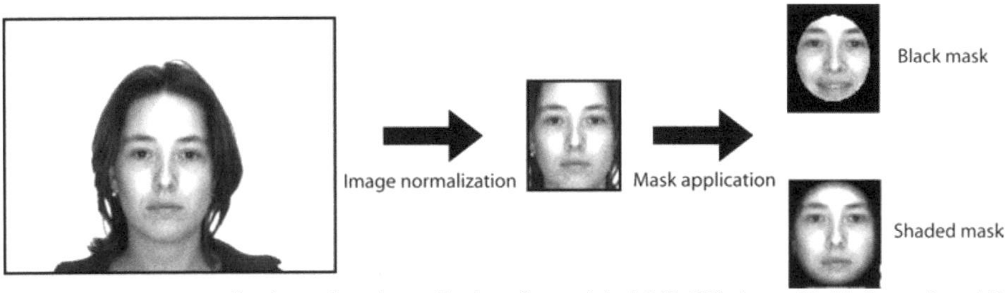

Figure 1. Image normalization and masks applications from original 768x576 size to 64x74. Image from AR Database.

Once we dispose of such a set of opportunely normalized images, it is possible to define an ellipse, centered in the centre of the image and with dimensions proportional to those of the image, that could enclose the face part in which we want to concentrate the process of face recognition. This part will remain unchanged, on the contrary outside the ellipse the brightness will be set to 0 for the black mask while it will decrease linearly down to 0 in function of the distance for the shaded mask.

Dealing with a grey scale image, such a shaded mask can be applied by defining a matrix *M(i,j)* of the same dimension of the image, with values varying from 1, in the completely transparent part inside the ellipse, to 0 in the blackened part, through a linear transition depending on the distance from the ellipse, and multiplying it scalarly with the image matrix. This means being (i, j) the generic pixel coordinates outside the ellipse:

$$M(i,j) = (l-d)/l \qquad d \le l$$
$$M(i,j) = 0 \qquad d > l$$

Where l is the width of the shaded area and d is the minimum distance between (i,j) and the ellipse. Dealing with just a simple scalar multiplication between matrix the application of the mask results being very fast once the image normalization has been realized.

2. Features extraction

The features extraction has been realized using the well known methods of eigenfaces and the DCT. Eigenfaces [6], also named principal component analisis (PCA), has been the first to attract the attention toward the holistic face recognition and is still object of refinement studies [7]. The DCT, which is at the base of coding image application as JPG and MPEG, has been used with success also in face recognition, due to its capacities for fast computation, dimensionality reduction and robustness to illumination variations [3] [8].

A deeper analysis and a comparison between these methods can be found in [9].

2.1 Eigenfaces

The eigenfaces method relies on the extraction of the eigenvectors of the covariance matrix of the set of the training face images. These eigenvectors can be thought of as a set of features that characterize the variation between the faces, referred to a mean face. For the testing process, the test faces are projected into the subspace computed previously and their coefficient (Ω) are compared with the ones of the training face. The number of possible eigenfaces is equal to the number of images in the training set. However, for computational efficiency, only the more significant eigenfaces, the ones that have the largest eigenvalues, can be used to approximate the faces. It has been suggested that by discarding the first most significant principal components, the variation due to lightening can be reduced [10]. In the case of the application of the face masks these components not only rely with illumination but also with the shape of the applied masks, so their discard leads to yet larger performance improvement.

2.2 The Discrete Time Cosine Transform

The DCT, which is at the base of coding image application as JPG and MPEG, has been used with success also in face recognition, due to its capacities of fast computation, dimensionality reduction and robustness to illumination variations [3].

The DCT [5] is a separable, linear transformation; this is, the two-dimensional transformation is equivalent to a one-dimensional DCT performed along a single dimension followed by a one-dimensional DCT on other side. The definition of a two-dimensional DCT for an input image A and a transformed image B is the following:

$$B_{pq} = \alpha_p \alpha_q \left\{ \sum_{m=0}^{M-1} \sum_{n=0}^{N-1} A_{mn} \cos\left(\frac{\pi(2m+1)p}{2M} \right) \cos\left(\frac{\pi(2n+1)q}{2N} \right) \right\}$$

Where:

$0 \leq p \leq M-1$ and $0 \leq q \leq N-1$

$$\alpha_p = \begin{cases} \frac{1}{\sqrt{M}}, & p = 0 \\ \sqrt{\frac{2}{M}}, & 1 \leq p \leq M-1 \end{cases} \qquad \alpha_q = \begin{cases} \frac{1}{\sqrt{N}}, & q = 0 \\ \sqrt{\frac{2}{N}}, & 1 \leq q \leq N-1 \end{cases}$$

M and *N* are respectively the row and column size of A.

The application of the DCT to an image (real data), produces a real result.

The DCT is used in many data compression applications in preference to the DFT because of its property that is frequently referred to as "energy compactation" [11]. It's easy to observe that most of the energy is concentrated around the origin (low frequency components located on the upper left corner) [Figure 2]. In order to classify the image it is consequently possible to give just a portion of the transformed image to the classifier, considering just a square windows of *N' x N' pixels*, located in the upper left corner.

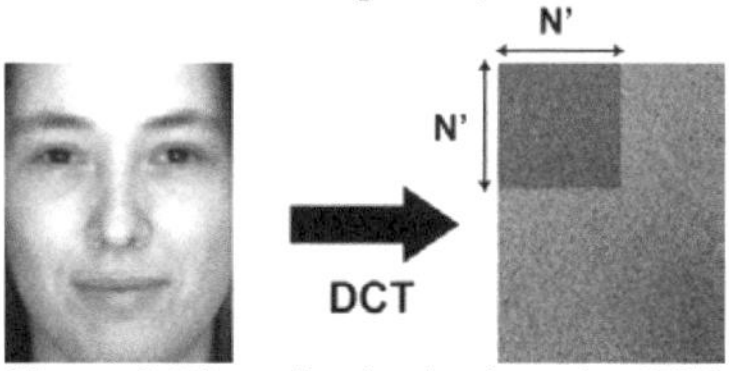

Figure 2. Dimensional reduction using DCT

The window's dimension is obviously the critic parameter for the system performances, determining both the recognition rate and the time of computation; so a good choice should grant fast computation without affecting too much the performance in terms of recognition.

3. Experimental Results

The AR database [12], used for the experiments, is a publicly available database (http://cobweb.ecn.purdue.edu/RVL/ARdatabase/ARdatabase.html) of 126 individuals, with 26 images for each, taken in two different sessions at a time distance of two weeks, varying the lighting and the facial expression. We have used 12 of the 26 images, excluding the ones in which the face was partially occluded by sunglasses or scarves, of 117 of the 126 individuals, those of the rest being either not complete or not available. All the images have been cropped and normalized to 64x74 grey ones [Figure 3] and two different masks have been applied for the experiments.

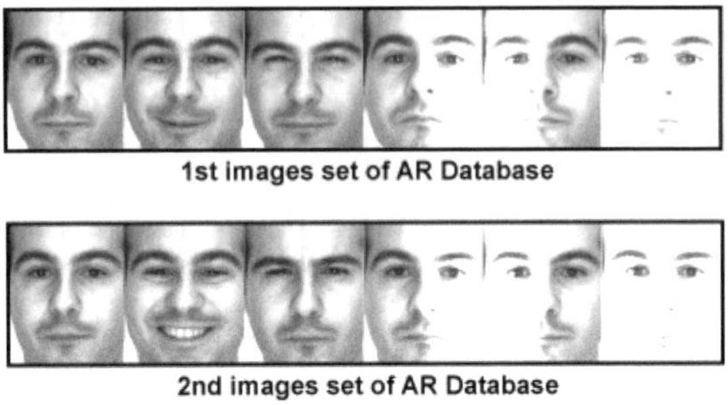

Figure 3. AR database samples of one person

In particular, a black mask and a shaded mask, defined by the same ellipse, have been used, as shown in figure 1, in order to analyze the effect of the introduction of the shade. The ellipse shape is related to the image dimension, by having the major and minor axes respectively equal to 4/5 of height and width of the image. The shaped area is 1/10 of the image width, which means that the shape extends up to the lateral boundaries of the image. Face images from 1 to 6 have been used for training, face from 7 to 12 for testing.

The performances in identification can simply be evaluated considering the percentage of successful identifications. On the contrary, in the verification task we assist to a trade-off between two different kinds of errors, missed detection (those situations where a user is incorrectly rejected) and false alarms (those situations where an impostor is accepted), which has usually to be established by adjusting a decision threshold. We have used the minimum value of the Detection Cost Function (DCF) for comparison purposes. The DCF is defined by [13]:

$$DCF = (C_{Miss} \times P_{Miss} \times P_{Target}) + (C_{FalseAlarm} \times P_{FalseAlarm} \times (1 - P_{Target}))$$

where C_{Miss} and $C_{FalseAlarm}$ represent respectively the cost of a missed detection and of a false alarm, P_{Miss} and $P_{FalseAlarm}$ represent respectively the Miss and the False Alarm probability and P_{Target} represents the a priori probability of the Target to be known.

An error type weighting of 1:1 and an equal probability that the target be known or unknown ($C_{Miss} = C_{FalseAlarm} = 1$; $P_{Target} = 0.5$) have been chosen.

3.1 Eigenfaces

Figure 4 shows a comparison between the recognitions rates obtained using eigenfaces method for features extraction with the original normalized images and with the images after the application of the masks in figure1. In order to optimize the results, for the original set of images the first three eigenfaces have been discarded, for the images with masks the first two.

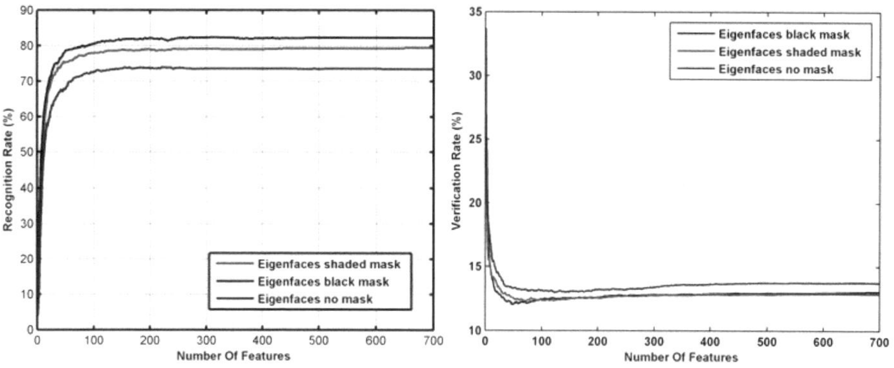

a) Identification b)Verification

Figure 4. a) Identification rates and **b)** Verification rates vs number of features using eigenfaces without mask, with the black mask and with the shaded mask.

The results highlight how the application of the elliptical masks leads to a significant improvement in recognition performances. In particular the black mask is the one that achieves the best results with a recognition rate stably over the 80% using a number of features superior to 95, with a maximum of 82,3% using 269 features. This is reasonable if we think that the eigenfaces method relies on the individualization of those features that characterize the variation between the faces, so the application of the black mask allows considering only features belonging to the most discriminant part of the face improving the recognition rate, while the application of the shade simply introduces further useless and misleading features that deteriorate the recognition rate.

3.2 DCT and Nearest Neighbours

Using the DCT for characteristics extraction, we obtain one model vector from each training image. During the testing each model input is compared against all the models in the database (117x6=702 in our case) and the model closer to the input image indicates the recognized person. A comparison between the different identification and verification rates, for the original set of normalized images and for the sets of images coming from the application of the black mask and the shaded mask, has been effectuated in function of the number of used DTC coefficients using a MAD nearest-neighbour classifier, that leads to better performances than the MSE [4].

Figure 5 shows a comparison of the results, in identification (a) and verification (b), using the original set of normalized images and the set of images coming from the application of the black and of the shaded mask. In the identification case, both the masks lead to a significant improvement of the performances. The black mask achieves better performances than the shaded one but only with an elevated number of coefficients, with a maximum identification rate of 86,2 %.

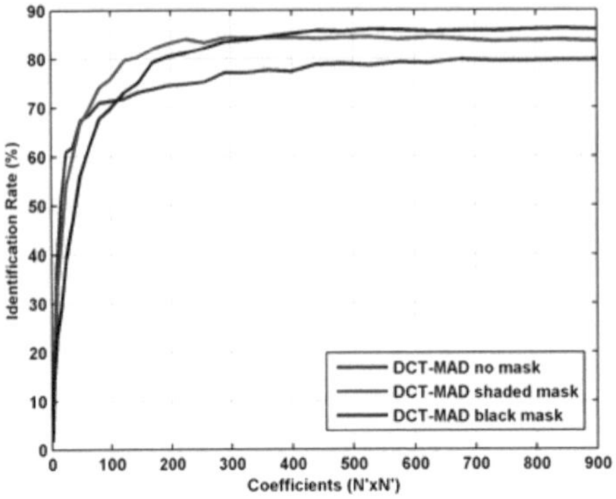

a) Identification

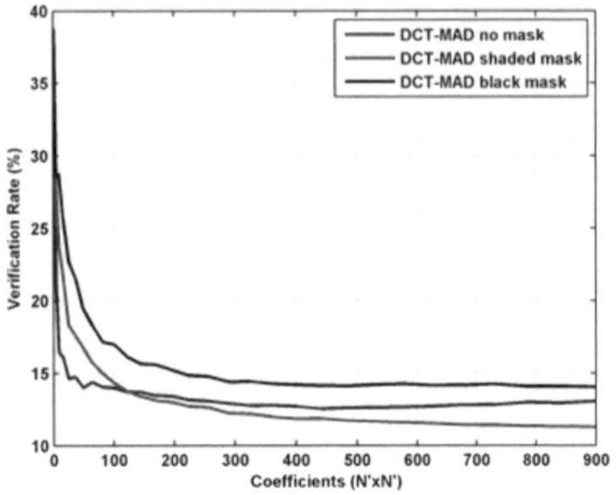

b) Verification

Figure 5. a) Identification rates and **b)** Verification rates vs. vector dimension using DCT-MAD Nearest Neighbour without mask, with the black mask and with the shaded mask

With a limited number of coefficients, minor to 18x18, the shaded mask is preferential, leading to a identification rate stably over the 80%, starting with a number of coefficients equal to 12x12. In verification the best performances are achieved using the shaded mask, with best verification rate with 30x30 coefficients (minDCF=11,2%), while the black one leads to a deterioration of the results.

3.3 DCT and RBF Neural Network

A further system performances refinement can be achieved using a Gaussian Radial Basis Neural Network as classifier. These nets can require more neurons than a standard feed-forward back-propagation networks, but they can be trained in a fraction of time needed by standard feed-forward networks [14]. A net with 117 output neurons has been used (one neuron for each person) with a spread (or scalar bias) of the Gaussian varying between 0 and 3. In the training phase the net receives as inputs the 702 training images, and the desired outputs will be vectors of 702 elements, one for every image, originating a 117x702 desired output matrix. The i-th output (i-th row) corresponds to the i-th person, the j-th columns to the j-th images. For every output the elements that identify the images of the corresponding person are set to 1, the others to -1. The net is then simulated with the 702 test images, and for every column of the originated output matrix (j-th test image), the element closer to 1 indicates the recognized person. If the image belongs to the person of the corresponding output the identification is successful.

Figure 6 shows the identification and verification rates versus the number of used coefficients for the original set of normalized images and for the set of images coming from the application of the black and of the shaded mask. Also in this case, system performances are improved using the shaded mask with a best identification rate of 92,75% with 25x25 coefficients and a best verification rate with 18x18 coefficients (minDCF=0,96%).

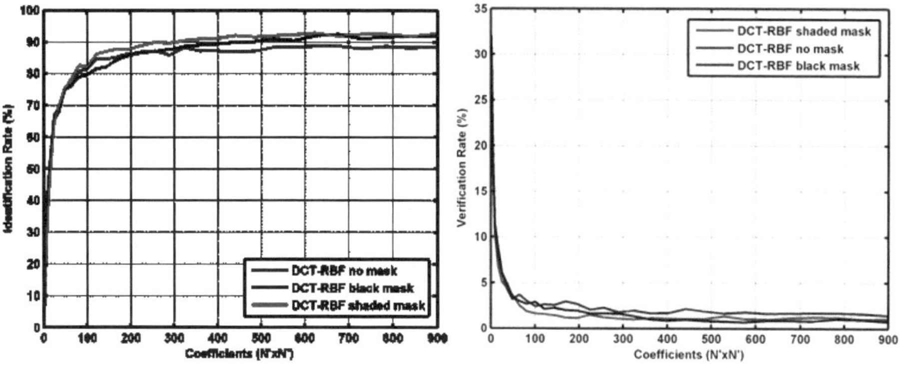

a) Identification b) Verification

Figure 6. a) Identification and **b)** verification rates vs vector dimension using the RBF Neural Network with and without the shaded mask

3.3 Results Comparison

In table 1 is presented a comparison between of the results obtained with the different tested methods and with those achieved using the Weighted PCA method [7] over the AR Database in the same test conditions, with only a little difference in the image format, that is of 64x64 pixels. This method is derived from the classic eigenfaces one and tries to assign a different weights to the different eigenfaces, in order to give more importance in the classification to the ones that result more significant for distinguishing faces of different individuals.

Table 1. Comparison between different methods

Method	Identification Rate % (100 features)	Identification Rate % (225 features)
Eigenfaces*	72,65	73,65
Eigenfaces black mask**	80,63	82,05
Eigenfaces shaded mask**	77,77	78,91
Weighted PCA	80,90	79,20
DCT + Nearest Neightbor (MAD)	71,36	74,78
DCT black mask + Nearest Neightbor (MAD)	69,94	81,19
DCT shaded mask + Nearest Neightbor (MAD)	75,92	84,04
DCT + RBF Neural Network	81,48	86,32
DCT black mask + RBF Neural Network	79,62	86,75
DCT shaded mask + RBF Neural Network	81,90	88,89

5. Conclusions

The results highlight how the application of the proposed image pre-processing method, consisting in the application of an elliptical linearly shaded mask, lead to a significant performances improvement in combination with all different tested methods and in particular used in combination with the DCT for the features extraction and a RBF neural network as classifier it leads to high recognition rates. This processing method is strictly related to the face normalization process and it's of very fast computation. In this way it allows a significant system performance improvement without modifying the computational weight. Futures efforts will be the practical implementation of an access control system, using the device we dispose, and the testing of the proposed method in this real situation.

References

[1] Stan Z. Li, Anil K. Jain (Eds.) "Face Recognition Handbook" Springer 2005
[2] M. Faundez-Zanuy, V. Espinosa-Duro, J. A. Ortega-Redondo "A low-cost webcam & personal computer opens doors". IEEE Aerospace and Electronic Systems Magazine. Vol.20, pp.23-26, November 2005.
[3] Meng Joo Er, Weilong Chen and Shiqian Wu "High-speed face recognition based on Discrete Cosine Transform and RBF Neural Networks" IEEE Transctions on neural networks, vol. 16, no.3, pp.679-691,May 2005
[4] Hamidreza Rashidy Kanan, Karim Faez and Mehdi Ezoji "An Efficient Face Recognition System Using a New Optimized Localization Method" 18th International Conference on Pattern Recognition 2006 pp. 564-567
[5] Ming-Hsuan Yang, David J. Kriegman and Narendra Ahuja "Detection Faces in Images: A Survey" IEEE Transaction On Pattern Analysis and Machine Inteligence, Vol. 24, No.1, pp. 34-58, January 2002
[6] M. Tuck, M & A Pentland "Eigenfaces for Recognition" Journal Cognitive Neuroscience, Vol.3, n°1 pp.71-86, Massachussets Institute of Thecnology, 1991
[7] Hui-Yuan Wang, Xiao-Juan Wu "Weighted PCA Space and Its Application In Face Recognition" Proceedings of The Fourth International Conference on Machine Learning and Cybernetics, Guangzhou, 18-21August 2005 pp.4522-4527
[8] Marcos Faundez-Zanuy, Joseph Ruore, Virginia Espinosa-Duro, Juan-Antonio Ortega "An Efficient Face Verification Method in a Transformed Domain" Accepted for Publication, Pattern Recognition Letters
[9] Marcos Faundez Zanuy "Face Recognition in a Transformed Domain" Proceedings 37th Annual International Carnahan Conference On Security Tecnology 2003 pp.290-297
[10] Peter N. Belhumeur , Joao P. Haspanha and David J. Kriegman "Eiegenfaces vs. Fisherfaces: Recognition Using Class Specific Linear Projection" IEEE Transictions On Pattern Analysis And Machine Intelligence, Vol.19, no.7, pp.711-720, July 1997
[11] Alan V. Oppenaim and Ronald W. Schafer "Discrete Time Signal Processing" 2nd Edition Prentice Hall 1999
[12] Aleix M. Martinez "Recognizing Imprecisely Localized, Partially Occluded, and Expression Variant Faces from a Single Sample per Class IEEE Transaction On Pattern Analysis and Machine Intelligence, Vol.24, N.6, pp. 748-763, June 2002
[13] A. Martin, G A. Martin, G. Doddington, T. Kamm, M. Ordowski, M. Przybocki "The DET curve in assessment of detection task performance" Vol.4, pp. 1895-1898, European Speech Processing Conference Eurospeech 1997
[14] Madan M. Gupta, Liang Jin, Noriyasu Homma "Static And Dynamic Neural Networks" John Wiley And Sons 2003

Section 4

Conversational Agents

From Research to Practice: The Technology, Psychology, and Pedagogy of Embodied Conversational Agents

Dominic W. MASSARO[1]
Department of Psychology, Santa Cruz, California, USA

Overview. The goal of the lectures is to provide an empirical and theoretical overview of a paradigm for inquiry on the Psychological and Biometrical Characteristics in Verbal and Nonverbal Communication. Baldi® actually consists of three branches of inquiry, which include the technology, the research, and the applications. Each of these lines of inquiry is described to give the reader an overview of the development, research findings, and applied value and potential of embodied conversational agents. We review various approaches to facial animation and visible speech synthesis, and describe the development and evaluation of Baldi, our embodied conversational agent. We then describe the empirical and theoretical research on the processing of multimodal information in face-to-face communication. A persistent theme of our approach is that humans are influenced by many different sources of information, including so-called bottom-up and top-down sources. Understanding spoken language, for example, is constrained by a variety of auditory, visual, and gestural cues, as well as lexical, semantic, syntactic, and pragmatic constraints. We then review our persistent goal to develop, evaluate, and apply animated agents to produce accurate visible speech, facilitate face-to-face oral communication, and to serve people with language challenges. Experiments demonstrating the effectiveness of Baldi and Timo for tutoring vocabulary learning for hearing-impaired, autistic, and English language learners are reviewed. We also demonstrate how the Baldi technology can be used to teach speech production. We hypothesize that these enhanced characters will have an important role as language tutors, reading tutors, or personal agents in human machine interaction.

Keywords: Embodied conversational agents, facial animation, speech synthesis

Technology

Visible speech synthesis is a sub-field of the general areas of speech synthesis and computer facial animation (see [18], Chapter 12, organizes the representative work that has been done in this area). The goal of visible speech synthesis in the Perceptual Science Laboratory (PSL) has been to develop a polygon (wireframe) model with realistic motions (but not to duplicate the musculature of the face to control the animation). We call this technique terminal analogue synthesis because its goal is to simply duplicate the facial articulation of speech (rather than necessarily simulate the physiological mechanisms that produce it). House et al. (this volume), Pelachaud (this

[1] Corresponding Author: Dominic W. Massaro, Perceptual Science Laboratory, Department of Psychology, University of California, Santa Cruz, Santa Cruz, CA. 95064, USA, e-mail: massaro@fuzzy.ucsc.edu .

volume), and Bailly (this volume) take a similar approach. This terminal analogue approach has also proven successful with audible speech synthesis. Other alternatives to facial animation and synthesis are actively being pursued. The goal of muscle models is to simulate the muscle and tissues during talking. One disadvantage of muscle models is that the necessary calculations are more computationally intensive than terminal analogue synthesis. Our software can generate a talking face in real time on a commodity PC, whereas muscle and tissue simulations are usually too computationally intensive to perform in real time [18]. Another technique is performance-based synthesis that tracks a live talker, but it does not have the flexibility of saying anything at any time in real time, as does our system [13].

More recently, image synthesis, which joins together images of a real speaker, has been gaining in popularity because of the realism that it provides. Image based models extract parameters of real faces using computer vision techniques to animate visual speech from video or image sequences. This data-driven approach can offer a high degree of realism since real facial images form the basis of the model. A typical approach is to use models as key frames and use image warping to obtain the rest of a sequence. Cosatto et. al [5] segmented the face and head into several distinct parts and collected a library of representative samples. New sequences are created by stitching together the appropriate mouth shapes and corresponding facial features based on acoustic triphones and expression. Brooke (cited in [5]) used an eigenspace model of mouth variation and modeled visual speech using acoustic triphone units. Ezzat et. al [11, 12] represented each discernible mouth shape with a static image and used an optical flow algorithm to morph between the images. The Video Rewrite system [4] automatically segmented video sequences of a person talking and, given sufficient training data, could reanimate a sequence with different speech by blending images from the training sequence in the desired order.

In addition to using graphics based models, one can also use image based models for representing visual speech. For example, researchers have developed Face Translation, a translation agent for people who speak different languages [36, 41]. The system can not only translate a spoken utterance into another language, but also produce an audio-visual output with the speaker's face and synchronized lip movement in the target language. The visual output is synthesized from real images based on image morphing technology. Both mouth and eye movements are generated according to linguistic and social cues. Image based models have the potential to implicitly capture all of the visual speech cues that are important to human visual speech perception because they are based on real data. However, for some applications such as language learning and psychological experiment tasks, parameter control of the model is also required. In addition, image based models can only represent visible cues from the face. They cannot synthesize tongue movements that are very important for language training tasks.

Our entry into the development of computer-animated talking heads as a research tool occurred well before we apprehended the potential value of embodied conversational agents (ECAs) in commerce, entertainment, and education. Learning about the many influences on our language perception and performance made us sensitive to the important role of the face in speech perception [22]. We had always used the latest technology in our research, employing speech synthesis in our studies of speech perception [6]. Our first research studies manipulating information in the face and the voice had used videotapes of natural speakers dubbed with synthetic speech. We realized that to delve into what information in the face is important and how it is processed, we would have to control it in the same manner that we were able to control synthetic auditory speech. A computer-animated face that could be controlled as one controls a puppet by a set of strings was the obvious answer. We learned about the seminal dissertation of Fred Parke [34] in which he created and controlled a three-dimensional texture-mapped wire frame face. This model was then implemented in GL and C code on a graphics computer by Pearce et al. [35].

A sobering realization surfaced: where do humble psychologists find a $100,000 computer to create stimuli for experiments? Luckily, at the time, one of my cycling colleagues was a mathematician with access to a heavy-duty Silicon Graphics 3030. We were given access to during the wee hours of the morning, which fit perfectly with my former student and now research colleague, Michael Cohen. Even given this expensive machine, it took 3 minutes to synthesize a single frame, but we succeeded in creating the syllables /ba/ and /da/ as well as 7 intermediate syllable varying the degree of /ba/-ness and /da/-ness. We used these stimuli in several speech perception studies using an expanded factorial design, which independently varied the audible and visible speech [23]. Given the success of this approach, we looked for funding to develop the technology to create an accurate and realistic talking head. We were stymied in this effort, however, and we had to piggyback this research on other much more mundane but funded research. After about 6 years of grant proposals, we were finally given funds to buy our computer (whose price had now dropped to about $60,000) to carry out the necessary speech science to create an intelligible synthetic talker.

The outcome of this research was the development of Baldi®, a 3-D computer-animated talking head. (Baldi is a trademark of Dominic W. Massaro), and his progeny Timo is the animated tutor used by Animated Speech Corporation http://animatedspeech.com/. This technology evolved from one of the first developments of fruitfully combining speech science and facial animation research [18]. The technology that emerged from this research was new in the sense that the facial animation produced realistic speech in real time. Other existing uses had not exploited the important of visible speech and emotion from the face, and from viewing the speech articulators inside the mouth.

Baldi provides realistic visible speech that is almost as accurate as a natural speaker [9, 18]. Our animated face can be aligned with either the output of a speech synthesizer or natural auditory speech. Our modifications over the last 15 years have included increased resolution of the model, additional and modified control parameters, a realistic tongue, a model of coarticulation, paralinguistic information and affect in the face, alignment with natural speech, text-to-speech synthesis, and bimodal (auditory/visual) synthesis. Most of our current parameters move vertices (and the polygons formed from these vertices) on the face by geometric functions such as rotation (e.g., jaw rotation) or translation of the vertices in one or more dimensions (e.g., lower and upper lip height, mouth widening). Other parameters work by scaling and interpolating different face subareas. Many of the face shape parameters—such as cheek, neck, or forehead shape, and also some affect parameters such as smiling—use interpolation.

We use phonemes as the unit of speech synthesis. Any utterance can be represented as a string of successive phonemes, and each phoneme is represented as a set of target values for the control parameters. Controls for the visible speech on the outside of the face include jaw rotation, lower lip f-tuck, upper lip raising, lower lip roll, jaw thrust, cheek hollow, philtrum indent, lip zipping, lower lip raising, rounding, and retraction. Because speech production is a continuous process involving movements of different articulators (e.g., tongue, lips, jaw) having mass and inertia, phoneme utterances are influenced by the context in which they occur. This coarticulation is implemented in the synthesis by dominance functions, which determine independently for each control parameter how much weight its target value carries against those of neighbouring segments. We have also developed the phoneme set and the corresponding target and coarticualtion values to allow synthesis of several other languages. These include Spanish (Baldero), Italian (Baldini), Mandarin (Bao), Arabic (Badr), French (Balduin), German (Balthasar), Japanese (Bushi), and Russian (Balda). Baldi, and his various multilingual incarnations are seen at: http://mambo.ucsc.edu/psl/international.html.

Baldi has a tongue, hard palate and three-dimensional teeth and his internal articulatory movements have been trained with natural speech. Baldi's synthetic tongue is constructed of a polygon surface defined by sagittal and coronal b-spline curves. The control points of these b-spline curves are controlled singly and in pairs by speech articulation control parameters. There are now 9 sagittal and 3 x 7 coronal parameters that are modified to mimic natural tongue movements. Two sets of observations of real talkers have been used to inform the appropriate movements of the tongue. These include 1) three dimensional ultrasound measurements of upper tongue surfaces and 2) EPG data collected from a natural talker using a plastic palate insert that incorporates a grid of about a hundred electrodes that detect contact between the tongue and palate. Minimization and optimization routines are used to create animated tongue movements that mimic the observed tongue movements [8].

Although many of the subtle distinctions among speech segments are not visible on the outside of the face, the skin of our talking head can be made transparent so that the inside of the vocal tract is visible, or we can present a cutaway view of the head along the sagittal plane. As an example, a unique view of Baldi's internal articulators can be presented by rotating the exposed head and vocal tract to be oriented away from the student. It is possible that this back-of-head view would be much more conducive to learning language production. The tongue in this view moves away from and towards the student in the same way as the student's own tongue would move. This correspondence between views of the target and the student's articulators might facilitate speech production learning. One analogy is the way one might use a map: We often orient the map in the direction we are headed to make it easier to follow (e.g. turning right on the map is equivalent to turning right in reality).

Another characteristic of the training is to provide additional cues for visible speech perception. Baldi can not only illustrate the articulatory movements, he can be made even more informative by embellishing of the visible speech with supplementary features. These features can distinguish phonemes that have similar visible articulations; for example, the difference between voiced and voiceless segments can be indicated by vibrating the neck. Nasal sounds can be marked by making the nasal opening red, and turbulent airflow can be characterized by lines emanating from the mouth during articulation. These embellished speech cues make the face more informative than it normally is.

A central and somewhat unique quality of our work is the empirical evaluation of the visible speech synthesis, which is carried out hand-in-hand with its development. The quality and intelligibility of Baldi's visible speech has been repeatedly modified and evaluated to accurately simulate naturally talking humans [9, 10, 18]. The gold standard we use is how well Baldi compares to a real person. Given that viewing a natural face improves speech perception, we determine the extent to which Baldi provides a similar improvement. We repeatedly modify the control values of Baldi in order to meet this criterion. We modify some of the control values by hand and also use data from measurements of real people talking. For examples, see http://mambo.ucsc.edu/psl/international.html. This procedure has been used successfully in English [9, 10, 18], Arabic [32], and Italian [10], and we now have proposed an invariant metric to describe the degree to which the embodied conversational agent approximates a real talker [33].

1. Psychology

I began my scientific research career as an experimental psychologist. The terms Cognitive Psychologist or Cognitive Scientist were not yet invented. My research on fundamental information-processing mechanisms expanded to include the study of reading and speech perception [16]. Previous experiments in speech perception, for example, had manipulated only a single variable in their studies, whereas our empirical work manipulated multiple sources of both bottom-up sensory and top-down contextual information. Learning about the many influences on our perception and performance made me sensitive to the important role of the face in speech perception. Although no one doubted the importance of the face in social interactions, speech perception was studied as a strictly auditory phenomenon. To compensate for this limitation, we carried out an extensive series of experiments manipulating information in the face and the voice within our experimental/theoretical framework. Our studies were successful in unveiling how perceivers are capable to process multiple sources of information from these two modalities to optimize their performance [17].

The face presents visual information during speech that is critically important for effective communication. While the auditory signal alone is adequate for communication, visual information from movements of the lips, tongue and jaws enhance intelligibility of the acoustic stimulus (particularly in noisy environments). The number of recognized words from a degraded auditory message can often be doubled by pairing the message with visible speech. In a series of experiments, we asked college students to report the words of sentences presented in noise [13]. On some trials, only the auditory sentence was presented. On other trials, the auditory sentence was aligned with our synthetic talking head which facilitated performance for each of the 71 participants. Performance was more than doubled for those participants performing relatively poorly given auditory speech alone. Moreover, speech is enriched by the facial expressions, emotions and gestures produced by a speaker [18]. The visual components of speech offer a lifeline to those with severe or profound hearing loss. Even for individuals who hear well, these visible aspects of speech are especially important in noisy environments. For individuals with severe or profound hearing loss, understanding visible speech can make the difference between effectively communicating orally with others and a life of relative isolation from oral society [37].

Empirical findings also show that speech reading is robust. Research has shown that perceivers are fairly good at speech reading even when they are not looking directly at the talker's lips. Furthermore, accuracy is not dramatically reduced when the facial image is blurred (because of poor vision, for example), when the face is viewed from above, below, or in profile, or when there is a large distance between the talker and the viewer [14, 18, 30]. These findings indicate that speech-reading is highly functional in a variety of non-optimal situations. Another example of the robustness of the influence of visible speech is that people naturally integrate visible speech with audible speech even when the temporal occurrence of the two sources is displaced by about a 1/5 of a second [24]. Given that light and sound travel at different speeds and that the dynamics of their corresponding sensory systems also differ, a cross-modal integration must be relatively immune to small temporal asynchronies [18]. This insensitivity to asynchrony can also be advantageous for transmitting bimodal speech across communication channels that cannot keep the two modalities in perfect synchrony.

Complementarity of auditory and visual information simply means that one of the sources is the most informative in those cases in which the other is weakest. Because of this, a speech distinction may be differentially supported by the two sources of information. That is, two segments that are robustly conveyed in one modality may be relatively ambiguous in the other modality. For example, the difference between /ba/ and /da/ is easy to see but relatively difficult to hear. On the other hand, the difference between /ba/ and /pa/ is relatively easy to hear but very difficult to discriminate visually. The fact that two sources of information are complementary makes their combined use much more informative than would be the case if the two sources were non-complementary, or redundant ([18], Chapter 14).

The final characteristic is that perceivers combine or integrate the auditory and visual sources of information in an optimally efficient manner [17, 27]. There are many possible ways to treat two sources of information: use only the most informative source, average the two sources together, or integrate them in such a fashion in which both sources are used but that the least ambiguous source has the most influence. Perceivers in fact integrate the information available from each modality to perform as efficiently as possible. One might question why perceivers integrate several sources of information when just one of them might be sufficient. Most of us do reasonably well in communicating over the telephone, for example. Part of the answer might be grounded in our ontogeny. Integration might be so natural for adults even when information from just one sense would be sufficient because, during development, there was much less information from each sense and therefore integration was all the more critical for accurate performance [15].

The Fuzzy Logical Model of Perception (FLMP), developed with Gregg Oden, has been shown to give the best description of this optimal integration process [31]. The FLMP addresses research questions asked by psycholinguists and speech and reading scientists. These questions include the nature of the sources of information; how each source is evaluated and represented; how the multiple sources are treated; whether or not the sources are integrated; the nature of the integration process; how decisions are made; and the time course of processing. Research in a variety of domains and tasks supports the predictions of the FLMP (for summary see [18]) that a) perceivers have continuous rather than categorical information from each of these sources; b) each source is evaluated with respect to the degree of support for each meaningful alternative; c) each source is treated independently of other sources; d) the sources are integrated to give an overall degree of support for each alternative; e) decisions are made with respect to the relative goodness of match among the viable alternatives; f) evaluation; integration; and decision are necessarily successive but overlapping stages of processing; and g) cross-talk among the sources of information is minimal. The fuzzy logical model of perception ([18] FLMP), which embodies these properties, is illustrated in Figure 1. One familiar example of language processing consistent with this model is the tendency to focus on the lips of the speaker when the listening conditions become difficult. Benjamin Franklin described how, during his tenure as Ambassador to France, the bifocal spectacles he invented permitted him to look down at his cuisine on the table and to look up at his neighbouring French dinner guest so he could adequately understand the conversation in French.

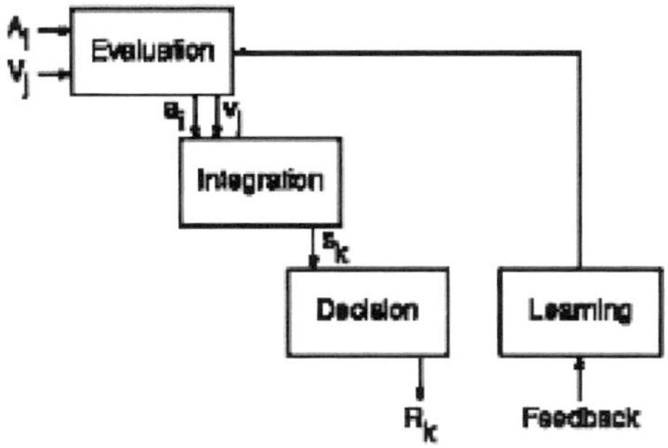

Figure 1. Schematic representation of the FLMP. The sources of information are represented by uppercase letters. Auditory information is represented by Ai and visual information by Vj. The evaluation process transforms these sources of information into psychological values ai and vj. These sources are then integrated to give an overall degree of support, sk, for each speech alternative k. The decision operation maps the outputs of integration into some response alternative, Rk. The response can take the form of a discrete decision or a continuous rating. The learning process is also included in the figure. Feedback at the learning stage is assumed to tune the psychological values of the sources of information used by the evaluation process.

Consistent with the FLMP, there is a large body of evidence that bimodal speech perception given the voice and the face is more accurate than unimodal perception given either one of these alone [18]. This result is found for syllables in isolation, as well as for words and sentences. The FLMP has also been able to account for a range of contextual constraints in both written and spoken language processing. Normally, we think of context effects occurring at a conscious and deliberate cognitive level but in fact context effects have been found at even the earliest stages of linguistic processing. The FLMP is also mathematically equivalent to Bayes theorem, an optimal method for combining several sources of evidence to infer a particular outcome. The fact that the model appears to be optimal and the empirical support from a plethora of empirical/theoretical studies provide an encouraging framework of the study of verbal and nonverbal communication [29].

2. Pedagogy

Since these early findings on the value of visible speech and the potential of animated tutors, our persistent goal has been to develop, evaluate, and apply animated agents to produce accurate visible speech, facilitate face-to-face oral communication, and to serve people with language challenges. These enhanced characters can function effectively as language tutors, reading tutors, or personal agents in human machine interaction.

Need For Language Tutoring and Computer-Animated Tutors

The need for language tutoring is pervasive in today's world. There are millions of pre-school and elementary school children who have language and speech disabilities, and these individuals require additional instruction in language learning. These include speech or language impaired children age 6-11 who were served in 2003 under the Individuals with Disabilities Education Act (U.S. Department of Education, 2003), an estimated 2.5 million elementary students who receive ELL services (U.S. Department of Education, May, 2003) and 6.6 million with low family socioeconomic status (SES) [28]. Hearing loss, mental retardation, physical impairments such as cleft lip or cleft palate, vocal abuse, autism, and brain injury are all factors that may lead to speech and language disorders (US Department of Health and Human Services, 2001, American Psychiatric Association, 1994).

These individuals require additional instruction in language learning. Currently, however, the important needs of these individuals are not being met. One significant problem faced by the people with these disabilities is the shortage of skilled teachers and professionals to give them the one on one attention that they need. While other resources, such as books or other media, may help alleviate this problem, these traditional interventions are not easily personalized to the students' needs, lack the engaging capability of a teacher, and are relatively ineffective.

Several advantages of utilizing a computer-animated agent as a language tutor are clear, including the popularity of computers and embodied conversational agents. An incentive to employing computer-controlled applications for training is the ease in which automated practice, feedback, and branching can be programmed. Another valuable component is the potential to present multiple sources of information, such as text, sound, and images in parallel. Instruction is always available to the child, 24 hours a day, 365 days a year. We have found that the students enjoy working with Baldi because he offers extreme patience, he doesn't become angry, tired, or bored, and he is in effect a perpetual teaching machine. Baldi has been successful in engaging foreign language and ESL learners, hearing-impaired, autistic and other children with special needs in face-to-face computerized lessons.

Vocabulary Learning

Vocabulary knowledge is critical for understanding the world and for language competence in both spoken language and in reading. There is empirical evidence that very young children more easily form conceptual categories when category labels are available than when they are not. Even children experiencing language delays because of specific language impairment benefit once this level of word knowledge is obtained. It is also well-known that vocabulary knowledge is positively correlated with both listening and reading comprehension [2]. It follows that increasing the pervasiveness and effectiveness of vocabulary learning offers a timely opportunity for improving conceptual knowledge and language competence for all individuals, whether or not they are disadvantaged because of sensory limitations, learning disabilities, or social condition.

Learning and retention are positively correlated with the amount of time devoted to learning. Our technology offers a platform for unlimited instruction, which can be initiated whenever and wherever the child and/or mentor chooses. Instruction can be tailored exactly to the student's need, which is best implemented in a one-on-one learning environment for the students. Other benefits of our program include the ability to seamlessly meld spoken and written language, and provide a semblance of a game-playing experience while actually learning. Given that education research has shown that children can be taught new word meanings by using direct instruction methods, we implemented these basic features in an application to teach vocabulary and grammar.

We have applied and evaluated the use of Baldi to teach vocabulary to children with language challenges. Vocabulary knowledge is critical for understanding the world and for language competence in both spoken language and in reading. Vocabulary knowledge is positively correlated with both listening and reading comprehension. It follows that increasing the pervasiveness and effectiveness of vocabulary learning offers a timely opportunity for improving conceptual knowledge and language competence for all individuals, whether or not they are disadvantaged because of sensory limitations, learning disabilities, or social condition.

Instruction can be tailored exactly to the student's need, which is best implemented in a one-on-one learning environment for the students. Other benefits of our program include the ability to seamlessly meld spoken and written language, and provide a semblance of a game-playing experience while actually learning. Given that education research has shown that children can be taught new word meanings by using direct instruction methods, we implemented these basic features in an application to teach vocabulary and grammar.

One of the principles of learning that we exploit most is the value of multiple sources of information in perception, recognition, learning, and retention. An interactive multimedia environment is ideally suited for learning. Incorporating text and visual images of the vocabulary to be learned along with the actual definitions and sound of the vocabulary facilitates learning and improves memory for the target vocabulary and grammar. Many aspects of these lessons enhance and reinforce learning. For example, the existing commercially available program, Timo Vocabulary [http://www.animatedspeech.com], which is derived directly from the Baldi technology, makes it possible for the students to 1) Observe the words being spoken by a realistic talking interlocutor, 2) Experience the word as spoken as well as written, 3) See visual images of referents of the words, 4) Click on or point to the referent or its spelling, 5) Hear themselves say the word, followed by a correct pronunciation, and 6) Spell the word by typing, and 7) Observe and respond to the word used in context.

To test the effectiveness of vocabulary instruction using an embodied conversational agent as the instructor, we developed a series of lessons that encompass and instantiate the developments in the pedagogy of how language is learned, remembered, and used. It is well known that children with hearing loss have significant deficits in both spoken and written vocabulary knowledge. To assess the learning of new vocabulary, we carried out an experiment based on a within-student multiple baseline design where certain words were continuously being tested while other words were being tested and trained [25]. Although the student's instructors and speech therapists agreed not to teach or use these words during our investigation, it is still possible that the words could be learned outside of the learning context. The single student multiple baseline design monitors this possibility by providing a continuous measure of the knowledge of words that are not being trained. Thus, any significant differences in performance on the trained words and untrained words can be attributed to the training program itself rather than some other factor.

We studied eight children with hearing loss, who needed help with their vocabulary building skills as suggested by their regular day teachers. The experimenter developed a set of lessons with a collection of vocabulary items that was individually composed for each student. Each collection of items was comprised of 24 items, broken down into 3 categories of 8 items each. Three lessons with 8 items each were made for each child. Images of the vocabulary items were presented on the screen next to Baldi as he spoke. Assessment was carried out on all of the items at the beginning of each lesson. It included identifying and producing the vocabulary item without feedback. Training on the appropriate word set followed this testing.

As expected, identification accuracy was always higher than production accuracy. This result is expected because a student would have to know the name of an item to pronounce it correctly. There was little knowledge of the test items without training, even though these items were repeatedly tested for many days. Once training began on a set of items, performance improved fairly quickly until asymptotic knowledge was obtained. This knowledge did not degrade after training on these words ended and training on other words took place. In addition, a reassessment test given about 4 weeks after completion of the experiment revealed that the students retained the items that were learned.

The tutoring application has also been used in evaluating vocabulary acquisition, retention and generalization in children with autism. Although the etiology of autism is not known, individuals diagnosed with autism must exhibit a) delayed or deviant language and communication, b) impaired social and reciprocal social interactions, and 3) restricted interests and repetitive behaviours. The language and communicative deficits are particularly salient, with large individual variations in the degree to which autistic children develop the fundamental lexical, semantic, syntactic, phonological, and pragmatic components of language. Vocabulary lessons were constructed, consisting of over 84 unique lessons with vocabulary items selected from the curriculum of two schools. The participants were eight children diagnosed with autism, ranging in age from 7 to 11 years [3].

The results indicated that the children learned many new words, grammatical constructions and concepts, proving that the application provided a valuable learning environment for these children. In addition, a delayed test given more than 30 days after the learning sessions took place showed that the children retained over 85% of the words that they learned. This learning and retention of new vocabulary, grammar, and language use is a significant accomplishment for autistic children (see also [40]).

Although all of the children demonstrated learning from initial assessment to final reassessment, it is possible that the children were learning the words outside of our learning program (for example, from speech therapists or in their school curriculum). Furthermore, it is important to know whether the vocabulary knowledge would generalize to new pictorial instances of the words. To address these questions, a second investigation used the single subject multiple probe design. Once a student achieved 100% correct, generalization tests and training were carried out with novel images. The placement of the images relative to one another was also random in each lesson. Assessment and training continued until the student was able to accurately identify at least 5 out of 6 vocabulary items across four unique sets of images. Although performance varied dramatically across the children and across the word sets during the pre-training sessions, training was effective for all words sets for all children. Given training, all of the students attained our criterion for identification accuracy for each word set and were also able to generalize accurate identification to four instances of untrained images. The students identified significantly more words following implementation of training compared to pre-training performance, showing that the program was responsible for learning. Learning also generalized to new images in random locations, and to new interactions outside of the lesson environment. These results show that our learning program is effective for children with autism, as it is for children with hearing loss.

We have several sources of evidence that application of this technology is making a contribution. One research study [20] was carried out with English Language Learners (ELL) and involved the use of our recently-released application, Timo Vocabulary (http://www.animatedspeech.com), which instantiated the pedagogy we found in our earlier research. Nine children ranging in age from 6-7 years were tested in the summer before first grade. Almost all of the children spoke Spanish in the home. The children were pre-tested on lessons in the application in order to find three lessons with vocabulary that was unknown to the children. A session on a given day included a series of three test lessons, and on training days, a training lesson on one of the three sets of words. Different lessons were necessarily chosen for the different children because of their differences in vocabulary knowledge. The test session involved the presentation of the images of a given lesson on the screen with Timo's request to click on one of the items, e.g., Please click on the oven. No feedback was given to the child. Each item was tested once in two separate blocks to give 2 observations on each item. Three different lessons were tested, corresponding to the three sets of items used in the multiple baseline design.

A training session on a given day consisted of just a single lesson in which the child was now given feedback on their response. Thus, if Timo requested the child to click on the dishwasher and the child clicked on the spice rack, Timo would say, "I asked for the dishwasher, you clicked on the spice rack. This is the dishwasher. The training session also included the Elicitation section in which the child was asked to repeat the word when it was highlighted and Timo said it, and the Imitation section in which the child was asked to say the item that was highlighted. Several days of pre-testing were required to find lessons with unknown vocabulary. Once the 3 lessons were determined, the pre-testing period was followed by the training days. The results indicated that training was effective in teaching new vocabulary to these English Language Learners.

We were gratified to learn that the same application could be used successfully with both autistic children and children with hearing loss, as well as children learning English as a new language. Specific interactions can be easily modified to accommodate group and individual differences. For example, autistic children are much more disrupted by negative feedback, and the lesson can be easily designed to instantiate errorless learning. Thus, the effectiveness and usability of our technology and pedagogy has been demonstrated with both hard of hearing and autistic children, and these results have meet the most stringent criterion of publication in highly regarded peer-reviewed refereed journals.

We appreciate the caveat that testimonials are not sufficient to demonstrate the effectiveness of an application or product: however, they are important to supplement controlled experimental evaluations. With caveat in mind, the testimonials we have received have illuminated some reasons for the effectiveness of our pedagogy. http://www.animatedspeech.com/Products/products_lessoncreator_testimonials.html

In a few instances, individuals reacted negatively to the use of synthetic auditory speech in our applications. Not only did they claim it sounded relatively robotic, they were worried that children may learn incorrect pronunciation or intonation patterns from this speech. However, we have learned that this worry is unnecessary. We have been using our applications for over 7 years at the Tucker-Maxon School of Oral Education (http://www.tmos.org/), and all of the children have succeeded or are succeeding in acquiring appropriate speech. In addition, the application has been successful in teaching correct pronunciation [25, 26].

The research and applications could serve as a model for others to evaluate. The intricate concern for innovation, effectiveness evaluation, and application value serves as an ideal model for technological, personal, and societal evolution. The original technology was initially developed for use in experimental and theoretical research but the potential for valuable applications became quickly transparent. There was a close relationship between research, evaluation, and application throughout the course of the work.

Although our computer-animated tutor, Baldi, has been successful in teaching vocabulary and grammar to hard of hearing and to autistic children, it is important to know to what extent the face facilitated this learning process relative to the voice alone. To assess this question, Massaro and Bosseler [21] created vocabulary lessons involving the association of pictures and spoken words. The lesson plan included both the receptive identification of pictures and the production of spoken words. A within-subject design with five autistic children followed an alternate treatment design in which each child continuously learned to criterion two sets of words with the face and two sets without the face. The rate of learning was significantly faster and the retention was better with than without the face. The research indicates that at least some autistic children benefit from the face in learning new language within an automated program centered on an embodied conversational agent, multimedia, and active participation.

In addition to Timo Vocabulary, teachers, parents, and even students can build original lessons that meet unique and specialized conditions. New lessons are made within a Lesson Creator application that allows personalized vocabulary and pictures to be easily integrated. This user-friendly application allows the composition of lessons with minimal computer experience and instruction. Although it has many options, the program has wizard-like features that direct the coach to explore and choose among the alternative implementations in the creation of a lesson.

The current application includes a curriculum of thousands of vocabulary words and images, and can be implemented to teach both individual vocabulary words and meta-cognitive awareness of word categorization and generalized usage. This application will facilitate the specialization and individualization of vocabulary and grammar lessons by allowing teachers to create customized vocabulary lists from words already in the system or with new words. If a teacher is taking her class on a field trip to the local Aquarium, for example, she will be able to create lessons about the marine animals the children will see at the museum. A parent could prepare lessons with words in the child's current reading, or names of her relatives, schoolmates, and teachers. Lessons can also be easily created for the child's current interests. Most importantly, given that vocabulary is essentially infinite in number, it is most efficient to instruct vocabulary just in time as it is needed.

Tutoring Speech Production

As noted earlier, Baldi can illustrate how articulation occurs inside the mouth, as well as provide supplementary characteristics about the speech segment. To evaluate the effectiveness of this information, Massaro and Light [26] tested seven students with hearing loss (2 male and 5 female). Children with hearing loss require guided instruction in speech perception and production. Some of the distinctions in spoken language cannot be heard with degraded hearing--even when the hearing loss has been compensated by hearing aids or cochlear implants. The students were trained to discriminate minimal pairs of words bimodally (auditorily and visually), and were also trained to produce various speech segments by observing visual information about how the inside oral articulators work during speech production. The articulators were displayed from different vantage points so that the subtleties of articulation could be optimally visualized. The speech was also slowed down significantly to emphasize and elongate the target phonemes, allowing for clearer understanding of how the target segment is produced in isolation or with other segments.

The students' ability to perceive and produce words involving the trained segments improved from pre-test to post-test. Intelligibility ratings of the post-test productions were significantly higher than pre-test productions, indicating significant learning. It is always possible that some of this learning occurred independently of our program or was simply based on routine practice. To test this possibility, we assessed the students' productions six weeks after training was completed. Although these productions were still rated as more intelligible than the pre-test productions, they were significantly lower than post-test ratings, indicating some decrement due to lack of continued use. This is evidence that at least some of the improvement must be due to our program. Although there were individual differences in aided hearing thresholds, attitude, and cognitive level, the training program helped all of the children.

The application could be used in new contexts. It is well-known that a number of children encounter in learning to read and spell. Dyslexia is a category used to pigeonhole children who have much more difficulty in reading and spelling than would be expected from their other perceptual and cognitive abilities. Psychological science has established a tight relationship between the mastery of written language and the child's ability to process spoken language. That is, it appears that many dyslexic children also have deficits in spoken language perception. The difficulty with spoken language can be alleviated through improving children's perception of phonological distinctions and transitions, which in turn improves their ability to read and spell. As mentioned in the previous section, the internal articulatory structures of Baldi can be used to pedagogically illustrate correct articulation. The goal is to instruct the child by revealing the appropriate movements of the tongue relative to the hard palate and teeth. Given existing 2D lessons such as the LiPs program by Lindamood Bell (http://www.ganderpublishing.com/lindamoodphonemesequencing.htm) and exploratory research, we expect that these 3D views would accelerate the acquisition of phonological awareness and learning to read.

References

[1] American Psychiatric Association. Diagnostic and Statistical Manual of Mental Disorders, DSM-IV (4th ed.). Washington, DC, 1994.

[2] I.L. Beck, M.G. McKeown, L. Kucan: Bringing Words to Life: Robust Vocabulary Instruction. New York: The Guilford Press., 2002.

[3] A. Bosseler, D.W. Massaro: Development and Evaluation of a Computer-Animated Tutor for Vocabulary and Language Learning for Children with Autism. *Journal of Autism and Developmental Disorders*, 33, 653-672, 2003.

[4] C. Bregler, S.M Omohundro: Learning Visual Models for Lipreading. In M. Shah, R. Jain (Eds.). *Motion-Based Recognition*, Volume 9 of Computational Imaging and Vision, Kluwer Academic, 301-320, 1997.

[5] E. Cosatto, H. Graf: Sample-Based Synthesis of Photo-Realistic Talking Heads. In Proceedings of *Computer Animation*, 103–110, 1998.

[6] M.M. Cohen, D.W. Massaro: Real-Time Speech Synthesis. *Behavior Research Methods and Instrumentation*, 8, 189-196, 1976.

[7] M.M. Cohen, D.W. Massaro. Synthesis of Visible Speech. *Behavior Research Methods, Instruments,& Computers*, 22, 260-263, 1990.

[8] M.M. Cohen, J. Besko, D.W. Massaro: Recent Developments In Facial Animation: An Inside View. In ETRW on *Auditory-Visual Speech Processing*. Terrigal-Sydney, Australia, 201-206, 1998.

[9] M.M. Cohen, D.W. Massaro, R. Clark: Training a Talking Head. In D.C. Martin (Ed.), Proceedings of the IEEE Fourth International Conference on *Multimodal Interfaces* (ICMI'02), Pittsburgh, PA, 499-510, 2002.

[10] P. Cosi, M.M. Cohen, D.W. Massaro: Baldini: Baldi Speaks Italian. In Proceedings of 7th International Conference on Spoken Language Processing, Denver, CO, 2349-2352, 2002.

[11] T. Ezzat, T. Poggio. MikeTalk: A Talking Facial Display Based on Morphing Visemes. In Proceedings of the *Computer Animation* Conference. Philadelphia, PA, June, 1998.

[12] T. Ezzat, G. Geiger, T. Poggio: Trainable Videorealistic Speech Animation. In Proceedings of ACM SIGGRAPH, 388-398, 2002.

[13] A. Jesse, N. Vrignaud, D.W. Massaro: The Processing of Information from Multiple Sources in Simultaneous Interpreting. *Interpreting* 5, 95-115, 2000.

[14] T. Jordan, P. Sergeant : Effects of distance on Visual and Audiovisual Speech Recognition. *Language and Speech*, 43, 107-124, 2000.

[15] D.J.Lewkowicz, K.S. Kraebel. The Value of Multisensory Redundancy in the Development of Intersensory Perception. In G. Calvert, C. Spence, B. E. Stein (Eds.), Handbook of Multisensory Processes, Cambridge, MA: MIT Press, 655-678, 2004.

[16] D.W. Massaro: Acoustic Features in Speech Perception. In D.W. Massaro (Ed.), *Understanding Language: An Information Processing Analysis of Speech Perception, Reading, and Psycholinguistics*, New York: Academic Press, 77-124, 1975.

[17] D.W. Massaro: *Speech Perception by Ear and Eye: A Paradigm for Psychological Inquiry*. Hillsdale, N.J.: Lawrence Erlbaum Associates, 1987.

[18] D.W. Massaro*: Perceiving Talking Faces: From Speech Perception to a Behavioral Principle*. MIT Press: Cambridge, MA, 1998.

[19] D.W. Massaro: Symbiotic Value of an Embodied Agent in Language Learning. In R.H. Sprague, Jr.(Ed.), Proceedings of 37th Annual Hawaii International Conference on System Sciences,(HICCS'04) (CD-ROM,10 pages). Los Alimitos, CA: IEEE Computer Society Press. Best paper in Emerging Technologies, 2004.

[20] D.W. Massaro. Embodied Agents in Language Learning for Children with Language Challenges. In K. Miesenberger, J. Klaus, W. Zagler, & A. Karshmer, (Eds.), Proceedings of the 10th International Conference on Computers Helping People with Special Needs, ICCHP 2006 (pp.809-816). University of Linz, Austria. Berlin, Germany: Springer, (2006).

[21] D.W. Massaro, A. Bosseler. Read My Lips: The Importance of the Face in a Computer-Animated Tutor for Autistic Children Learning Language. *Autism: The International Journal of Research and Practice*,10(5),495-510, 2006.

[22] D.W. Massaro, M.M. Cohen: Evaluation and Integration of Visual and Auditory Information in Speech Perception. *Journal of Experimental Psychology: Human Perception and Performance*, 9(5), 753-771, 1983.

[23] D.W. Massaro, M.M. Cohen: Perception of Synthesized Audible and Visible Speech. *Psychological Science*, 1, 55-63, 1990.

[24] D.W. Massaro, M.M. Cohen: Perceiving Asynchronous Bimodal Speech in Consonant-Vowel and Vowel Syllables. *Speech Communication*, 13, 127-134, 1993.

[25] D.W. Massaro, J. Light; Improving the Vocabulary of Children with Hearing Loss. *Volta Review*, 104(3), 141-174, 2004.

[26] D.W. Massaro, J. Light: Using Visible Speech for Training Perception and Production of Speech for Hard of Hearing Individuals. *Journal of Speech, Language, and Hearing Research*, 47(2), 304-320, 2004.
[27] D.W. Massaro, D.G. Stork: Sensory Integration and Speechreading by Humans and Machines. *American Scientist*, 86, 236-244, 1998.
[28] D. Meyer, D. Princiotta, L. Lanahan: The Summer After Kindergarten: Children's Activities and Library Use by Household Socioeconomic Status, National Center for Education Statistics, Institute of Education Sciences, U.S. Dept. of Education, Education Statistics Quarterly, 6:3, 2004.
[29] J. Movellan & J.L. McClelland: The Morton-Massaro Law of Information Integration: Implications for Models of Perception. *Psychological Review*, 108, 113-148, 2001.
[30] K. Munhall & E. Vatikiotis-Bateson. Spatial and Temporal Constraints on Audiovisual Speech Perception. In G. Calvert, C. Spence, B. E. Stein (Eds.), *Handbook of Multisensory Processes*, Cambridge, MA: MIT Press, 177-188, 2004.
[31] G.C. Oden, D.W. Massaro. Integration of Featural Information in Speech Perception. *Psychological Review*, 85, 172-191, 1978.
[32] S. Ouni, M.M. Cohen, D.W. Massaro: Training Baldi to be Multilingual: A Case Study for an Arabic Badr. *Speech Communication, 45(2),* 115-137, 2005.
[33] S. Ouni, M.M. Cohen, H. Ishak, D.W. Massaro: Visual Contribution to Speech Perception: Measuring the Intelligibility of Animated Talking Heads. *EURASIP Journal on Audio, Speech, and Music Processing*, 1, 1-13, 2006.
[34] F.I. Parke: A Model for Human Faces That Allows Speech Synchronized Animation. *Computers and Graphics Journal*, 1, 1-4, 1975.
[35] A. Pearce, B. Wyvill, G. Wyvill, D. Hill: Speech and Expression: A Computer Solution to Face Animation. In Proceedings of Graphics Interface '86, 1986.
[36] M. Ritter, U. Meier, J. Yang, A. Waibel: Face Translation: A Multimodal Translation Agent. In Proceedings of AVSP 99, 1999.
[37] S. Trychin.: Coping with Hearing Loss. In R. Cherry, T. Giolas (Eds), *Seminars in Hearing*, vol 18, #2, New York: Thieme Medical Publishers, Inc., 1997.
[38] U.S. Department of Education. Table AA9, Number of Children Served Under IDEA by Disability and Age Group, 1994 Through 2003, Office of Special Education Programs, Washington, DC. http://www.ideadata.org/tables2aa94th/ar_.htm, 2003.
[39] U.S. Department of Education. Table 10. Number and percentage of public school students participating in selected programs, by state: School year 2001–02, "Public Elementary/Secondary School Universe Survey," 2001–02 and "Local Education Agency Universe Survey," 2001–02, National Center for Education Statistics, Institute of Education Sciences, Washington, DC, 2003.
[40] J.H.G. Williams, D.W. Massaro, N.J. Peel, A. Bosseler, T. Suddendorf: Visual-Auditory Integration During Speech Imitation in Autism. *Research in Developmental Disabilities*, 25, 559-575, 2004.
[41] J. Yang, J. Xiao, M. Ritter: Automatic Selection of Visemes for Image-based Visual Speech Synthesis. In Proceedings of First IEEE International Conference on Multimedia IEEE ME2000, 2000.

A. Esposito et al. (Eds.)
IOS Press, 2007

Virtual Talking Heads and Ambiant Face-to-Face Communication

Gérard BAILLY[1], Frédéric ELISEI, Stephan RAIDT
Institut de la Communication Parlée, Grenoble - France

Abstract. We describe here our first effort for developing a virtual talking head able to engage a situated face-to-face interaction with a human partner. This paper concentrates on the low-level components of this interaction loop and the cognitive impact of the implementation of mutual attention and multimodal deixis on the communication task.

Keywords: Talking heads, virtual conversational agents, audiovisual speech synthesis, face-to-face interaction.

Introduction

Building Embodied Conversational Agents (ECA) able to engage a convincing face-to-face conversation with a human partner is certainly one of the more challenging Turing test one can imagine [1]. The challenge is far more complex than the experimental conditions of the Loebner Prize[2] where dialog is conducted via textual information and of the perception test conducted by Ezzat et al [2] where a non interactive ECA was evaluated. Features of situated face-to-face communication including mixed initiative, back channeling, sense of presence, rules for turn taking should be implemented. The interaction loop should not only rely on a convincing animation but also requires a detailed scene analysis: the analysis and comprehension of an embodied interaction is deeply grounded in our senses and actuators and we do have strong expectations on how dialogic information is encoded into multimodal signals.

Appropriate interaction loops have to be implemented. They have to synchronize at least two different perception/action loops. On the one hand there are low-frequency dialogic loops. They require analysis, comprehension and synthesis of dialog acts with time scales of the order of a few utterances. On the other hand there are interaction loops of higher frequency. These include the prompt reactions to the scene analysis such as involved in eye contact, or exogenous saccades. The YTTM model [3] of turn-taking possesses three layered feedback loops (reactive, process control and content). The intermediate process control loop is responsible for the willful control of the social interaction (starts and stops, breaks, back-channeling, etc). In all interaction models,

[1] Corresponding Author:: Institut de la Communication Parlée, 46 av. Félix Viallet, 38031 Grenoble – France, e-mail : bailly@icp.inpg.fr.

[2] The Loebner Prize for artificial intelligence wards each year the computer programs that delivers the most human-like responses to questions given by a panel of judges over a computer terminal.

information- and signal-driven interactions should then be coupled to guarantee efficiency, believability, trustfulness and user-friendliness of the information retrieval.

The work described here is dedicated to the analysis, modeling and control of multimodal face-to-face interaction between a virtual ECA and a user. We particularly study here the impact of facial deictic gestures of the ECA on user performance in simple search and retrieval tasks.

1. Eye gaze and human-computer interaction

1.1. Face-to-face interaction, attention and deixis

Eye gaze is an essential component of face-to-face interaction. Eyes constitute a very special stimulus in a visual scene. Gaze and eye-contact are important cues for the development of social activity and speech acquisition [4]. In conversation it is involved in the regulation of turn taking, accentuation and organization of discourse [5, 6]. We are also very sensitive to the gaze of others when directed towards objects of interest within our field of view or even outside [7]. In the Posner cueing paradigm [8, 9], observers' performance in detecting a target is typically better in trials in which the target is present at the location indicated by a former visual cue than in trials in which the target appears at the uncued location. The outstanding prominence of the human face in this respect was shown by Langton et al. [10, 11], who have shown that observers react more quickly when the cue is an oriented face than when it is an arrow. Driver et al. [12] have shown that a concomitant eye gaze also speeds reaction time.

Eye gaze is thus capable of attracting visual attention whereas visual features associated with the objects themselves such as highlighting or blinking are not given so much attention, unless they convey important information for the recognition of a scene. As an example the work of Simons and Chabris [13] suggests that attention is essential to consciously perceive any aspect of a scene. Major changes to objects or scenes may be ignored ('change blindness') and objects may not even be perceived ('inattentional blindness') if they are not in our focus of attention. Perceptual salience is thus not the only determinant of interest. The cognitive demand of a task has a striking impact on the human audiovisual analysis of scenes and their interpretation. Yarbus [14] showed notably that eye gaze patterns are influenced by the instructions given to the observer during the examination of pictures. Similarly Vatikiotis-Bateson et al [15] showed that eye gaze patterns of perceivers during audiovisual speech perception are influenced both by environmental conditions (audio signal-to-noise ratio) and by the recognition task (identification of phonetic segments vs. the sentence's modality).

1.2. Interacting with humanoids and avatars

The faculty of interpreting eye gaze patterns of the others is thus crucial for humans and machines interacting with humans. For the "theory of mind" (TOM) as described by Baron-Cohen [16], the perception of gaze direction is an important element of the set of abilities that allow an individual, on the basis of the observation of his behavior, to infer the hidden mental states of another. Several TOM have been proposed [17, 18]. Baron-Cohen proposes an Eye Direction Detector (EDD) and an Intentionality Detector (ID) as basic components of a Shared Attention Mechanism (SAM) that is essential to the TOM's bootstrap. The actual implementation of these modules requires the coordination of a large number of perceptual, sensorimotor, attentional, and cognitive processes.

Scassellati [19] applied the "theory of mind" concept to humanoid robots developing an "embodied theory of mind" to link high-level cognitive skills to the low-level motor and perceptual abilities of such a robot. The low-level motor abilities comprised coordinated eye, head and arm movements for pointing. The low-level perceptual abilities comprised essentially detection of salient textures and motion for monitoring pointing and visual attention. This work still inspires a lot of works on humanoid robots where complex behaviors emerge from interaction with the environment and users despite the simple tasks to be fulfilled by the robot such as expressing empathy for Kismet [20] or following turn-taking for Robita [21, 22].

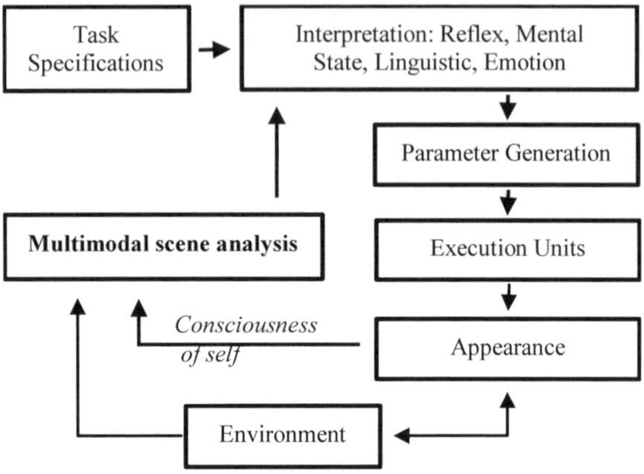

Figure 1: ECA-Human interaction scheme

2. Interacting with an ECA

Most ECAs derive their "theory of mind" from high-level linguistic information gathered during the dialog. These virtual agents are generally not equipped with the means of deriving meaning from the implicit and explicit communicational gestures of a human interlocutor and are also not generally equipped to generate such gestures for communication purposes. Although no real evaluation has been performed, ECA eye gaze can be generated without grounding these gestures in the scene by simply reproducing statistical properties of saccadic eye movements [23]. Note however that Itti et al [24] propose a model that couples physical scene analysis and control of eye gaze of a virtual ECA while preserving cognitive permeability of the analysis strategy thanks to use of a so-called pertinence map.

In situations where context-aware face-to-face interaction is possible, an ECA should be able to give direct and indirect signs that it actually knows about *where* the interaction is taking place, *who* is its interlocutor and *what* service it may provide to the user considering the given environment. By signalling its ability to interpret human behavior, the system encourages the interlocutor to show the appropriate natural activity. Such a complex face-to-face interaction requires intensive collaboration between an elaborate scene analysis and the specification of the task to be performed in order to generate appropriate and convincing actions of the ECA (see Figure 1).

Our perspective is to develop an embodied TOM for an ECA that will link high-level cognitive skills to the low-level motor and perceptual abilities and to demonstrate that such a TOM will provide the information system with enhanced user satisfaction, efficient and robust interaction. The motor abilities are principally extended towards speech communication i.e. adapting content and speech style to pragmatic needs (e.g. confidentiality), speaker (notably age and possible communication handicaps) and environmental conditions (e.g. noise). If the use of a virtual talking head instead of a humanoid robot limits physical actions, it extends the domain of interaction to the virtual world. The user and the ECA can thus involve both physical and virtual objects - such as icons surrounding the virtual talking head – in their interaction.

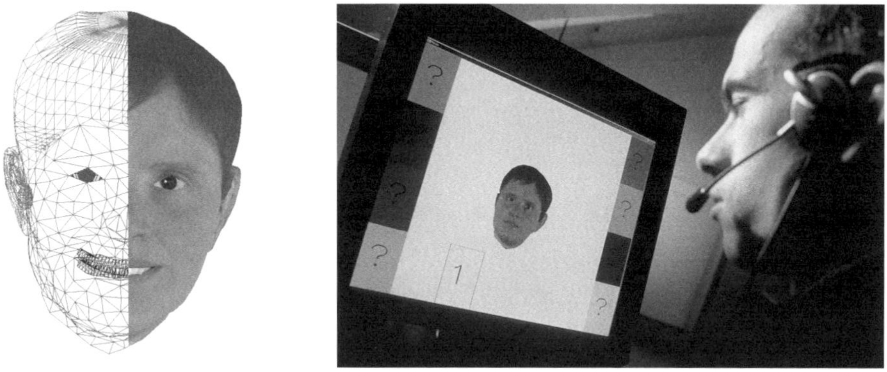

Figure 2: Left. Animated 3D clone with independent head and eye movements. Right: face-to-face interaction platform with a 3D clone

3. Experimental Setup

Our experimental platform aims at implementing context-aware face-to-face interaction between a user and an ECA. Adequate processing of the activity of human partners and changes of ambient environment delivers information to the ECA that should properly condition its behavior.

3.1. Hardware

Sensors. The core element for the experiments described here is a Tobii 1750 eye-tracker[3] that consists of a standard-looking flat screen that discretely embeds infrared lights and a camera (see Figure 2). It monitors at up to 60Hz, the eye gaze of the user whose head can move and rotate freely within a 40cm square cube centered at 60cm away from the screen. A short calibration procedure typically leads to a mean accuracy of 0.5 degrees i.e. 5mm when eyes 50cm away from the 17" screen.

During interaction with the system, the user sits in front of the eye-tracker where our 3D talking head is displayed, as shown in Figure 2. Hardware and software specificities allow the user to interact with the system using eye gaze, mouse and keyboard. The Tobii eye tracker also delivers eye positions relative to its camera and

[3] Please consult http://www.tobii.se/ for technical details.

we use this information for head tracking. Additional data input from a video camera and speech recognition is available for other experimental setups.
Actuators. The visual representation of our ECA is implemented as the cloned 3D appearance and articulation gestures of a real human [25, 26], (see Figure 2). The eye gaze can be controlled independently to look at 2D objects on the screen or spots in the real world outside the screen. Hereby the vergence of the eyes is controlled and provides a crucial cue for inferring spatial cognition. The virtual neck is also articulated and can accompany the eye gaze movements. Standard graphic hardware with 3D acceleration allows real-time rendering of the talking head on the screen. The ECA has also the gift of speech: audiovisual utterances can either be synthesized from text input or mimic pre-recorded human stimuli. We expect that the proper control of these capabilities will enable the ECA to maintain mutual attention - by appropriate eye saccades towards the user or his/her points of interest – and actively draw the user's attention.

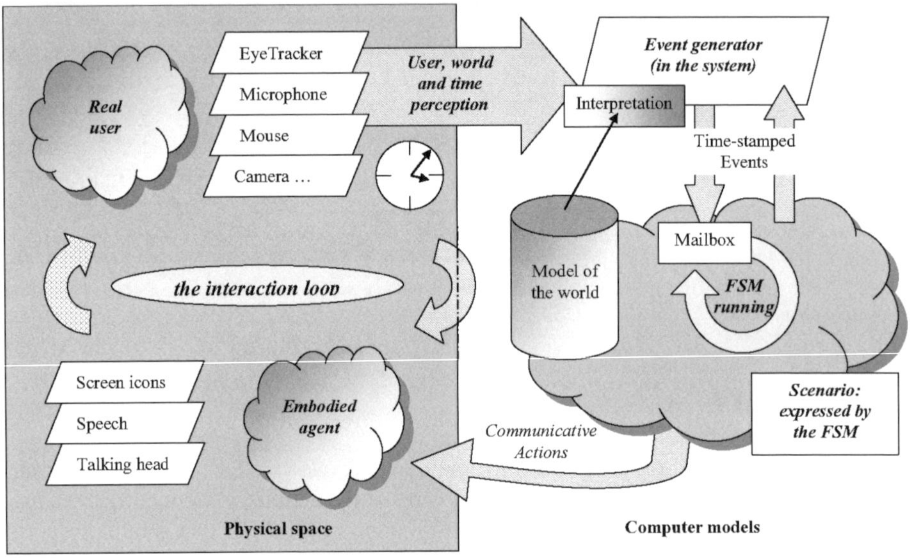

Figure 3: The finite state machine and event handler controlling the interaction.

3.2. Software: scripting multimedia scenarios

The perception-action control loop is described by an event-based language. This language allows the simple description and modification of multimedia scenarios. We developed a corresponding C++ code generator that permits to compile any scenario into an executable binary file. The C++ benefits, like variables, procedural and complex algorithms remain accessible through code inclusion inside any script. Compilation of the generated source code allows accurate recording of the involved events and timings.

In our event-based language a finite state machine (FSM) describes each scenario as a series of states with pre-conditions and post-actions.

Each input device emits events according to user action and a constantly refreshed internal model of the space of interaction (see Figure 3). Triggerable areas on the screen, such as selectable icons or parts of the talking head are defined and surveyed by the eye tracker. Each time the user looks at such a zone, the system posts new events, such as "entering zone" and "quitting zone" and may emit additional "zone fixation duration" events. The FSM is called each time an event is generated or updated.

As the user progresses in the scenario, the FSM specifies which states are waiting on events. Pre-conditions consist of conjunctions or successions of expected multimodal events as for instance recognized keywords, mouse clicks or displacements, eye movements or gaze to active objects. Each of these events is time-stamped.

Pre-conditions can include tests on intervals between time-stamps of events. This allows, for example to associate speech items in terms of words that are identified as a sub product of speech recognition with a certain focus of attention.

Post-actions typically consist of the generation of multimodal events. Time-stamps of these events can be used to delay their actual instantiation in the future. Post-actions can also generate phantom events, to simulate multimodal input or to share information. These phantom events are potential triggers for pre-conditions inside the following state of the FSM.

4. Experiments

4.1. The interaction scenario

To follow up Langton and Driver experiments on visual orienting in response to social attention [10, 11], we designed an interaction scenario where an ECA should direct the user's attention in a complex virtual scene. Our aim was to investigate the effect of multimodal deictic gestures on the user's performance during a search and retrieval task. We chose an on-screen card game, where the user is asked to locate the correct target position of a played card.

4.1.1. Design of the card game

The card game consists of eight cards, the numbers of which are revealed once the played card at the lower middle of the screen is selected with a mouse click. The played card has then to be put down on one of the eight possible target cards placed on the sides of the screen. The correct target card is the one with the same digit as the played card. To anticipate memory effects, the numbers on the cards are shuffled before each turn. The target position is alternated randomly and uniformly distributed amongst the eight possibilities provided that the number of cycles is a multiple of eight. The background color for each position is not changed and thus not correlated with numbers.

4.1.2. Interaction loop

The ECA utter spoken commands and cue directions with an eye saccade combined with a head turn. The ECA alternates between mutual attention and deixis. His gaze and head orientation focus on three regions of interest: the face of the user, his current focus of interest and the target card. In the experiments described below, spoken instructions are not allowed and the ECA gaze alternates between the user's eyes when fixating his face. When the user speaks, the ECA gaze pattern includes the speaker's mouth in his attention loop [15].

4.1.3. Experimental conditions

We tested several experimental conditions corresponding to different levels of assistance given by the ECA. Screenshots of the game interface are given in Figure 4. Each experimental condition comprises three training cycles to allow the subjects to become accustomed to the task, which are followed by 24 measurement cycles. The characteristics of the upcoming condition are described as text on the screen before the training cycles and thus inform the user about the expected gaze behavior of the clone. General information explaining the task is given as text on the screen at the beginning of the experiment. The user is instructed to put down the played card on the target position as fast as possible but no strategy is suggested.

4.1.4. Data acquisition

For all experiments the reaction time and the gaze behavior are monitored. The reaction time is measured as the time span between the first mouse click on the played card and the click on the correct target position. As the card game is displayed on the monitor with embedded eye-tracking, the visual focus of the user on the screen can be recorded. We thus compute which regions on the screen are looked at and how much time users spend on them. Monitored regions of interest are the eight cards on the sides and the ECA. Eye gaze towards the played card was not monitored, as it was constantly moving during the experiment.

At the end of the experiment, which lasted about 15 minutes, participants answer a questionnaire. They rank various subjective aspects of the experiment on a five-point MOS scale, and chose which condition they consider as most appropriate and fastest.

4.2. Experiment I : does our clone have cues to direct social attention?

This experiment aims at evaluating the capacity of our ECA for attracting user's attention using facial cues and quantifying the impact of good and bad hints on the user's performance. This work builds on the psychophysical experiments on visual priming conducted by Langton et al [10, 11].

4.2.1. Conditions

The first series of experiments consists of four different conditions, screenshots of which are displayed in Figure 4. For condition 1, no ECA is displayed. For condition 2, the ECA is visible and provides bad hints: it indicates randomly one of the non-matching positions with a facial gesture as soon as the user selects the played card. In condition 3, it provides good hints: it indicates the correct target position. For condition 4, cards remain upside down and the correct visual cues provided by the ECA are the only ones to find the correct target position.

In conditions 2,3 and 4, the ECA rewards the user with randomly chosen utterances alternating between motivation and congratulation. The utterances are generated off-line to avoid computation delays.

We have strong expectations about the data to be collected: we expect a negative influence on the test person's performance when the clone gives misleading cues and a positive influence when giving good hints. The condition where no clone is displayed serves as a reference. From the fourth condition, we expect to measure the precision with which the gaze direction of the ECA could be perceived. As we expect the ECA to strongly influence the users' attention, we keep the order of conditions as described above for all subjects awaiting the mentioned behavior to be free from learning effects.

Ten users (six male and four female) take part in the first series of experiments. Participants range from 23 to 33 years and most were students. All regularly use a computer mouse and none reported vision problems. The dominant eye is the right eye for all but one subject. Each user had to play the game with the four successive experimental conditions as described above.

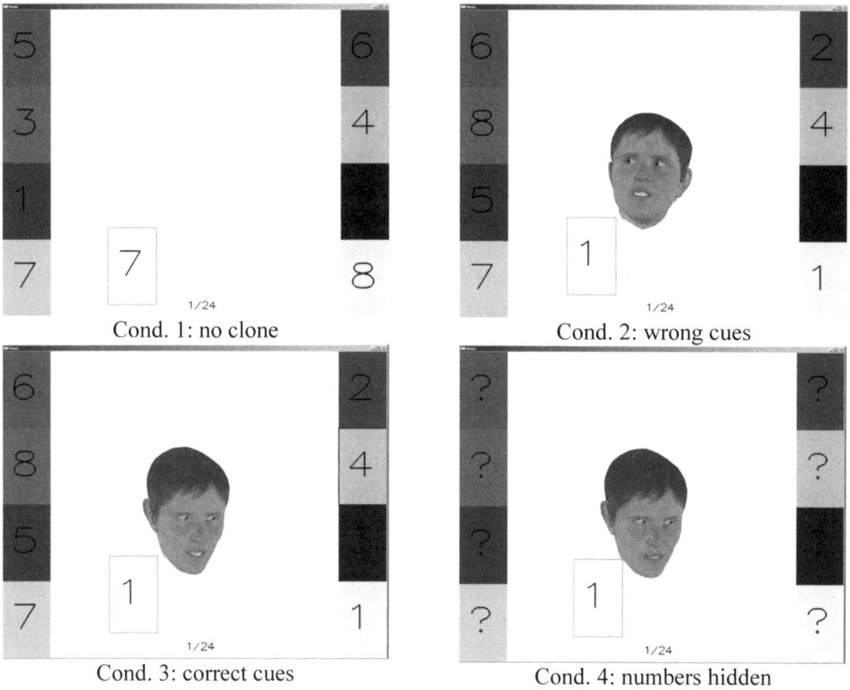

Cond. 1: no clone — Cond. 2: wrong cues — Cond. 3: correct cues — Cond. 4: numbers hidden

Figure 4: Experimental conditions: The experiment is divided into four conditions with different levels of help and guidance by the clone.

4.2.2. Data processing

Reaction-times. Before evaluating the measured reaction times, extreme outliers (distance from median > 5 times inter quartile range) are detected and replaced by a mean value computed from the remaining valid data. Such outliers may be due to the dual screens setup chosen for the experiment. The mouse pointer may in fact leave the screen on one side to appear on the other screen. This happens occasionally when users overshoots the target card and loose time while moving the mouse pointer back into view. The distribution of reaction time is log-normal. We thus analyse the logarithms of the reaction times within each experiment and check with an ANOVA for significance at p=0.05. The significant differences between pairs of distributions are indicated in Figure 5 and Figure 7 with stars.

Number of cards inspected. The number of possible target positions looked at while searching for the correct target position was computed in order to analyse the search strategy of the subjects. This data depends heavily on the data delivered by eye tracker. If less than 60% of all cycles of a condition are not valid (eye gaze not detected, strong deviations between left and right eyes, etc), the data of this condition is entirely rejected. We characterize the log-normal distribution of the number of cards inspected during a game. To avoid invalid operations (log of 0) an offset of one was added to all observed values before statistical analysis. An ANOVA analysis is then performed on valid data and significant differences between pairs of distributions are indicated in Figure 6 and Figure 8 with stars.

4.2.3. Results

Errors. For the conditions where the target cards are turned up, only one wrong selection occurred (condition 3). The pairing task can therefore be considered as accomplished successfully. Numerous errors occur during condition 4 where users could only rely on the deictic gestures of the ECA. In total there are 34 cases in which subjects clicked on a wrong card before finding the correct target position (15% error). Only one subject accomplished the task without any errors. This indicates that users have difficulties to precisely interpret the gaze direction of the ECA. Nevertheless, as all of these errors occurred between neighbouring cards, we consider the assistance given by the facial gestures as sufficient since the user benefits from additional information to localize the target during the other conditions.

Reaction times. The results are displayed in Figure 5. The mean reaction times are sorted for increasing difference between the compared conditions. Subjects are represented by their order of participation. Significance is marked with a red star above the subject number on the x-coordinate. The diagram shows that 5 out of 10 subjects show significantly shorter reaction times for condition 3 (with correct cues) compared to condition 2 (with wrong cues). Three subjects behave the same compared to the condition 1 (without the ECA). These users gain a substantial amount of 200 milliseconds (~10% of the mean duration of a game) at each drawing. Conditions 1 and 2 lead in fact to similar results: comparing the conditions without the ECA and the ECA giving wrong hints, one subject out of 10 show significant shorter reaction times whereas one show longer ones. As several selection errors occurred during the condition 4 (with cards remaining hidden until selection), it is obvious that this entails longer reaction times for most of the subjects.

Number of cards inspected. The results are displayed in Figure 6. Due to the verification of the reliability of the eye tracker data, the data of subject 7 are excluded completely. Probably the subject changed his head position considerably during the experiment. For subject 8 the data of condition four, where no digits appear on the cards, is also excluded. Analysis of the means with an ANOVA at p=0.05 evidences a clear advantage for the condition 3 (with correct hints given by the ECA): 6 of the remaining 9 users check significantly fewer cards compared to condition 2 with misleading hints while 5 users also behave the same way when compared to the condition 1 without the ECA. On average these users inspect in fact 1,5 cards less with a correct gaze than with a wrong or no deictic gaze. We interpret this as a clear decrease of cognitive load since less cognitive resources are used for matching cards. Again no obvious interpretation emerges when comparing the conditions 1 and 2. The condition 4 (where the cards are only shown when selected) doubles the number of cards inspected. This is statistically significant for all except one subject.

Questionnaire. 4 of the 10 subjects think they are faster with the helpful assistance of the ECA and prefer this condition when playing.

° no Clone; * misleading hints; + correct hints; ◀ Cards remain hidden; * $p = 0.05$, × not valid

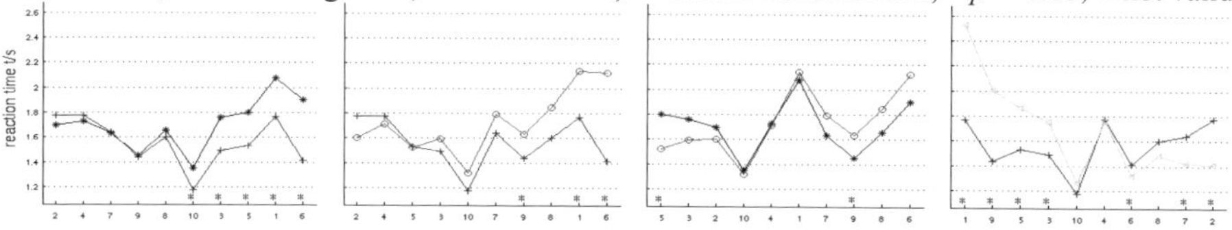

Figure 5: Comparing reaction times for four pairs of conditions. From left to right: condition 2 vs. condition 3; condition 1 vs. condition 3; condition 1 vs. condition 2; condition 4 vs. condition 3. Mean reaction times for each user and for each session are displayed together with the statistical significance of the underlying distributions (stars displayed at the bottom when p<0.05).

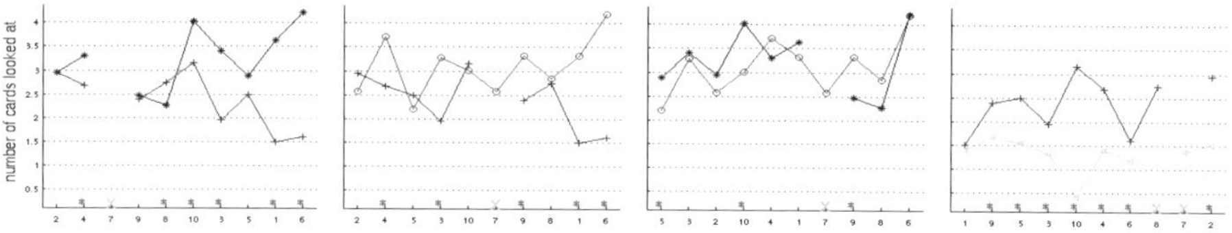

Figure 6: Same as Figure 5 but for nb. of cards inspected. The user order is the one of Figure 5.

4.3. Experiment II: multimodal deixis

This experiment aims at evaluating the benefit of multimodal deixis in drawing user's attention using facial cues together with a spoken instruction.

4.3.1. Conditions

This second series of experiments consists likewise of four different conditions. They resembled the conditions of experiment I. As a major difference the head and gaze movements of the clone are accompanied by the uttering of the demonstrative adverb "là!" (engl.: "there!"). Condition 1 with no clone was replicated for reference. In conditions 2 (wrong cues) and 3 (good cues) speech onset is initiated 100ms after the onset of the deictic gestures: this delay corresponds to the average duration of the eye saccade towards the target position implemented in our ECA. All other rewarding utterances are now omitted. Condition 4 of experiment I is replaced by a condition with correct hints, where an additional delay of 200 ms is introduced between the gestural and the following acoustic deictic gestures in order to comply with data on speech and gesture coordination [27]. We expect this natural coordination to enhance the ability of the ECA to attract user attention. The data collection and treatment is identical to that used for the experiment I.

Fourteen users (ten male and four female) took part to this experiment. They range from 21 to 48 years and most are students. All regularly use a computer mouse and none report vision problems. The dominant eye is the right eye for 8 subjects and the left eye for the other 6 subjects.

As the influence of the clone is not as strong as expected in experiment I when providing bad hints, it was not clear if the order of presentation might have a major influence. Therefore the conditions are here presented in random orders.

4.3.2. Results

Errors. Only one click error between neighbouring cards occurred (subject six in the condition 2 with misleading hints).
Reaction times. The analysis of the reaction time evidences now a clear advantage for 7 subjects out of 14 during the condition 3 (with correct hints) against the condition 2 (with misleading hints), and for 8 subjects of 14 compared to the condition 1 (without the ECA). These users now gain on average a substantial amount of almost 400 milliseconds (~20% of the mean duration of a game) at each drawing. The proportion of users benefiting from this advantage and the amount of benefit are both more important than for Experiment I. When comparing the condition 1 versus conditions 2 and 3, subjects show faster reaction times in condition 2 while 3 other subjects just behave the opposite way. When comparing delayed vs. synchronized spoken instructions, 2 subjects show shorter reaction time for condition 3 while 3 show longer reaction times.
Number of cards inspected. Due to insufficient monitoring of the data collected with the eye tracker, the data of subject 2 is completely excluded from evaluation and data of subjects 1 and 9 only partly. Analysing the remaining data with an ANOVA for significance at p=0.05 it was found that 11 of the 13 subjects with usable data look at fewer cards for condition 3 compared to condition 2, and 10 of the 12 subjects with usable data when compared to condition 1 (without the ECA). On average these users have in fact to inspect 1,5 cards less with a correct gaze than with a wrong or no deictic gaze. These numbers are in line with the data of experiment I. Again data between condition 1 and 2 were statistically significant for only 1 subject. No clear tendency can be reported when considering influence of delay on performance except that the delayed stimuli cause 2 subjects to look at more cards than for the synchronous condition 3.
Questionnaire. 11 of the 14 subjects estimate that they have the best reaction times when correct hints are given by the ECA. Most of the subjects declare that they glance a lot at the ECA giving correct hints and discard gestures in condition 2 but that these cues have poor influence on their reaction time. The movements of the ECA are judged realistic.

° no Clone; * misleading hints; + correct hints; ◄ delayed correct hints; * $p = 0.05$, ; × not valid

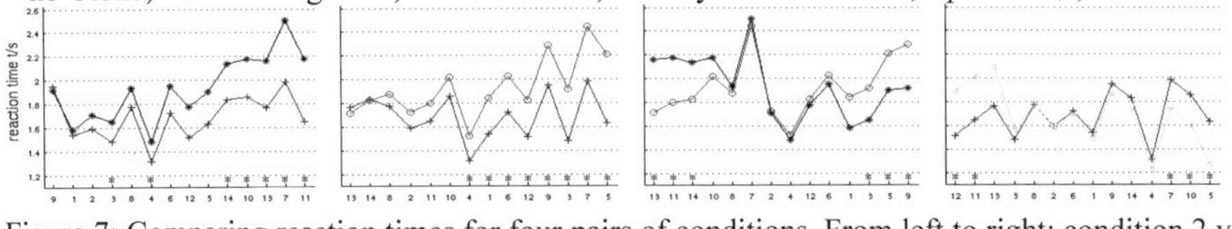

Figure 7: Comparing reaction times for four pairs of conditions. From left to right: condition 2 vs. condition 3; condition 1 vs. condition 3; condition 1 vs. condition 2; condition 4 vs. condition 3. Same conventions as for Figure 5.

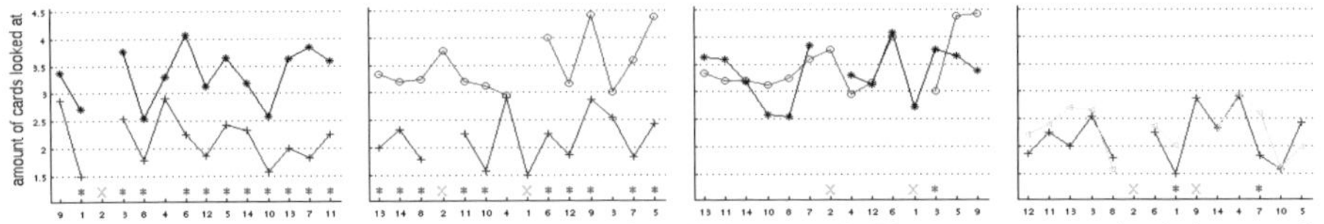

Figure 8. Same as Figure 7 but for number of cards inspected. The same user order is kept as for Figure 7.

5. Discussion and Perspectives

When considering the number of cards inspected and the number of wrong selections in condition 4 of experiment I, the current control and rendering of deixis gestures of the ECA are sufficient to localize objects as long as there is information available at the target position to take the final decision. Without such additional information the gestures of the ECA are not precise enough to decide between close neighboring objects. Apart from the limitations of 3D rendering on a 2D screen, this may be due to the synchronization between gaze and head orientation that are not yet derived from empirical data. An additional limitation is the poor rendering of the facial deformations around the eyes of the ECA when eye gaze deviates from head direction: eyelids should be notably enlarged to widen the aperture available for the iris. These additional cues may contribute significantly to the estimation of eyes direction.

When considering reaction time, 30% to 50% of the users benefit from the assistance given by the ECA. When considering the number of cards a user had to check visually to find the correct target position, the percentage is slightly higher. When misleading or no assistance is given by the ECA, no major differences are observed. The influence of the ECA when giving misleading hints is however less strong than expected and most users seem able to willingly ignore its gesturing. No clear correlations between the data emerge that would enable a more detailed comprehension of the individual strategies followed to fulfill the task.

Several subjects complained for being disturbed by rewarding utterances in experiment I. Therefore these utterances fail as means to maintain attention and make the interaction more natural. A more appropriate feedback should be short and clear according to the instruction given to subjects as to react fast. Furthermore it should contribute to attract the attention to the object of interest. No subject complains effectively for spoken instructions in experiment II.

The results of experiment II show that the benefit in reaction time from the assistance of the ECA using multimodal deixis could still be improved (up to 90%). A more important finding is the reduced number of looked at cards for more than 80% of the subjects. The majority of participants manage to complete the task looking at significantly less cards when the ECA is giving helpful assistance. This means that even if they do not improve their reaction time, the search process is more efficient and probably more relaxed. We conclude that this helpfully diminishes the cognitive load of the task. The answers to the questionnaire confirm this finding as the good ratings for naturalness of the ECA and the preference of the condition where it is giving correct hints are outlined more clearly for experiment II compared with experiment I.

The experimental scenario presented here can probably be further improved by displaying more objects on the screen and using smaller digits. This should enhance the benefit of ECA assistance since this would require a closer examination of the objects and increase the number of objects to check in order to find the correct one without the assistance of the ECA. However, we consider the results with the current implementation as sufficient confirmation of our assumptions and encouraging motivation to study further possibilities to enhance the capabilities of the ECA.

6. Conclusions

Our first implementation of an embodied conversational agent able to maintain face-to-face interaction with a human interlocutor proves here its capability to direct user attention using multimodal deictic gestures. We demonstrate that users can largely benefit from a very basic implementation even in a rather simple selection task. ECA guidance results in reduced reaction time and lower cognitive load for the given search and retrieval task. Subjects benefiting from ECA guidance have a substantial gain of 200ms (~10% of the task) and 1.5 cards compared with improper or no guidance in experiment I. The impact could be enhanced (up to 400ms in reaction time) by appropriate and well timed speech commands which especially entail reduction of the cognitive load by reducing the search space and number of matches. We confirm here that the rather modest impact of visual cues found in psychophysical experiments [the 20ms benefit in up/down directions found by 10] is enhanced by more ecological interactions.

We believe that the study, modeling and implementation of the components of human face-to-face interaction are crucial elements to obtain an intuitive, robust and reliable communication interface able to establish an effective and efficient interaction loop. While most experimental data on speech and gaze examine attention of the listener using recorded videos [15], only few experimental data is currently available on gaze patterns when speaking [5, 28] and during face-to-face interaction. We recently identify and track facial degrees-of-freedom involved in face-to-face conversation [29] and conduct similar experiments on eye gaze.

Such interaction platforms involving actual real-time interaction between a user and autonomous (or semi-autonomous in Wizard-of-Oz experiments animating the ECA with control movements captured on a human operator) ECA is highly valuable for recording characteristic control signals, investigating the influence of embodiment and assessing the benefit of enhanced strategies on performance, learning and acceptability.

7. Acknowledgements

We gratefully acknowledge the patience of Hélène Lœvenbruck, the human model for our clone. We thank Alain Arnal and Christophe Savariaux for their technical assistance with the audiovisual capture platform, Denis Beautemps for helping us with statistical analysis, as well as Matthias Odisio, Pauline Welby and the reviewers for their helpful comments. We are also grateful to all participants of the experiments.

8. References

[1] Cassell, J., Sullivan, J., Prevost, S., and Churchill, E. (2000) *Embodied Conversational Agents*. Cambridge: MIT Press.

[2] Geiger, G., Ezzat, T., and Poggio, T. (2003) *Perceptual evaluation of video-realistic speech*. Massachusetts Institute of Technology: Cambridge, MA.

[3] Thórisson, K. (2002) *Natural turn-taking needs no manual: computational theory and model from perception to action*, in *Multimodality in language and speech systems*, B. Granström,

D. House, and I. Karlsson, Editors. Kluwer Academic: Dordrecht, The Netherlands. p. 173–207.

[4] Carpenter, M. and Tomasello, M. (2000) *Joint attention, cultural learning and language acquisition: Implications for children with autism*, in *Communicative and language intervention series. Autism spectrum disorders: A transactional perspective*, A.M. Wetherby and B.M. Prizant, Editors. Paul H. Brooks Publishing: Baltimore. p. 30–54.

[5] Argyle, M. and Cook, M. (1976) *Gaze and mutual gaze*. London: Cambridge University Press.

[6] Kendon, A. (1967) *Some functions of gaze-direction in social interaction.* Acta Psychologica, **26**: p.22-63.

[7] Pourtois, G., Sander, D., Andres, M., Grandjean, D., Reveret, L., Olivier, E., and Vuilleumier, P. (2004) *Dissociable roles of the human somatosensory and superior temporal cortices for processing social face signals.* European Journal of Neuroscience, **20**: p.3507-3515.

[8] Posner, M. and Peterson, S. (1990) *The attention system of the human brain.* Annual Review of Neuroscience, **13**: p.25-42.

[9] Posner, M.I. (1980) *Orienting of attention.* Quarterly Journal of Experimental Psychology, **32**: p.3-25.

[10] Langton, S. and Bruce, V. (1999) *Reflexive visual orienting in response to the social attention of others.* Visual Cognition, **6**(5): p.541-567.

[11] Langton, S., Watt, J., and Bruce, V. (2000) *Do the eyes have it ? Cues to the direction of social attention.* Trends in Cognitive Sciences, **4**(2): p.50-59.

[12] Driver, J., Davis, G., Riccardelli, P., Kidd, P., Maxwell, E., and Baron-Cohen, S. (1999) *Shared attention and the social brain : gaze perception triggers automatic visuospatial orienting in adults.* Visual Cognition, **6**(5): p.509-540.

[13] Simons, D.J. and Chabris, C.F. (1999) *Gorillas in our midst: sustained inattentional blindness for dynamic events.* Perception, **28**: p.1059-1074.

[14] Yarbus, A.L. (1967) *Eye movements during perception of complex objects*, in *Eye Movements and Vision'*, L.A. Riggs, Editor. Plenum Press: New York. p. 171-196.

[15] Vatikiotis-Bateson, E., Eigsti, I.-M., Yano, S., and Munhall, K.G. (1998) *Eye movement of perceivers during audiovisual speech perception.* Perception & Psychophysics, **60**: p.926-940.

[16] Premack, D. and Woodruff, G. (1978) *Does the chimpanzee have a theory of mind?* Behavioral and brain sciences, **1**: p.515-526.

[17] Baron-Cohen, S., Leslie, A., and Frith, U. (1985) *Does the autistic child have a "theory of mind"?* Cognition, **21**: p.37-46.

[18] Leslie, A.M. (1994) *ToMM, ToBY, and Agency: Core architecture and domain specificity*, in *Mapping the Mind: Domain specificity in cognition and culture*, L.A. Hirschfeld and S.A. Gelman, Editors. Cambridge University Press: Cambridge. p. 119–148.

[19] Scassellati, B. (2001) *Foundations for a theory of mind for a humanoid robot*, in *Department of Computer Science and Electrical Engineering*. MIT: Boston - MA. 174 pages.

[20] Breazeal, C. (2000) *Sociable machines: expressive social exchange between humans and robots*. Sc.D. dissertation, in *Department of Electrical Engineering and Computer Science*. MIT: Boston.

[21] Fujie, S., Fukushima, K., and Kobayashi, T. (2005) *Back-channel feedback generation using linguistic and nonlinguistic information and its application to spoken dialogue system*. in *Interspeech*. Lisbon, Portugal. p.889-892.

[22] Matsusaka, Y., Tojo, T., and Kobayashi, T. (2003) *Conversation Robot Participating in Group Conversation.* IEICE Transaction of Information and System, **E86-D**(1): p.26-36.

[23] Lee, S.P., Badler, J.B., and Badler, N. (2002) *Eyes alive.* ACM Transaction on Graphics, **21**(3): p.637-644.

[24] Itti, L., Dhavale, N., and Pighin, F. (2003) *Realistic avatar eye and head animation using a neurobiological model of visual attention*. in *SPIE 48th Annual International Symposium on Optical Science and Technology*. San Diego, CA. p.64-78.

[25] Bailly, G., Bérar, M., Elisei, F., and Odisio, M. (2003) *Audiovisual speech synthesis.* International Journal of Speech Technology, **6**: p.331-346.
[26] Revéret, L., Bailly, G., and Badin, P. (2000) *MOTHER: a new generation of talking heads providing a flexible articulatory control for video-realistic speech animation.* in *International Conference on Speech and Language Processing*. Beijing - China. p.755-758.
[27] Castiello, U., Paulignan, Y., and Jeannerod, M. (1991) *Temporal dissociation of motor responses and subjective awareness.* Brain, **114**: p.2639-2655.
[28] Vertegaal, R., Slagter, R., van der Veer, G., and Nijholt, A. (2001) *Eye gaze patterns in conversations: There is more to conversational agents than meets the eyes.* in *Conference on Human Factors in Computing Systems*. Seattle, USA. p.301 - 308.
[29] Bailly, G., Elisei, F., Badin, P., and Savariaux, C. (2006) *Degrees of freedom of facial movements in face-to-face conversational speech.* in *International Workshop on Multimodal Corpora*. Genoa - Italy. p.33-36.

Fundamentals of Verbal and Nonverbal Communication and the Biometric Issue
A. Esposito et al. (Eds.)
IOS Press, 2007

Analyzing and Modelling Verbal and Non-verbal Communication for Talking Animated Interface Agents

David HOUSE[1] and Björn GRANSTRÖM
Centre for Speech Technology, CSC, KTH, Stockholm, Sweden

Abstract. The use of animated talking agents is a novel feature of many multimodal spoken dialogue systems. The addition and integration of a virtual talking head has direct implications for the way in which users approach and interact with such systems. The focus of this contribution is to look into the communicative function of the agent, both the capability to increase intelligibility of the spoken interaction and the possibility to make the flow of the dialogue smoother, through different kinds of communicative gestures, such as visual prosodic gestures (e.g. focal accent and emphatic stress) and gestures for different expressive states, turntaking and negative or positive system feedback.

Keywords. ECA, animated agent, audiovisual speech, non-verbal communication, visual prosody

1. Introduction

In our interaction with others, we easily and naturally use all of our sensory modalities as we communicate and exchange information. Our senses are exceptionally well adapted for these tasks, and our brain enables us to effortlessly integrate information from different modalities fusing data to optimally meet the current communication needs. As we attempt to take advantage of the effective communication potential of human conversation, we see an increasing need to embody the conversational partner using audiovisual verbal and non-verbal communication in the form of animated talking agents. The use of animated talking agents is currently a novel feature of many multimodal experimental spoken dialogue systems. The addition and integration of a virtual talking head has direct implications for the way in which users approach and interact with such systems [1]. However, techniques for analyzing, modeling and testing this kind of interaction are still in a stage of rapid development.

Effective interaction in dialogue systems involves both the presentation of information and the flow of interactive dialogue. A talking animated agent can provide the user with an interactive partner whose goal is to take the role of the human agent. An effective agent is one who is capable of supplying the user with relevant

[1] Corresponding Author: Centre for Speech Technology, Dept. of Speech, Music and Hearing, School of Computer Science and Communication, KTH, Lindstedtsvägen 24, 10044, Stockholm, Sweden. Email: davidh@speech.kth.se .

information, can fluently answer questions concerning complex user requirements and can ultimately assist the user in a decision-making process through the flow of conversation. One way to achieve believability is through the use of a talking head where information is transformed into speech, articulator movements, speech-related gestures and conversational gestures. Useful applications of talking heads include aids for the hearing impaired, educational software, audiovisual human perception experiments [2], entertainment, and high-quality audiovisual text-to-speech synthesis for applications such as news reading. The use of the talking head aims at increasing effectiveness by building on the user's social skills to improve the flow of the dialogue [3]. Visual cues to feedback, turntaking and signalling the system's internal state are key aspects of effective interaction.

The talking head developed at KTH is based on text-to-speech synthesis. Audio speech synthesis is generated from a text representation in synchrony with visual articulator movements of the lips, tongue and jaw. Linguistic information in the text is used to generate visual cues for relevant prosodic categories such as prominence, phrasing and emphasis. These cues generally take the form of eyebrow and head movements which we have termed "visual prosody" [4]. These types of visual cues with the addition of e.g. a smiling or frowning face are also used as conversational gestures to signal such things as positive or negative feedback, turntaking regulation, and the system's internal state. In addition, the head can visually signal attitudes and emotions. More recently, we have been exploring data-driven methods to model articulation and facial parameters of major importance for conveying social signals and emotions.

The focus of this chapter is to look into the communicative function of the agent, both the capability to increase intelligibility of the spoken interaction and the possibility to make the flow of the dialogue smoother, through different kinds of communicative gestures such as gestures for emphatic stress, emotions, turntaking and negative or positive system feedback. The chapter reviews state-of-the-art animated agent technologies and some of their possible applications primarily in spoken dialogue systems. The chapter also includes some examples of methods of evaluating communicative gestures in different contexts.

2. KTH Multimodal Speech Synthesis

Animated synthetic talking faces and characters have been developed using a number of different techniques and for a variety of purposes for more than two decades. Historically, our approach is based on parameterised, deformable 3D facial models, controlled by rules within a text-to-speech framework [5]. The rules generate the parameter tracks for the face from a representation of the text, taking coarticulation into account [6]. We employ a generalized parameterisation technique to adapt a static 3D-wireframe of a face for visual speech animation [7]. Based on concepts first introduced by Parke [8], we define a set of parameters that will deform the wireframe by applying weighted transformations to its vertices. One critical difference from Parke's system, however, is that we have de-coupled the model definitions from the animation engine. The animation engine uses different definition files that are created for each face. This greatly increases flexibility, allowing models with different topologies to be animated from the same control parameter program, such as a text-to-speech system.

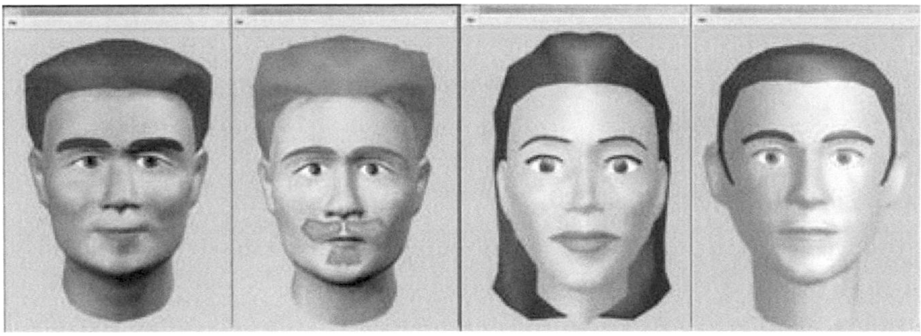

Figure 1. Some different versions of the KTH talking head

The models are made up of polygon surfaces that are rendered in 3D using standard computer graphics techniques. The surfaces can be articulated and deformed under the control of a number of parameters. The parameters are designed to allow for intuitive interactive or rule-based control. The display can be chosen to show only the surfaces or the polygons for the different components of the face. The surfaces can be made (semi) transparent to display the internal parts of the model, including the tongue, palate, jaw and teeth [9]. The internal parts are based on articulatory measurements using MRI (Magnetic Resonance Imaging), EMA (ElectroMagnetic Articulography) and EPG (ElectroPalatoGraphy), in order to assure that the model's movements are realistic. This is of importance for language learning situations, where the transparency of the skin may be used to explain non-visible articulations [10] [11] [12] [13]. Several face models have been developed for different application, some of them can be seen in Figure 1. All can be parametrically controlled by the same articulation rules.

For stimuli preparation and explorative investigations, we have developed a control interface that allows fine-grained control over the trajectories for acoustic as well as visual parameters. The interface is implemented as an extension to the WaveSurfer application (www.speech.kth.se/wavesurfer) [14] which is a freeware tool for recording, playing, editing, viewing, printing, and labelling audio data.

The interface makes it possible to start with an utterance synthesised from text, with the articulatory parameters generated by rule, and then interactively edit the parameter tracks for F0, visual (non-articulatory) parameters as well as the durations of individual segments in the utterance to produce specific cues. An example of the user interface is shown in Figure 2. In the top box a text can be entered in Swedish or English. The selection of language triggers separate text-to-speech systems with different phoneme definitions and rules, built in the Rulsys notation [5]. The generated phonetic transcription can be edited. On pushing "Synthesize", rule generated parameters will be created and displayed in different panes below. The selection of parameters is user controlled. A talking face is displayed in a separate window. The acoustic synthesis can be exchanged for a natural utterance and synchronised to the face synthesis on a segment by segment basis by running the face synthesis with phoneme durations from the natural utterance. The combination of natural and synthetic speech is useful for different experiments on multimodal integration and has been used in the Synface project (see section 3). In language learning applications this feature could be used to add to the naturalness of the tutor's voice in cases when the acoustic synthesis is judged to be inappropriate.

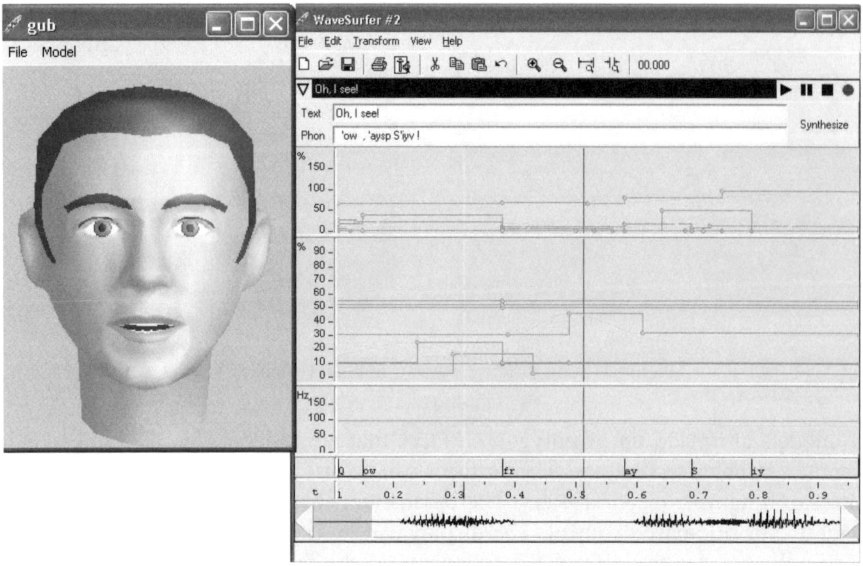

Figure 2. The WaveSurfer user interface for parametric manipulation of the multimodal synthesis.

More recently, we have begun experimenting with data-driven visual synthesis using a newly developed MPEG-4 compatible talking head [15]. A data-driven approach enables us to capture the interaction between facial expression and articulation. This is especially important when trying to synthesize expressive speech [16].

To automatically extract important facial movements we have employed a motion capture procedure. We want to be able to obtain articulatory data as well as other facial movements at the same time, and it is crucial that the accuracy in the measurements is good enough for resynthesis of an animated head. The Qualisys optical motion tracking system that we use has an accuracy of better than 1 mm with a temporal resolution of 60 Hz. The data acquisition and processing is very similar to earlier facial measurements carried out by [17]. The recording set-up can be seen in Figure 3.

The subject could either pronounce sentences presented on the screen outside the window or be engaged in a (structured) dialogue with another person as shown in Figure 3. In the present set-up, the second person cannot be recorded with the Qualisys system but is only video recorded. In the illustrated case we used 30 markers to register lip movements as well as other facial movements such as eyebrows, cheek, chin and eyelids. Additionally we placed three markers on the chest to register head movements with respect to the torso. A pair of spectacles with four markers attached was used as a reference to be able to factor out head and body movements when looking at the facial movements specifically.

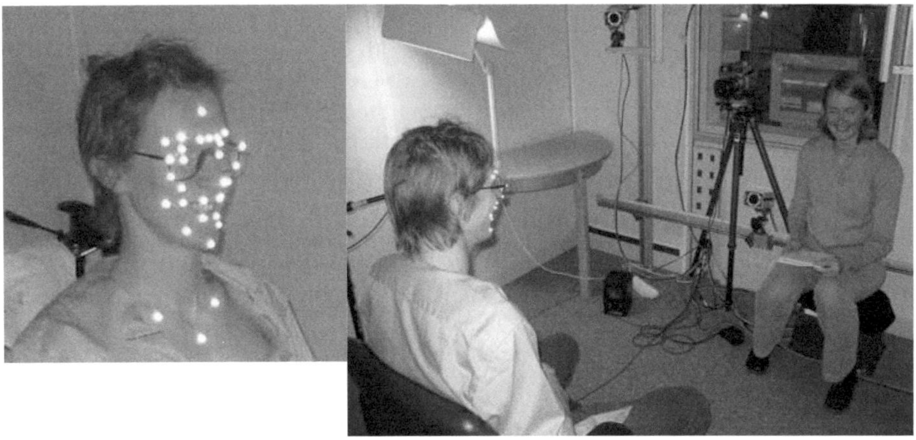

Figure 3. Data collection setup with video and IR-cameras, microphone and a screen for prompts (right) and test subject with the IR-reflecting markers glued to the face (left).

The databases we have thus far collected have enabled us to analyze speech movements such as articulatory variation in expressive speech [16] in addition to providing us with data with which to develop data-driven visual synthesis. The data has also been used to directly drive synthetic 3D face models which adhere to the MPEG-4 Facial Animation (FA) standard [18] enabling us to perform comparative evaluation studies of different animated faces within the EU-funded PF-Star project [19][20]. Examples of the new head displaying different emotions taken from the database are shown in Figure 8.

3. Improving Intelligibility and Information Presentation

One of the more striking examples of improvement and effectiveness in speech intelligibility is taken from the Synface project which aims at improving telephone communication for the hearing impaired [21]. A demonstrator of the system for telephony with a synthetic face that articulates in synchrony with a natural voice has now been implemented as a result of the project.

Evaluation studies within this project were mainly oriented towards investigating differences in intelligibility between speech alone and speech with the addition of a talking head. These evaluation studies were performed off-line, e.g. the speech material was manually labelled so that the visible speech synthesis always generated the correct phonemes rather than being generated from the Synface recognizer which can introduce recognition errors. The results of a series of tests using non-sense VCV words and hearing-impaired subjects showed a significant gain in intelligibility when the talking head was added to a natural voice. With the synthetic face, consonant identification improved from 29% to 54% correct responses. This compares to the 57% correct response result obtained by using the natural face. In certain cases, notably the consonants consisting of lip movement (i.e. the bilabial and labiodental consonants), the response results were in fact better for the synthetic face than for the natural face. This points to the possibility of using overarticulation strategies for the talking face in these kinds of applications. Results indicate that a certain degree of overarticulation can be advantageous in improving intelligibility [22]. The average preferred hyper articulation was found to be 24%, given the task to optimise the subjective ability to lip-read. The highest and lowest preferred values were 150 and 90% respectively with a standard deviation of 16%. The option of setting the articulation strength to the user's subjective preference could be included in the Synface application.

Intelligibility tests have also been run using normal hearing subjects with similar results where the audio signal was degraded by adding white noise [21]. For example, for a synthetic male voice, consonant identification improved from 31% without the face to 45% with the face. A different type of application aimed at normal hearing persons is the use of a talking head in combination with targeted audio. To transmit highly directional sound, targeted audio makes use of a technique known as "parametric array" [23]. This type of highly directed sound can be used to communicate a voice message to a single person within a group of people (e.g. in a meeting situation or at a museum exhibit), minimally disturbing the other people. Within the framework of the EU project CHIL, experiments have shown that targeted audio combined with a talking head directed at the listener provides an efficient method of decreasing the disturbance by further lowering the audio intensity with retained and even increased intelligibility for the targeted listener [24].

4. Analyzing and Modelling Visual Cues for Prominence

4.1. Perception Experiments

Another quite different example of the contribution of the talking head to information presentation is taken from the results of perception studies in which the percept of emphasis and syllable prominence is enhanced by eyebrow and head movements. In an experiment investigating the contribution of eyebrow movement to the perception of prominence in Swedish [25], a test sentence was created using our audiovisual text-to-speech synthesis in which the acoustic cues and lower-face visual cues were the same for all stimuli. Articulatory movements were created by using the text-to-speech rule system. The upper face cues were eyebrow movement where the eyebrows were raised on successive words in the sentence. The movements were created by hand editing the eyebrow parameter. The degree of eyebrow raising was chosen to create a subtle movement that was distinctive although not too obvious. In the stimuli, the acoustic signal was always the same, and the sentence was synthesized as one phrase. Six versions were included in the experiment: one with no eyebrow movement and five where eyebrow raising was placed on one of the five content words in the test sentence. The words with concomitant eyebrow movement were generally perceived as more prominent than words without the movement. This tendency was even greater for a subgroup of non-native (L2) listeners. The mean increase in prominence response following an eyebrow movement was 24 percent for the Swedish native (L1) listeners and 39 percent for the L2 group. Similar results have also been obtained for Dutch in [26].

In another study [27] both eyebrow and head movements were tested as potential cues to prominence. The goal of the study was two-fold. First of all we wanted to see if head movement (nodding) is a more powerful cue to prominence than is eyebrow movement by virtue of a larger movement. Secondly, we wanted to test the perceptual sensitivity to the timing of both eyebrow and head movement in relationship to the syllable.

As in the previous experiment, our rule-based audiovisual synthesizer was used for stimuli preparation. The test sentence used to create the stimuli for the experiment was the same as that used in an earlier perception experiment designed to test acoustic cues only. The sentence, *Jag vill bara flyga om vädret är perfekt* (I only want to fly if the weather is perfect) was synthesized with focal accent rises on both *flyga* (fly) and *vädret* (weather) and was ambiguous as to prominence when subjects were presented with the audio stimulus only. For the face, the eyebrows were raised to create a subtle movement that was distinctive although not too obvious. In quantitative terms the movement comprised 4% of the total possible movement. The head movement was a slight vertical lowering comprising 3% of the total possible vertical head rotation. Statically, the displacement is difficult to perceive, while dynamically, the movement is quite distinct.

Two sets of stimuli were created: set one in which both eyebrow and head movement occurred simultaneously and set two in which the movements were separated and potentially conflicting with each other. In set one, six stimuli were created by synchronizing the movement in stimulus 1 with the stressed vowel of *flyga*. This movement was successively shifted in intervals of 100 ms towards *vädret* resulting in the movement in stimulus 6 being synchronized with the stressed vowel of *vädret*. In set two, stimuli 1-3 were created by fixing the head movement to synchronize with the stressed vowel of *vädret* and successively shifting the eyebrow movements from the stressed vowel of *flyga* towards *vädret* in steps of 100 ms. Stimuli 4-6 were created by fixing the eyebrow movement to *vädret* and shifting the head movement from *flyga* towards *vädret*. The acoustic signal and articulatory movements were the same for all stimuli. A schematic illustration of the stimuli is presented in Figure 4.

The results from stimulus set 1 where eyebrow and head movements occurred simultaneously clearly reflect the timing aspect of these stimuli as can be seen in Figure 5 where percent votes for *vädret* increase successively as movement is shifted in time from *flyga* to *vädret*.

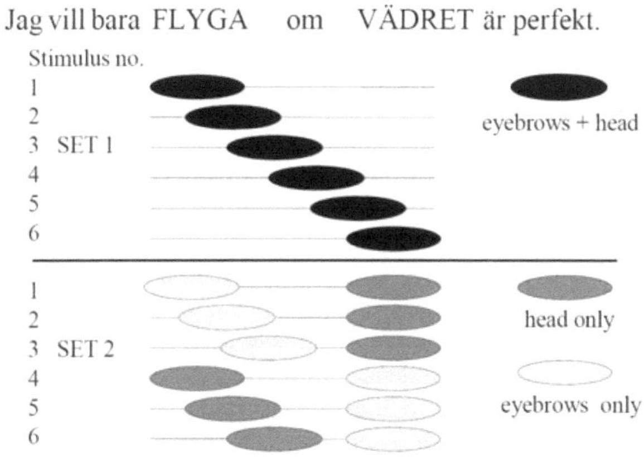

Figure 4. Schematic illustration of face gesture timing

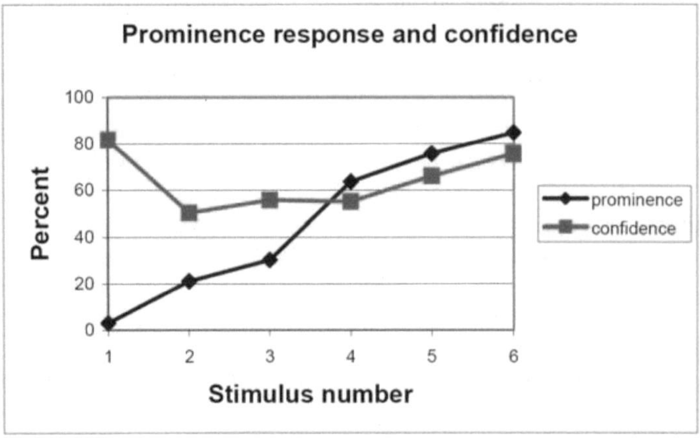

Figure 5. Results for stimulus set one showing prominence response for "vädret" and confidence in percent.

It is clear from the results that combined head and eyebrow movements of the scope used in the experiment are powerful cues to prominence when synchronized with the stressed vowel of the potentially prominent word and when no conflicting acoustic cue is present. The results demonstrate a general sensitivity to the timing of these movements at least on the order of 100 ms as the prominence response moves successively from the word *flyga* to the word *vädret*. However, there is a tendency for integration of the movements to the nearest potentially prominent word, thus accounting for the jump in prominence response between stimulus 3 and 4 in set 1. This integration is consistent with the results of similar experiments using visual and auditory segmental cues [28].

As could be expected, the results from set 2, where eyebrow and head movement were in conflict, showed more stimulus ambiguity. Head movement, however, demonstrated a slight advantage in signalling prominence. This advantage can perhaps be explained by the fact that the movement of the head may be visually more salient than the relatively subtle eyebrow movement. The advantage might even be increased if the head is observed from a greater distance. In an informal demonstration, where observers were further away from the computer screen than the subjects in the experiment, head-movement advantage was quite pronounced.

A number of questions remain to be answered, as a perception experiment of this type is necessarily restricted in scope. Amplitude of movement was not addressed in this investigation. If, for example, eyebrow movement were exaggerated, would this counterbalance the greater power of head movement? A perhaps even more crucial question is the interaction between the acoustic and visual cues.

Duration of the eyebrow and head movements is another aspect which was not tested here. It seems plausible that similar onset and offset durations (100 ms) combined with substantially longer static displacements would serve as conversational signals rather than as cues to prominence. In this way, non-synchronous eyebrow and head movements can be combined to signal both prominence and e.g. feedback giving or seeking. Some of the subjects also commented that the face seemed to convey a certain degree of irony in some of the stimuli in set 2, most likely in those stimuli with non-synchronous eyebrow and head movement.

4.2. Production Studies with Several Expressive Modes

The types of studies reported on in the previous section have established the perceptual importance of eyebrow and head movement cues for prominence. These experiments do not, however, provide us with quantifiable data on the exact timing or amplitude of such movements used by human speakers such as can be found in e.g. [29][30]. Nor do they give us information on the variability of the movements in communicative situations. This kind of information is important if we are to be able to implement realistic facial gestures and head movements in our animated agents. In this section we present measurement results obtained from a speech corpus in which focal accent was systematically varied in a variety of expressive modes.

The speech material used for the study consisted of 39 short, content neutral sentences such as "*Båten seglade förbi*" (The boat sailed by) and "*Grannen knackade på dörren*" (The neighbor knocked on the door), all with three content words which could each be focally accented. To elicit visual prosody in terms of prominence, these short sentences were recorded with varying focal accent position, usually on the subject, the verb and the object respectively, thus making a total of 117 sentences. The utterances were recorded in a variety of expressive modes including certain, confirming, questioning, uncertain, happy, angry and neutral. With the exception of angry, these were all expressions which are considered to be relevant in a spoken dialogue system scenario. The corpus is presented in more detail in [31]. An earlier study using the material is reported in [32], and the current study can be found in more detail in [33].

For recording the database, a total of 29 IR reflective markers were attached to the speaker's face, of which 4 markers were used as reference markers (on the ears and on the forehead). The marker setup, similar to that shown in Figure 3, largely corresponds to MPEG-4 feature point (FP) configuration. Audio data was recorded on DAT-tape, and video was recorded using a mini-DV digital video camera. A synchronisation signal from the Qualisys system was fed into one audio channel of the DAT and DV to facilitate post-synchronisation of the data streams.

In the present study, we chose to base our quantitative analysis of facial movement on the MPEG-4 Facial Animation Parameter (FAP) representation, because it is a compact and standardised scheme for describing movements of the human face and head. Specifically, we chose a subset of 31 FAPs out of the 68 FAPs defined in the MPEG-4 standard, including only the ones that we were able to calculate directly from our measured point data (discarding e.g. parameters for inner lip contour, tongue, ears and eyes).

We wanted to obtain a measure of how (in what FAPs) focus was realised by the recorded speaker for the different expressive modes. In an attempt to quantify this, we introduce the Focal Motion Quotient, FMQ, defined as the standard deviation of a FAP parameter taken over a word in focal position, divided by the average standard deviation of the same FAP in the same word in non-focal position. This quotient was then averaged over all sentence-triplets for each expressive mode separately.

As a first step in the analysis the FMQs for all the 31 measured FAPs were averaged across the 39 sentences. These data are displayed in Figure 6 for the analyzed expressive modes, i.e. Angry, Happy, Confirming, Questioning, Certain, Uncertain and Neutral. As can be seen, the FMQ mean is always above one, irrespective of which facial movement (FAP) that is studied. This means that a shift from a non-focal to a focal pronunciation on the average results in greater dynamics in all facial movements for all expressive modes. It should be noted that these are results from only one speaker

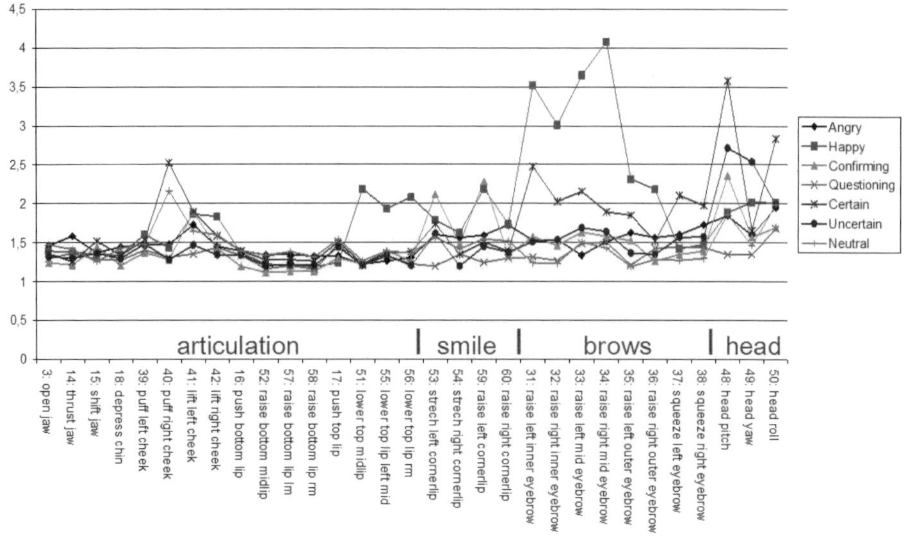

Figure 6. The focal motion quotient, FMQ, averaged across all sentences, for all measured MPEG-4 FAPs for several expressive modes (see text for definitions and details).

and averages across the whole database. It is however conceivable that facial movements will at least reinforce the perception of focal accent. The mean FMQ taken over all expressive modes is 1.6. The expressive mode yielding the largest mean FMQ is happy (1.9) followed by confirming (1.7), while questioning has the lowest mean FMQ value of 1.3. If we look at the individual parameters and the different expressive modes, some FMQs are significantly greater, especially for the Happy expression, up to 4 for parameter 34 ”raise right mid eyebrow”.

While much more detailed data on facial movement patterns is available in the database, we wanted to show the strong effects of focal accent on basically all facial movement patterns. Moreover, the results suggest that the specific gestures used for realization of focal accent are related to the intended expressive mode.

5. Visual Cues for Feedback

The use of a believable talking head can trigger the user's social skills such as using greetings, addressing the agent by name, and generally socially chatting with the agent. This was demonstrated by the results of the public use of the KTH August system [34]. These results led to more specific studies on visual cues for feedback such as [35] in which smile, for example, was found to be the strongest cue for affirmative feedback.

The stimuli used in [35] consisted of an exchange between a human, who was intended to represent a client, and the face, representing a travel agent. An observer of these stimuli could only hear the client's voice, but could both see and hear the agent. The human utterance was a natural speech recording and was exactly the same in all exchanges, whereas the speech and the facial expressions of the travel agent were synthetic and variable. The fragment that was manipulated always consisted of the following two utterances:

Human: "Jag vill åka från Stockholm till Linköping."
("I want to go from Stockholm to Linköping.")
Head: "Linköping."

The stimuli were created by orthogonally varying six visual and acoustic parameters, using two possible settings for each parameter: one which was hypothesised to lead to affirmative feedback responses, and one which was hypothesised to lead to negative responses. The task was to respond to this dialogue exchange in terms of whether the agent signals that he understands and accepts the human utterance, or rather signals that he is uncertain about the human utterance. A detailed description of the experiment and the analysis can be found in [35]. Here, we would only like to highlight the strength of the different acoustic and visual cues. In Figure 7 the mean difference in response value (the response weighted by the subjects' confidence ratings) is presented for negative and affirmative settings of the different parameters. The effects of Eye_closure and Delay are not significant, but the trends observed in the means are clearly in the expected direction. There appears to be a strength order with Smile being the most important factor, followed by F0_contour, Eyebrow, Head_movement, Eye_closure and Delay.

This study clearly shows that subjects are sensitive to both acoustic and visual parameters when they have to judge utterances as affirmative or negative feedback signals. One obvious next step is to test whether the fluency of human-machine interactions is helped by the inclusion of such feedback cues in the dialogue management component of a system.

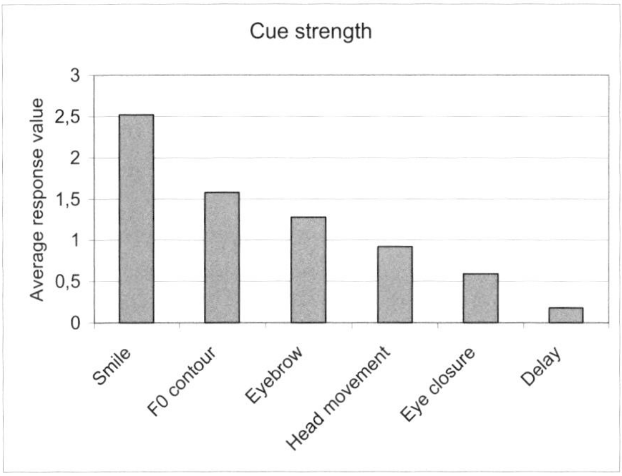

Figure 7. The mean response value difference for stimuli with the indicated cues set to their affirmative and negative value.

6. Modelling Agent Expressiveness and Attitude

In conjunction with the development of data-driven visual synthesis as reported in section 2, two different evaluation studies have been carried out. One was designed to evaluate expressive visual speech synthesis in the framework of a virtual language tutor [36]. The experiment, reported in detail in [37], used a method similar to the one reported on in section 5 above. The talking head had the role of a language tutor who was engaged in helping a student of Swedish improve her pronunciation. The test subjects hear the student's pronunciation and both see and hear the tutor's response. The task was to judge what emotion the talking head expressed in its final turn of the interaction.

Visual synthesis derived from a happy, angry, sad and neutral database was used to drive the new MPEG-4 compatible talking head as described in section 2. For the audio part of the stimuli, a pre-recorded human voice was use to portray the three emotions since we have not yet developed suitable audio data-driven synthesis with different emotions. All possible combinations of auditory and visual stimuli were tested. The results indicated that for stimuli where the auditory and visual emotions matched, listener recognition of each emotion was quite good: 87% for neutral and happy, 70% for sad and 93% for angry. For the mismatched stimuli, the visual elements seemed to have a stronger influence than the auditive elements. These results point to the importance of matching audio and visual emotional content and show that subjects attend to the visual element to a large degree when judging agent expressiveness and attitude.

In another experiment reported on in [38], the new talking head was evaluated in terms of degrees of friendliness. Databases of angry, happy and neutral emotions were used to synthesize the utterance, "*Vad heter du?*" (What is your name?). Samples of the three versions of the visual stimuli are presented in Figure 8. The three versions of the visual synthesis were combined with two audio configurations: low, early pitch peak and high, late pitch peak resulting in six stimuli. Previous experiments showed that the high, late pitch peak elicited more friendly responses [39]. A perception test using these six stimuli was carried out by asking 27 subjects to indicate on a four-point scale (0-3) how friendly they felt the agent was.

The results showing mean scores are presented in Figure 9. It is quite clear that the face synthesized from the angry database elicited the lowest friendliness score. However, there is still evidence of interaction from the audio, as the angry face with the late, high peak received a higher friendliness score than did the angry face with the early, low peak. The faces from the other databases (happy and neutral) elicited more friendliness responses, but neither combination of face and audio received a particularly high friendliness score. The happy face did not elicit more friendliness responses than did the neutral face, but the influence of the audio stimuli remained consistent for all the visual stimuli. Nevertheless, the results show that the visual modality can be a powerful signal of attitude. Moreover, the effects of the auditory cues for friendliness indicate that subjects make use of both modalities in judging speaker attitude. These results stress the need to carefully consider both the visual and audio aspects of expressive synthesis.

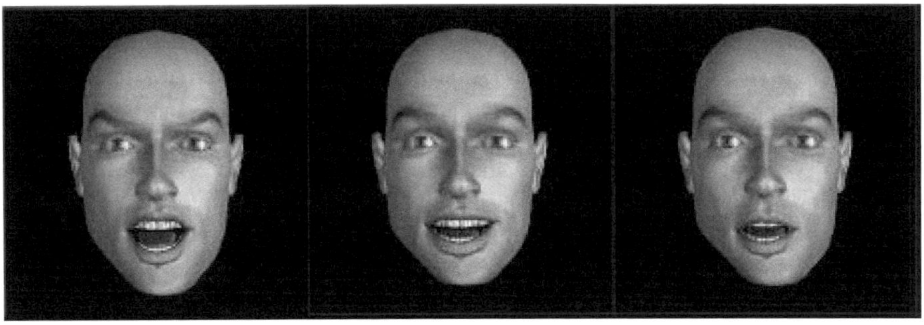

Figure 8. Visual stimuli generated by data-driven synthesis from the angry database (left) the happy database (middle) and the neutral database (right). All samples are taken from the middle of the second vowel of the utterance "*Vad heter du?*" (What is your name?).

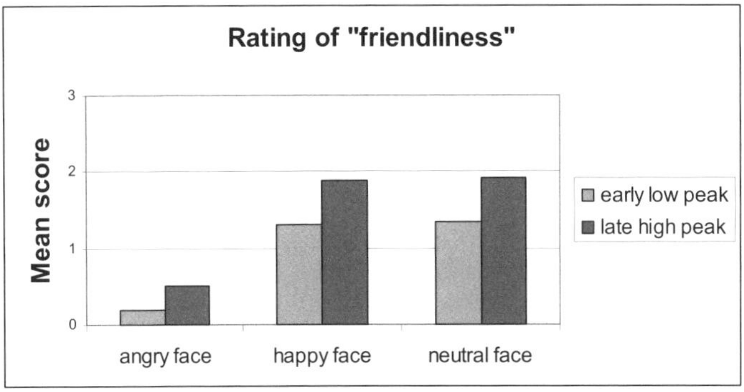

Figure 9. Results from the data-driven synthesis test showing the mean "friendliness score" for each stimulus.

7. Future Challenges

In this chapter, we have presented an overview of some of the recent work in audiovisual synthesis, primarily at KTH, regarding data collection methods, modeling and evaluation experiments, and implementation in animated talking agents for dialogue systems. From this point of departure, we can see that many challenges remain before we will be able to create a believable, animated talking agent based on knowledge concerning how auditory and visual signals interact in verbal and non-verbal communication. In terms of modelling and evaluation, there is a great need to explore in more detail the coherence between audio and visual prosodic expressions especially regarding different functional dimensions. As we demonstrated in the section on prominence above, head nods which strengthen the percept of prominence tend to be integrated with the nearest candidate syllable resulting in audiovisual coherence. However, head nods which indicate dialogue functions such as feedback or turntaking may not be integrated with the audio in the same way. Visual gestures can even be used to contradict or qualify the verbal message, which is often the case in e.g. ironic expressions. On the other hand, there are other powerful visual communicative cues such as the smile which clearly affect the resulting audio (through articulation) and must by definition be integrated with the speech signal. Modelling of a greater number of parameters is also essential, such as head movement in more dimensions, eye movement and gaze, and other body movements such as hand and arm gestures. To model and evaluate how these parameters combine in different ways to convey individual personality traits while at the same time signalling basic prosodic and dialogue functions is a great challenge.

8. Acknowledgements

The work at KTH reported here was carried out by a large number of researchers at the Centre for Speech Technology which is gratefully acknowledged. The work has also been supported by the EU/IST projects SYNFACE and PF-Star.

References

[1] Cassell J., Bickmore T., Campbell L., Hannes V. and Yan H. "Conversation as a System Framework: Designing Embodied Conversational Agents". In: Cassell J., J. Sullivan, S. Prevost & E. Churchill (Eds.). *Embodied Conversational Agents*, 29-63, Cambridge MA: The MIT Press. 2000.

[2] Massaro D.W. *"Perceiving Talking Faces: From Speech Perception to a Behavioural Principle"*, Cambridge, MA: The MIT Press. 1998.

[3] Bickmore T. and Cassell J. "Social Dialogue with Embodied Conversational Agents" In J. van Kuppevelt, L. Dybkjaer, and N. O. Bernsen, (eds.) *Advances in Natural Multimodal Dialogue Systems*, 23-54, Springer, Dordrecht, The Netherlands. 2005.

[4] Granström B., House D., Beskow J. and Lundeberg M. "Verbal and visual prosody in multimodal speech perception." In W. von Dommelen and T. Fretheim eds. *Nordic Prosody: Proc of the VIIIth Conference, Trondheim 2000* . Frankfurt am Main: Peter Lang, 77-88. 2001.

[5] Carlson R. and Granström B. "Speech Synthesis", In Hardcastle W and Laver J (Eds.). *The Handbook of Phonetic Sciences*, 768-788, Oxford: Blackwell Publishers Ltd. 1997.

[6] Beskow J. "Rule-based Visual Speech Synthesis", In *Proceedings of Eurospeech '95*, 299-302, Madrid, Spain. 1995.

[7] Beskow, J. "Animation of Talking Agents", In: Proceedings of AVSP'97, *ESCA Workshop on Audio-Visual Speech Processing*, 149-152, Rhodes, Greece. 1997.

[8] Parke F. I. "Parameterized models for facial animation", IEEE Computer Graphics, 2(9), 61-68, 1982.

[9] Engwall O. "Combining MRI, EMA and EPG measurements in a three-dimensional tongue model," *Speech Communication, 41(2-3),* 303-329. 2003.

[10] Cole R., Massaro D.W., de Villiers J., Rundle B., Shobaki K., Wouters J., Cohen M., Beskow J., Stone P., Connors P., Tarachow A. and Solcher D. "New tools for interactive speech and language training: Using animated conversational agents in the classrooms of profoundly deaf children". In *Proceedings of ESCA/Socrates Workshop on Method and Tool Innovations for Speech Science Education (MATISSE)*, 45-52, London: University College London. 1999.

[11] Massaro D.W., Bosseler A. and Light J. "Development and Evaluation of a Computer-Animated Tutor for Language and Vocabulary Learning", *15th International Congress of Phonetic Sciences (ICPhS '03)*, Barcelona, Spain, 143-146. 2003.

[12] Massaro D.W. and Light J. "Read My Tongue Movements: Bimodal Learning To Perceive And Produce Non-Native Speech /r/ and /l/", *Eurospeech 2003* Geneva, Switzerland, 2249-2252. 2003

[13] Bälter O., Engwall O., Öster A.-M. and Kjellström H., "Wizard-of-Oz Test of ARTUR - a Computer-Based Speech Training System with Articulation Correction," In *Proceedings of the Seventh International ACM SIGACCESS Conference on Computers and Accessibility,* pp. 36-43, October 9-12, 2005, Baltimore, MD. 2005.

[14] Sjölander K. and Beskow J. "WaveSurfer - an Open Source Speech Tool", Proc of ICSLP 2000, Vol. 4, 464-467, Beijing. 2000.

[15] Beskow, J. and Nordenberg, M. "Data-driven Synthesis of Expressive Visual Speech using an MPEG-4 Talking Head", In *Proceedings of INTERSPEECH 2005*, Lisbon, Portugal, 793-796. 2005.

[16] Nordstrand M., Svanfeldt G., Granström B. and House D. "Measurements of articulatory variation in expressive speech for a set of Swedish vowels." *Speech Communication, 44,* 187-196. 2004.

[17] Beskow J., Engwall O. and Granström, B. "Resynthesis of facial and intraoral articulation from simultaneous measurements". *Proc of ICPhS 2003,* Barcelona, Spain, 431-434. 2003.

[18] Pandzic I.S. and Forchheimer, R. *MPEG Facial animation – the standard, implementation and applications,* Chichester, England: John Wiley & Sons. 2002.

[19] Beskow J., Cerrato L., Cosi P., Costantini E., Nordstrand M., Pianesi F., Prete M. and Svanfeldt G. "Preliminary Cross-cultural Evaluation of Expressiveness in Synthetic Faces." In *Proc Affective Dialogue Systems (ADS) 2004*, 301-304, Kloster Irsee, Germany. 2004.

[20] Beskow J., Cerrato L., Granström B., House D., Nordenberg M., Nordstrand M. and Svanfeldt G. (2004). "Expressive Animated Agents for Affective Dialogue Systems ." In *Proc Affective Dialogue Systems (ADS) 2004*, 240-243, Kloster Irsee, Germany. 2004.

[21] Agelfors E., Beskow J., Dahlquist M., Granström B., Lundeberg M., Spens K.-E. and Öhman T. "Synthetic faces as a lipreading support". In: *Proceedings of ICSLP'98*, Sydney, Australia, 3047-3050. 1998.

[22] Beskow J., Granström B. and Spens K.-E. "Articulation strength - Readability experiments with a synthetic talking face" *TMH-QPSR vol. 44,* Stockholm: KTH, 97-100. 2002.

[23] Westervelt P.J. "Parametric acoustic array*", J. Acoust. Soc. Amer. 35*, 535-537. 1963.

[24] Svanfeldt, G. and Olszewski, D. "Perception experiment combining a parametric loudspeaker and a synthetic talking head", In *Proceedings of INTERSPEECH 2005*, Lisbon, Portugal, 1721-1724, 2005.

[25] Granström B, House D. and Lundeberg M. "Prosodic Cues in Multimodal Speech Perception", In *Proceedings of the International Congress of Phonetic Sciences (ICPhS99),* 655-658, San Francisco. 1999.

[26] Krahmer E., Ruttkay Z., Swerts M. and Wesselink W. "Perceptual evaluation of audiovisual cues for prominence." In: *Proceedings 7th International Conference on Spoken Language Processing* (ICSLP 2002), Denver, Colo., USA. 1933-1936. 2002.

[27] House D., Beskow J. and Granstrom B. "Timing and interaction of visual cues for prominence in audiovisual speech perception*". Proc. Eurospeech 2001*, 387-390, Aalborg, Denmark. 2001.

[28] Massaro D.W., Cohen M.M. and Smeele P.M.T. "Perception of asynchronous and conflicting visual and auditory speech". *J. Acoust. Soc. Am. 100.* 1777-1786. 1996.

[29] Keating P., Baroni M., Mattys S., Scarborough R., Alwan A., Auer E. and Bernstein L. "Optical Phonetics and Visual Perception of Lexical and Phrasal Stress in English", In *Proc. 15th International Congress of Phonetic Sciences*, 2071-2074. 2003.

[30] Dohen M. *Deixis prosodique multisensorielle: Production et perception audiovisuelle de la focalisation contrastive en Français.* PhD thesis, Institut de la Communication Parlée, Grenoble. 2005.

[31] Beskow J., Cerrato L., Granström B., House D., Nordstrand M. and Svanfeldt G. "The Swedish PF-Star Multimodal Corpora." In *Proc LREC Workshop, Multimodal Corpora: Models of Human Behaviour for the Specification and Evaluation of Multimodal Input and Output Interfaces*, 34-37. Lisbon, Portugal. 2004.

[32] Cerrato L. and Svanfeldt G. "A method for the detection of communicative head nods in expressive speech", In *Proceedings of The Second Nordic Conference on Multimodal Communication,* 153-165, Gothenburg University. 2006.

[33] Beskow J., Granström B. and House D. "Visual correlates to prominence in several expressive modes." In *Proceedings of Interspeech 2006*, 1272-1275. Pittsburg, PA. 2006.

[34] Bell L. and Gustafson J. "Interacting with an animated agent: an analysis of a Swedish database of spontaneous computer directed speech." *Proc of Eurospeech '99*, 1143-1146. 1999.

[35] Granström B., House D. and Swerts M.G. "Multimodal feedback cues in human-machine interactions." In Bernard Bel & Isabelle Marlien (Eds.), *Proceedings of the Speech Prosody 2002 Conference*, 347-350. Aix-en-Provence: Laboratoire Parole et Langage. 2002.

[36] Granström B. "Towards a virtual language tutor", *Proc InSTIL/ICALL2004 NLP and Speech* Technologies in Advanced Language Learning Systems, 1-8, Venice. 2004.

[37] Beskow J. and Cerrato L. "Evaluation of the expressivity of a Swedish talking head in the context of human-machine interaction", In: *Proceedings of Gruppo di Studio della Comunicazione Parlata (GSCP) '04*, Padova, Italy. 2006.

[38] House D. "On the interaction of audio and visual cues to friendliness in interrogative prosody", In *Proceedings of The Second Nordic Conference on Multimodal Communication,* 201-213, Gothenburg University. 2006.

[39] House D. "Phrase-final rises as a prosodic feature in wh-questions in Swedish human–machine dialogue", Speech *Communication 46*: 268-283. 2005.

Fundamentals of Verbal and Nonverbal Communication and the Biometric Issue
A. Esposito et al. (Eds.)
IOS Press, 2007

Towards a Socially and Emotionally Attuned Humanoid Agent

Christopher PETERS[a] Catherine PELACHAUD[1a], Elisabetta BEVACQUA[a]
Magalie OCHS[a,b], Nicolas Ech CHAFAI [a,b], Maurizio MANCINI[a]
[a]LINC University of Paris 8
[b]France Telecom R&D

Abstract. The ability for AI controlled humanoid agents to interact and resonate with the user and with each other in a social and emotional manner is of the utmost importance for creating a sense of plausibility, immersion and ultimately enhancing user experience. In order to express themselves effectively, agents will need to be adept in a broad range of communicative verbal and nonverbal behaviors based on internal models that are not incoherent with their human counterparts. In this work, we present a number of avenues of research that we feel help enhance agents abilities to communicate socially and emotionally. The core of this work consists of models of attention and emotion which are augmented with communication management and behavior expressivity control. The work presented here has particular potential as a basis for supporting, at a low-level, individual agent behavior driven by high-level specifications from dialog systems in the next generation of human-user interfaces.

Introduction

Human-user Interaction have become an increasingly useful, if not challenging domain, for the testing and application of human-level artificial intelligence techniques. Virtual agents populate web applications, interactive systems, video games, etc. A common difficulty in each of these spheres is the control of the animation of the agents: Pre-scripted animation is simply not adequate for truly interactive systems. Autonomy in terms of deciding which action to take and which behavior to show is a must for plausibly animated agents. In interactive systems that contain a social aspect an essential role of virtual agents is to communicate with other agents, both user and computer controlled.

It is our view that humanoid virtual agents should be endowed with capabilities to allow them to perceive their physical and social context, to feel and express emotions, to dialog and communicate with other agents. In current technology, agents often behave as if being in an empty world: they are not aware of their surrounding, they do not pay attention to whom they are talking to, they do not try to grasp the attention of other agents, etc. In our view, what really matters in interaction is not that agents look like a human, but rather that they are able to resonate with the users through naturalistic communicative and emotional capabilities.

---

[1] Corresponding Author: Catherine Pelachaud, Universite de Paris 8, France, e-mail: **c.pelachaud@iut.univ-paris8.fr**.

Communication involves complex processes. It is fundamentally multi-modal. Nonverbal behaviors do have a major role in the interaction between people. They have several functions such as: they may be used to process the environment and others; they may indicate our mental and emotional state; they may be used as signals to be read by others. For a virtual character to be encompassed with these functions requires that it can: (1) generate multi-modal communicative behavior; (2) have emotions and display expressive behaviors; (3) perceive the environment and the other's gaze and behaviors; (4) manage the conversation by showing feedback.

In this paper we describe how we model these various capabilities. After describing our model of perceptual attention, we present our emotional model. In the following section we explain how we model communicative behaviors. We then turn our attention to our model of communication management that encompass two types of interactants, namely speaker and listener. We end our paper going through a scenario where we detailed our emotion computational model.

Perceptual Attention

The ability to sense information from the environment in a manner congruent with real humans is an important prerequisite for plausible agent behavior. We provide agents with real-time synthetic vision, attention and memory capabilities (see [23]). All three capabilities interact with each other in order to collect information and orient the agents senses with respect to the environment. The visual sensing module is monocular with multiple resolutions; it operates in a snapshot manner by taking frequent updates of the visible region of the scene from the point of view of the agent and passing them to the visual attention module. The purpose of the visual attention module is to choose a set of visible objects or locations in the scene for enhanced processing by the agent and is based on a popular real-time model from cognitive engineering [14]. We use it with virtual scenes by passing full scene renderings from the synthetic vision module in order to compute the saliency map: a 2D gray-scale spatial ranking of the scene in terms of areas that `pop-out' and are thought to attract visual attention. A memory system [25] stores uncertainty levels for each object to ensure the focus of attention moves throughout the scene; values are combined into an uncertainty map, with which the saliency map is modulated in order to provide the final attention map. Artificial regions of interest are enumerated from the attention map and are used to create artificial scan-paths for the agent.

In a gaming environment, these capabilities provide autonomous and human-like looking behaviors and help drive more subtle movements, such as eye saccades and blinking. Such systems serve as a basis for implementing higher-level cognitive and behavioral mechanisms: for example, we have used these capabilities, along with theory of mind, to control autonomous interaction initiation between agents (see the Communication Management Section).

Emotion

Emotion is a crucial element in any gaming experience: virtual characters who are able to display emotions are more likely to be able to invoke an emotional reaction in the user and thus add to the users' experience. An affective virtual agent should be able to adopt an emotional behavior to convey its emotions and be able to represent its environment in terms of its own emotions and those of other characters [5]. Three emotional processes have to be simulated [27]: the elicitation of emotion, expression of emotion and experience of emotion. These allow an agent to identify the circumstances in which an emotion is triggered, display emotions and manifest emotional influence on their reasoning and actions, respectively. Appraisal theory [34] is one way in which we may give an agent the ability to identify the emotional meaning of a situation. According to this theory, emotions are elicited by a subjective interpretation of an event, which depends both on situational and cultural factors and on a particular individual's features. Major determinants in emotion elicitation include the person's beliefs and goals. The interpretation of an event corresponds to the appraisal of a set of variables called appraisal variables. Particular combinations of these variables lead to emotion elicitation (using, in many cases, the OCC model of emotion [20]). Appraisal variables are often hard-coded [8, 32] or based on the agent's plan [11]. In contrast, we aim to create a domain-independent model of emotion elicitation that enables the virtual agent to assess its own emotions and those of other characters or users without having prior access to their plans. We propose to represent emotion elicited-events based on a BDI approach [31]. The mental state of a BDI agent is composed of mental attitudes such as beliefs and intentions. It corresponds to its cognitive representation of the world at a given instant. It includes a representation of the events perceived in the environment. Accordingly, an occurred emotion elicited-event is also represented through mental attitudes. Emotion elicited-events can then be represented by a particular mental state. Based on the OCC model, we have described values of appraisal variables through beliefs and intentions. For instance, a desirable event corresponds to the belief that an occurred event enables an agent to achieve one of its goals. According to appraisal theory, we have represented emotion elicited-events by combinations of values of appraisal variables; that is, by a combination of mental attitudes. From this formalization of emotion elicited-events, a virtual agent can identify in real-time its own emotions triggered by a situation. By knowing other characters intentions, their emotions can also be identified. More details on this computational model of emotion elicitation and associated model of facial expressions of emotions can be found in [19].

Communicative behaviors

In social settings, the study of proper communicative behaviors are of importance for believable and interactive characters. Humans communicate in a rich variety of ways through multiple channels [16, 32, 7, 17]: through our choice of words, facial expressions, body postures, gaze, gestures... Nonverbal behaviors accompany the flow of speech and are synchronized at the verbal level, punctuating accented phonemic segments and pauses. They may have several communicative functions [28, 4], such as managing the dialog flow, emphasizing a word, or even providing information on our mental and emotional state.

In our model, we follow the taxonomy of communicative functions proposed in [28] where four main categories are defined. Behaviors are gathered depending if they provide information about the speaker's beliefs, intentions and affective state. The fourth category corresponds to behaviors that provide meta-cognitive information about speaker's mental state (the speaker may try to remember or recall an information). Each communicative function is defined as a (meaning, signal) pair. Each function may be associated with different signals; that is for a given meaning, there may be several ways to communicate it. Vice versa, a same signal may be used to convey different meanings. To control the agent's behavior we are using a representation language, called `Affective Presentation Markup Language' (APML) where the tags of this language are these communicative functions [7]. In APML, text to be spoken by the agent is annotated with tags denoting emotion and interaction information (see Section entitles Scenario).

Individual styles through expressivity and multi-modal behavior modeling

When animating humanoid characters, it is important that their behavior should be consistent with their personality, mood, emotional state and their individual styles [35]. To express an emotional or mental state, one uses given communicative behaviors. We model individual styles through the choice of behaviors and their manner of execution (that is, multi-modal behaviors and expressivity).

In terms of expressivity modeling, many researchers (see, for example, Wallbott and Scherer [36], Pollick [30]) have investigated human motion characteristics and encoded them into categories, for example, slow / fast and weak / energetic. We define the *expressivity* of behavior as the "quality" of the communicated information through the execution of some physical behavior. We have defined and implemented expressivity [12] as a set of parameters that affect the quality of execution of behaviors, as performed by arms or head movements. These parameters are: speed of arms / head, spatial volume taken by the arms / head, energy and fluidity of arms / head movement, and the number of gesture stroke repetitions.

For multi-modal behavior, a social agent needs to decide which of its available modalities of expression to use: speech, gestures, gaze, facial expressions, body movements and body posture [20, 35]. We have implemented a mechanism to choose between these different available modalities during conversation based on a hierarchy of preferences for modalities. For example, an agent with the tendency to use gestures during communication will have a higher value associated with the gesture modality.

Communicative Management

Communication management deals with the agents ability to autonomously start, maintain and end interactions. Furthermore, communication should not stop when the agent is listening, but the agent should provide feedback to the speaker.

Start of interaction and Perceptual Theory of Mind

Interaction initiation behaviors such as gaze orienting, waving and name utterances are of great significance for adding realism to interactive agent systems and virtual communities where visual social interaction behaviors are desired between agents and avatars [15]. In real environments, people must routinely deploy their senses in order to scan for possible contacts, must gain the attention of one they wish to interact with, must signal interest in communicating both verbally and non-verbally and must seek the cooperation, as well as gauge the willingness, of the other to reciprocate in conversation. For a human viewer, in behavioral plausibility terms, sometimes it is not the destination that counts (i.e. conversation), so much as getting there (i.e. the acts involved in initiating the conversation). We implement an automatic non-verbal interaction initiation system based primarily on direction of attention and perceptual theory of mind.

This allows for special social perception by [24] incorporating a perceptual theory of mind model based on evolutionary psychology and supported by neurophysiological evidence [22]. In our model, an intentionality detection (ID) mechanism filters perceived agents into this special pathway and segments them into eye, head and body subparts. A direction of attention detector (DAD) computes the direction of the eyes, head and body of the other agent with respect to the self. Orientations are merged into an *attention level* metric through a subpart weighting process, which is boosted if mutual attention exists between the agents, as signaled by a mutual attention mechanism (MAM). Results are stored in short-term memory. An agent may integrate these metrics over a time interval on demand in order to formulate theories of whether the other has seen it, whether the other has seen this agent see it and the interest that the other agent appears to have in this agent.

These theories drive a hierarchical finite state machine between various interactional states, such as monitoring the environment, grabbing the attention of the other and gauging the reaction of the other. Thus, unlike other systems, agent behavior is based not only on the goals of the agent, but also on its theory of the intentions of the other in order to avoid the social embarrassment of engaging in conversation with an unwilling participant [10]. Our evaluation of conversation initiation behaviors seems to suggest that users may be as sensitive to subtle non-verbal social cues in the virtual environment as in the real situation [26].

Feedback

During a conversation, two flows of information are established between speaker and listener, one concerning the effective content of the speaker's speech and the other regarding the listener's reaction. Conversation does not happen as a unique flow of information (the speaker providing information to the listener) but it is a continuous exchange of information between interacts. The listener informs at all time whether she is engaged and interested in the conversation, whether she understands and agrees, and furthermore which emotional and attitudinal reactions are elicited in her by the speaker's speech [1, 29]. These signals are called *feedback*; without them a communication becomes difficult and frustrating.

We distinguished two main clusters of feedback signals: cognitive feedbacks and reactive feedbacks. The *cognitive feedbacks* embed signals emitted after a cognitive evaluation of the speaker's actions and says: the listener may show her agreement/disagreement, belief/disbelief, that she understands or she does not, or even that she accepts or not what is being said. The other cluster, the *reactive feedback*, gathers the signals the listener emits unconsciously [1].

A complete feedback model requires to develop a computational model of both feedback types. Implementing the cognitive feedback requires to have a semantic representation of the discourse as well as a representation of the mental state of the listener (her knowledge, her choices, her beliefs,...). We have left aside this difficult task for the moment and have started implemented the reactive model. The listener's behaviors is defined as a set of rules based on corpus analysis [17, 9, 3]. Feedback signals (like head nods, verbal responses) often appear at a pitch variation in the speaker's voice; frowns, body movements and gaze shifts are produced when the speaker shows uncertainty; facial expressions, postural and gaze shifts are provided to reflect those made by the speaker (mimicry).

Interaction Maintenance

In applications featuring conversational interaction with a user, the use of gestures may enhance the communication and help to maintain the interaction. Gestures help the speaker formulate his thoughts and create a link with the speaker's thoughts and have a pragmatic value in the sense that they illustrate his communicative efforts. To maintain a listener in a conversational interaction, we need to study this pragmatic dimension of gestures. The aim of our study is to point out the importance of the pragmatic value of gestures and their rhetorical relations when trying to create applications in which agents are able to maintain the user's interest during conversation interaction.

While previous works [12, 2] have focused on gestures that elicit focal attention through the collection of objective data from an eye tracker, we have been collecting subjective data from a corpus of conversational interactions in 2D traditional animations. From this data, we have been considering the pragmatic intention of the speaker from the expressive parameters of gestures (see previous Section on Communicative Behaviors) in order to establish rhetorical relationship (relations of similarity or contrast) between the elements of the verbal or non-verbal utterance.

More particularly, we have gathered a corpus of 2D animations depicting conversation scene. We have defined a schema for the annotation of gesture expressivity at a low-level of segmentation that we used to annotate the cartoons. Annotations took place at the gesture phase level. Interested in the communicative value of gestures, we looked at the modulations of gesture expressivity over time rather than their values. This led us to introduce two notions of expressivity breaks: irregularities, and discontinuities. In irregularities, within a sequence of gestures with a constant expressivity, only one gesture phase has a different value. On the other hand, discontinuities refer to sudden variation in the expressivity values; that is two successive sequences of gestures are performed with different expressivity. These types of gesture expressivity breaks suggest that they act as a relation of similarity with a function of segmentation, for the irregularities, and as a relation of contrast that derives from the speaker's effort to communicate her intention, for the discontinuities.

We have started to investigate how these breaks may be used to trigger an increase in user's attention, for example, by creating a contrasting gesture in an otherwise similar sequence by modulating the gesture expressivity parameters.

Scenario

We have constructed a scenario featuring two agents in dialog in order to demonstrate our research ideas, particularly in terms of feedback and emotion models.

The scenario has been constructed using the Torque game engine [36] and a real-time version of the Greta agent. In APML, text to be spoken by the agent is annotated with tags denoting emotion and communication function [6].

In the scenario, two agents, Lucy and Greta, meet. The following dialog takes place:

Greta : Hi Lucy

Lucy : Hi Greta ; How are you ?
G : I am fine and you ?
L : Oh I had a terrible day :(
G : (surprise - a little concern) oh really ?
L : Yes. I had to do a demo for a conference. But it is over now (relief, positive emotion)
G : (shows relief) 'phew'

Each sentence in the dialog is represented by a single APML file and is annotated accordingly. For example, the APML file for Greta's reply to Lucy with "Oh really" is as follows:

```
<?xml version="1.0"?>
<!DOCTYPE apml SYSTEM "apml.dtd" []>
<apml>
<performative type="inform">
<theme affect="concern">
Oh really
<boundary type="LH"/>
</theme>
</performative>
<pause sec="2.0"/>
</apml>
```

We explain now the different steps our computational model undertake when computing which emotions arise within this dialog for a particular conversational situation. We describe how the mental states of the agents trigger different emotions.

The emotion elicitation of Lucy

During the demo of the conference, Lucy's mental state is:

- She has the goal that her demo operates well:
  Goal(operationalDemo)
- She has the belief that it is possible that her demo fails :
  Belief(∃ event(Feasible(event, ¬operationalDemo)))

So, Lucy has an emotion of **fear** because she has the following mental state:

Goal(p) ∧ Belief(∃ event(Feasible(event, ¬p))) where p = operationalDemo

Where Feasible(event, p) means that event can take place and if it does p will be true after event

After the demo of the conference, Lucy's mental state is:

- She has the goal that her demo operates well :
  Goal(operationalDemo)
- She believed before the end of the conference (noted Event2) that it is possible that her demo fails :
  Belief(Done(Event2,Belief(∃ event(Feasible(¬operationalDemo)))))
- She believes now that her demo cannot fail anymore :
  Belief(¬(∃ event(Feasible(¬operationalDemo))))

So, Lucy has an emotion of **relief** because she has the following mental state :

$$Goal(p) \wedge Belief(Done(Event2, Belief(\exists\, event(Feasible(\neg p))))) \wedge Belief(\neg(\exists\, event(Feasible(\neg p))))$$

where p = operationalDemo and Done(event, p) means that event has just taken place and p was true before event.

Moreover, Lucy changes her mental state after the demo of the conference because :

- She has the goal during the conference that her demo operates well : Goal(operationalDemo)
- She has achieved her goal at the end of the conference: Belief(operationalDemo)

So, Lucy has an emotion of **joy** because she has the following mental state:

$$Belief(Done(Event2, Goal(p)) \wedge p) \text{ where } p = operationalDemo$$

Let us turn now our attention to the second agent, Greta.

During the conversation, Greta feels empathic emotions of **relief** and **joy** because:

- She knows the emotion elicitation mental state. In this case, she knows the mental states that trigger the emotion of joy and relief as described above.
- Lucy told Greta her goal ("I had to do a demo for a conference"), and Greta can suppose Lucy's beliefs. So, she knows the mental state (goals and beliefs) of Lucy.
- From this knowledge, she can infer Lucy's emotions.

Such an elicitation is an important way to tie mental states, defined by agent's beliefs and goals, to emotion modulation, and hence drive the emotional expression of the agent. Figures 1 and 2 illustrates the facial expressions Greta displayed depending on her emotional states and communicative intent.

Although work is still in progress in enumerating new annotation tags relevant for feedback, this scenario is a proof-of-concept of what we are trying to achieve in terms of feedback during conversation between two agents. We are also using the scenario as an example case for emotion elicitation based on mental state, as outlined above.

Conclusion

This paper has presented some avenues of research that we have been following with a view to contribute to the social and emotional interaction skills of the next generation of real-time interactive agents. Our work focuses on low-level high-detail interaction scenarios between a small number of agents and is particularly suited to applications that require a social aspect. Among many challenges is the effort to extend our models beyond scenarios involving only two interactants, to handle more agents. This is especially challenging due to the predominance of existing theory for low number of interactants, making formulation of appropriate models difficult.

Fig.1. Feedback of surprise from the Greta agent on the left (close-up shown on inset) based on the dialog of the speaking agent of the right. In the scenario shown here, the agent also provides feedback by nodding at the speaker.

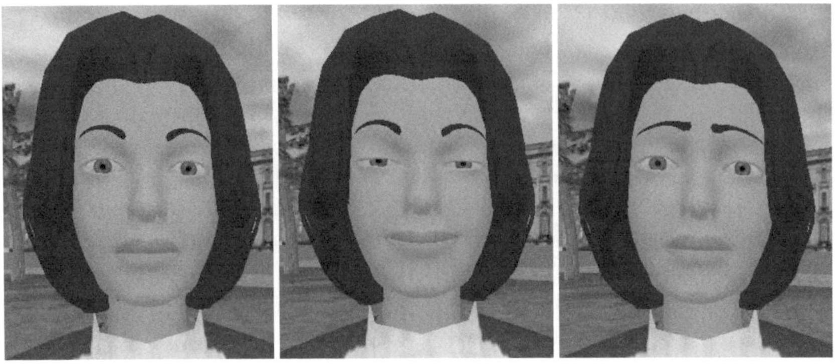

Fig.2. A number of different facial expressions generated by the agent as feedback in her role of listener. Controlled by APML, these are just a few examples of the wide range of facial expressions the Greta agent is capable of.

Acknowledgement

This work has been funded by the Network of Excellence Humaine, IST-2002-2.3.1.6 Contract no. 507422 http://emotion-research.net/

Reference

[1] J. Allwood, J. Nivre, and E. Ahlsn. On the semantics and pragmatics of linguistic feedback. Semantics, 9(1), 1993.
[2] G. Barrier, J. Caelen, and B. Meillon. La visibilité des gestes: Paramètres directionnels, intentionnalité du signe et attribution de pertinence. In Workshop Français sur les Agents Conversationnels Animés, pages 113-123, Grenoble, France, 2005.
[3] J. Cassell, C. Pelachaud, N. Badler, M. Steedman, B. Achorn, T. Becket, B. Douville, S. Prevost, M. Stone, Animated Conversation: Rule-Based Generation of Facial Expression, Gesture and Spoken Intonation for Multiple Conversational Agents. Proceedings of SIGGRAPH'94, pp. 413-420, ACM Special Interest Group on Graphics, 1994.
[4] N. Chovil. Social determinants of facial displays. Journal of Nonverbal Behavior, 15(3):141-154, Fall 1991.
[5] C. Crawford. On Game Design. New Riders Games, 2003.
[6] B. DeCarolis, C. Pelachaud, I. Poggi, and M. Steedman. APML, a markup language for believable behavior generation. In Helmut Prendinger and Mitsuru Ishizuka, editors, Life- Like Characters, Cognitive Technologies. Springer, 2004.
[7] P. Ekman. About brows: Emotional and conversational signals. In M. von Cranach, K. Foppa, W. Lepenies, and D. Ploog, editors, Human ethology: Claims and limits of a new discipline: contributions to the Colloquium, pages 169-248. Cambridge University Press, Cambridge, England; New-York, 1979.
[8] C. Elliot. The Affective Reasoner: A process model of emotions in a multi-agent system. Computer science, Northwestern University, 1992.
[9] D. Friedman and M. Gillies. Teaching virtual characters to use body language. In Intelligent Virtual Agents, Lecture Notes in Artificial Intelligence. Springer-Verlag, 2005.
[10] E. Goffman. Behaviour in public places: notes on the social order of gatherings. The Free Press, New York, 1963.
[11] J. Gratch. Emile: Marshalling passions in training and education. In The Fourth International Conference on Autonomous Agents, pages 325-332, Barcelona, Catalonia, Spain, 2000.
[12] M. Gullberg and K. Holmqvist. Keeping an eye on gestures: Visual perception of gestures in face-to-face communication. Pragmatics and Cognition, 7:35-63, 1999.
[13] B. Hartmann, M. Mancini, and C. Pelachaud. Towards affective agent action: Modelling expressive ECA gestures. In Proceedings of the IUI Workshop on Affective Interaction, San Diego, CA, January 2005.
[14] L. Itti. Models of Bottom-Up and Top-Down Visual Attention. PhD thesis, California Institute of Technology, Jan 2000.
[15] K. Isbister. Better Game Characters by Design: A Psychological Approach. Elsevier Science and Technology Books, 2006.
[16] A. Kendon. Movement coordination in social interaction: Some examples described. In S. Weitz, editor, Nonverbal Communication. Oxford University Press, 1974.
[17] R. M. Maatman, J. Gratch, and S. Marsella. Natural behavior of a listening agent. In 5th International Conference on Interactive Virtual Agents. Kos, Greece, 2005.
[18] D. McNeill. Hand and Mind: What Gestures Reveal about Thought. University of Chicago, 1992.
[19] M. Ochs, R. Niewiadomski, C. Pelachaud, and D. Sadek. Intelligent expressions of emotions. In Jianhua Tao, Tieniu Tan, and Rosalind W. Picard, editors, The 1st International Conference on Affective Computing and Intelligent Interaction (ACII05), pages p. 707-714, Beijing, China, 2005. Springer.
[20] A. Ortony, G.L. Clore, and A. Collins. The cognitive structure of emotions. Cambridge University Press, 1988.
[21] C. Pelachaud, V. Carofiglio, B. De Carolis, and F. de Rosis. Embodied contextual agent in information delivering application. In First International Joint Conference on Autonomous Agents and Multi-Agent Systems (AAMAS), Bologna, Italy, July 2002.
[22] D.I. Perrett and N.J. Emery. Understanding the intentions of others from visual signals: Neurophysiological evidence. Current Psychology of Cognition, 13:683-694, 1994.
[23] C. Peters. Bottom-Up Visual Attention for Autonomous Virtual Human Animation. PhD thesis, Department of Computer Science, Trinity College Dublin, 2004.

[24] C. Peters. Direction of attention perception for conversation initiation in virtual environments. In International Working Conference on Intelligent Virtual Agents, pages 215-228, Kos, Greece, September 2005.
[25] C. Peters. Designing Synthetic Memory Systems for Supporting Autonomous Embodied Agent Behaviour, Proceedings of the 15th International Symposium on Robot and Human Interactive Communication RO-MAN), pp. 14–19, September 2006.
[26] C. Peters. Evaluating perception of interaction initiation in virtual environments using humanoid agents. In Proceedings of the 17th European Conference on Artificial Intelligence, pages 46-50, Riva Del Garda, Italy, August 2006.
[27] R. Picard. Affective Computing. MIT Press, 1997.
[28] I. Poggi. Mind markers. In N. Trigo M. Rector, I. Poggi, editor, Gestures. Meaning and use. University Fernando Pessoa Press, Oporto, Portugal, 2003.
[29] I. Poggi. Backchannel: from humans to embodied agents. In Conversational Informatics for Supporting Social Intelligence and Interaction - Situational and Environmental Information Enforcing Involvement in Conversation workshop in AISB05. University of Hertfordshire, Hatfield, England, 2005.
[30] F. E. Pollick. The features people use to recognize human movement style. In A. Camurri and G. Volpe, editors, Gesture-Based Communication in Human-Computer Interaction - GW 2003, number 2915 in LNAI, pages 10-19. Springer, 2004.
[31] A. S. Rao and M.P. Georgeff. Modeling rational agents within a BDI architecture. In Proceedings of International Conference on Principles of Knowledge Representation and Reasoning (KR), pages 473-484, San Mateo, CA, USA, 1991.
[32] S. Reilly. Believable Social and Emotional Agents. Computer science, University of Carnegie Mellon, 1996.
[33] A.E. Scheflen. The significance of posture in communication systems. Psychiatry, 27, 1964.
[34] K. Scherer. Introduction to Social Psychology: A European perspective, chapter Emotion. Oxford, 2000.
[35] Z. Ruttkay and C. Pelachaud. Exercises of style for virtual humans. In Symposium of the AISB02 Convention, Volume Animating Expressive Characters for Social Interactions, London, 2002.
[36] The Torque Game Engine (TGE), GarageGames, http://www.garagegames.com.
[37] H. G. Wallbott and K. R. Scherer. Cues and channels in emotion recognition. Journal of Personality and Social Psychology, 51(4):690-699, 1986.

Fundamentals of Verbal and Nonverbal Communication and the Biometric Issue
A. Esposito et al. (Eds.)
IOS Press, 2007

Nonverbal and Bodily Interaction in Ambient Entertainment

Anton NIJHOLT[1], Dennis REIDSMA, Zsofia RUTTKAY, Herwin van WELBERGEN, Pieter BOS
University of Twente, Center of Telematics & Information Technology, the Netherlands

Abstract. In future Ambient Intelligence environments we assume intelligence embedded in the environment and its virtual, sometimes visualized agents (virtual humans). These environments support the human inhabitants or visitors in their activities and interactions by perceiving them through their sensors. In this paper we look at our research on bodily and gestural interaction with environments equipped with simple sensors, application-dependent intelligence, and an embodied virtual agent employed in the display of reactive and pro-active activity. The virtual humans we discuss play roles such as dance partner, conductor or trainer. All of them require the perception and the generation of bodily activity and other display of nonverbal communication. The role of affect and persuasion in these ambient entertainment environments is touched upon.

Keywords. Ambient Intelligence, entertainment, nonverbal communication, embodied agents, bodily interaction, behavioral processes, affect

1. Introduction

In future Ambient Intelligence (AmI) environments we assume intelligence embedded in the environment, its objects (furniture, mobile robots) and its virtual, sometimes visualized agents (virtual humans). These environments support the human inhabitants or visitors of these environments in their activities and interactions by perceiving them through their sensors (proximity sensors, cameras, microphones, etc.). Support can be reactive, but also and more importantly, pro-active and unobtrusive, anticipating the needs of the inhabitants and visitors by sensing their behavioral signals and being aware of the context in which they act [8].

Health, recreation, sports and playing games are among the needs inhabitants and visitors of smart environments will have. Sensors in these environments can detect and interpret bodily activity and can give multimedia feedback to invite, stimulate, guide and advise on bodily activity. Rather than aiming at improving user task efficiency, in the environments we investigate the aim is to improve physical and mental health (well-being) through exercise and through play. Exercises can be done in order to improve fitness, to prevent certain injuries (e.g., RSI), or to recover from an accident (e.g., physiotherapy exercises). Other exercises may aim at improving certain capabilities related to a profession (ballet, etc.), some kind of recreation (juggling, etc.),

[1] Corresponding Author: University of Twente, Center of Telematics & Information Technology, PO Box 217, 7500 AE Enschede, the Netherlands, e-mail: A.Nijholt@ewi.utwente.nl.

or sports (fencing, etc.). Fun, just fun, achieved from interaction (e.g. dancing or physical gaming) can be another aim of such environments.

In this paper we look at our research on bodily and gestural interaction with environments equipped with some simple sensors (cameras, microphones, dance pads), some application-dependent intelligence (allowing reactive and pro-active activity), and an embodied virtual agent employed in the display of reactive and pro-active activity. The activities we look at are dance movements (both from a virtual and a human dancer), music conducting (a virtual conductor conducting human musicians) and fitness, aerobics or physiotherapy exercises (to be performed by a human trainee or patient and to monitored and demonstrated by a virtual trainer).

The organization of this paper is as follows. In section 2 we look at application domains of our research on nonverbal and bodily interaction in smart environments. Section 3 is devoted to some of our virtual agents performing in these application domains. We show both practical and theoretical design considerations. In section 4 we look at shortcomings of our applications and topics of future research.

2. Bodily Interaction: Entertainment, Health, and Others

Entertainment, health, sports, and leisure applications using information and communication technology often require and encourage physical body movements and often applications are designed for that reason. In our research we look at bodily and gestural interaction with game and leisure environments that are equipped with sensors (cameras, microphones, touch, and proximity sensors) and application-dependent intelligence (allowing reactive and proactive activity). Interpretation of the bodily interaction, requiring domain-dependent artificial intelligence, needs to be done by the environment and the agents that maintain the interaction with the human partner. In the display of reactive and pro-active activity embodied virtual agents play an important role. Virtual agents can play the role of teacher, coach, partner or buddy. One underlying assumption is that emphasis on activities in which the experience rather than the result will guide the design of social and intelligent systems that will become part of ambient intelligence home environments [7].

3. Inviting and Maintaining Interaction by Virtual Humans

Three applications have been designed in which our ideas about nonverbal and bodily interaction have been implemented. The implementations are available, but they are certainly not final. We will have a look at a virtual dancer that invites a visitor to her environment to dance with her, a conductor that guides musicians in its environment to play according the score designed by a composer, and a virtual trainer (e.g. in the role of fitness trainer or physiotherapist) that knows about exercises that need to be performed by a user or patient. In all these applications there is a continuous interaction between embodied agent and its human partner. Moreover, rather than have the more traditional verbal interaction supported by nonverbal communication, here the main interaction that takes place is nonverbal, and speech and language, when present at all, take the supporting role. External signals like music being played can also have a role in addition to the multimodal communication.

3.1 A Virtual Dancer

Our Virtual Dancer [10] is an interactive dancing agent that dances together with the user to the beat of the music (cf. [13] for related work). The dancer adapts its performance to whatever the human user is doing. The moves of the virtual dancer, to be chosen from a database obtained from motion-capturing, are aligned to the beats of the music. These beats are detected real-time. The system observes the movements of the human dancer using computer vision software. The system extracts global characteristics about the movements of the human dancer, such how much (s)he moves around or how much (s)he waves with the arms. Thee characteristics are used to select moves from the database that are in some way "appropriate" to the dancing style of the human dancer. Finally, there is a dance pad that registers feet activity. By alternating patterns of following the user with taking the lead with new moves, the system attempts to achieve a mutual dancing interaction where both human and virtual dancer influence each other. Finding the appropriate nonverbal interaction patterns that allow us to have a system that establishes rapport with its visitors is one of the longer term issues that is being addressed in this research.

3.2 A Virtual Conductor

We have designed and implemented a virtual conductor [2] that is capable of leading, and reacting to, live musicians in real time. The conductor possesses knowledge of the music to be conducted, and it is able to translate this knowledge to gestures and to produce these gestures. The conductor extracts features from the music and reacts to them, based on information of the knowledge of the score. The reactions are tailored to elicit the desired response from the musicians.

Clearly, if an ensemble is playing too slow or too fast, a (human) conductor should lead them back to the correct tempo. She can choose to lead strictly or more leniently, but completely ignoring the musicians' tempo and conducting like a metronome set at the right tempo will not work. A conductor must incorporate some sense of the actual tempo at which the musicians play in her conducting, or else she will lose control. If the musicians play too slowly, the virtual conductor will conduct a little bit faster than they are playing. When the musicians follow him, he will conduct faster yet, till the correct tempo is reached again.

The input of the virtual conductor consists of the audio from the human musicians. From this input volume and tempo are detected. These features are evaluated against the original score (currently stored in MIDI) to determine the conducting style (lead, follow, dynamic indications, required corrective feedback to musicians, etc) and then the appropriate conducting movements of the virtual conductor are generated. Computer vision has not yet been added to the system. That is, musicians can only interact with the conductor through their music. In a future implementation we can look at the possibility to have the conducting behavior directed to (the location of) one or more particular instruments and their players.

3.3 A Virtual Trainer

A Virtual Trainer is a virtual human capable of presenting physical exercises that are to be performed by a user, monitoring the user's performance and providing feedback accordingly at different levels [1,3]. Our virtual trainer [11] fulfills most of the functions of a real trainer: not only performs the exercises to be followed, provides professionally and psychologically sound, human-like coaching. Depending on the motivation and the application context, the exercises may be general ones of fitness to improve the user's physical condition, special exercises to be performed from time to time during work to prevent for example RSI, or physiotherapy exercises with medical indications.

Currently, the trainer is in its design phase, but parts of its intended behavior are already implemented. In the design the focus is on the reactivity of the trainer, manifested in natural language comments on readjusting the tempo, pointing out mistakes or rescheduling the exercises. When choosing how to react, the static and dynamic characteristics of the user and the objectives to be achieved are taken into account, and evaluated with respect to biomechanical knowledge and psychological considerations of real experts. Hence if the user is just slowing down, the trainer will urge him in a friendly way to keep up with the tempo, and acknowledge with cheerful feedback good performance and introduce small talk every now and then to keep the user motivated.

Figure 1. Virtual dancer, virtual conductor, and virtual trainer.

3.4 Aligning Interaction with Goals and Events

In Figure 1 we have illustrated our three applications. In our applications there is a strong alignment or integration to external channels or events over which our embodied agents and the other partners in the interaction hardly have control. For example, whatever the bodily and nonverbal interaction is that takes place between virtual dancer and human dancer, the virtual dancer is programmed to follow the music and the human dancer is assumed to cooperate. Hence, their dance movements focus on the alignment of the dance behavior to the music. Similar observations can be made about the virtual conductor and the virtual trainer or therapist. The conductor has the score of the music as basis of his conducting behavior and its behavior aims at making the musicians follow the music. Our therapist has not yet been fully implemented, but one of the aims is to have its gestures, body movements, and verbal expressions aligned with music that supports the fitness, aerobics and physiotherapy exercises it offers to its human partners. It should be mentioned that similar kinds of alignment need to be observed in every application where we have rules and constraints on interaction, not only in dancing, conducting and training applications, but also in sports (tennis, fencing, chess) and in dialogues or conversations. For example, also in a dialogue there are constraints, there are dialogue dynamics and there often is an aim that guides the progress of the dialogue and the conversational partners align their conversation (and turn taking) behavior to conversational principles and progress towards a joint goal or towards individual goals.

To define the alignment of the multimodal interaction channels (like speech, gesture, motion and sounds) with each other and events like the beat of the music, we make use of our own extension of the multimodal synchronization language BML [6]. To make behavior, like body movement, fit external events, we might have to stretch or skew its timing. Currently we are researching which parameters can be used to speed up and slow down movement [12]. We are analyzing captured motion of humans to develop models of modifying movements. These models will be used to drive a virtual human, and the resulting motion will be evaluated by user tests. We made use of a time warping technique similar to [14] to align dance or conducting movement to the beat of music. However, this technique allows only relatively small variations in timing. Movements that are played back too slow look like a slow-motion movie. Playing motion too fast produces physically impossible movement. Our preliminary motion capture analysis indicates that amplitude reduction (in our test defined as the maximum amount of distance between the hands in a clapping motion) of a movement can be used as a speed up mechanism.

4. Shortcomings: Intelligence, Social Influence and Persuasion

A serious shortcoming of our agents (conductor, physiotherapist, and dancer) is an explicit modeling of their intelligence and their reasoning behavior. We certainly need to look at BDI&E (Beliefs, Desires, Intentions & Emotions) agents in order to model both the aims and the knowledge of our virtual humans as well as their ability to plan interaction behavior. This planning requires reasoning capabilities that take into account the role of emotions and persuasive strategies. The latter aim at inducing changes in actions and behavior of the partners of the virtual humans.

It is well-known that people respond to computers as if they are social actors [9]. This is even truer if the computer is hidden behind a virtual human that has been designed to be a social actor in its interactions with a human partner. Such a virtual human can exercise social influence and therefore it can persuade and motivate.

Persuasion plays a role in our applications. We do want interaction between trainer and user, between conductor and musicians and between virtual and human dancer. For example, the virtual dancer is not really subtle in inviting you to dance with her and she is really disappointed when you leave. In [4] a comprehensive survey of persuasive technologies is presented. Persuasive agents are discussed in [5]. It is certainly useful, but the length of this paper does not allow this, to look at our applications from the point of view of persuasion. That is, analyze our designs and implementations, hopefully including evaluations, and redesign them from the point of view of computers as social actors that can persuade and from the point of view of persuasive environments where we can simulate desired, expected and inspiring behavior with the aim to improve physical performance or enhance entertaining experience.

References

[1] S. Babu et al. Virtual Human Physiotherapist Framework for Personalized Training and Rehabilitation. Proceedings Graphics Interface 2005, Victoria, Canada, 2005.

[2] P. Bos, D. Reidsma, Z. Ruttkay & A. Nijholt. Interacting with a Virtual Conductor. Proceedings International Conference on Entertainment Computing (ICEC 2006), Cambridge, UK, LNCS 4161, Springer, 2006, 25-30.
[3] J.W. Davis & A.F. Bobick. Virtual PAT: A Virtual Personal Aerobics Trainer. MIT Media Laboratory, TR 436, 1998.
[4] B.J. Fogg. *Persuasive Technology*. Morgan Kaufmann Publishers, San Francisco, CA, 2003.
[5] M. Guerini. Persuasion Models for Multimodal Message Generation. Ph.D. thesis, University of Trento, Trento, Italy, March 31, 2006.
[6] S. Kopp, B. Krenn, S. Marsella, A.N. Marshall, C. Pelachaud, H. Pirker, K.R. Thórisson & H.H. Vilhjálmsson. Towards a Common Framework for Multimodal Generation: The Behavior Markup Language. Proceedings Intelligent Virtual Agents (IVA 2006), Marina del Rey, USA, LNAI 4133, Springer, 2006, 205-217.
[7] A. Nijholt. Google Home: Experience, Support and Re-experience of Social Home Activities. Information Sciences. Elsevier, Amsterdam, 2007, to appear.
[8] M. Pantic, A. Pentland, A. Nijholt & T. Huang. Human Computing and Machine Understanding of Human Behavior: A Survey. ACM SIGCHI Proceedings of the Eight International Conference on Multimodal Interfaces (ICMI'06), Banff, Alberta, Canada, ACM, New York, NY, 239-248.
[9] B. Reeves & C. Nass. *The Media Equation*. Cambridge University Press, Cambridge, MA., 1996.
[10] D. Reidsma, H. van Welbergen, R. Poppe, P. Bos, & A. Nijholt. Towards Bi-directional Dancing Interaction. Proceedings International Conference on Entertainment Computing (ICEC 2006), Cambridge, UK, LNCS 4161, Springer, 2006, 1–12.
[11] Z. Ruttkay, J. Zwiers, H. Welbergen van, & D. Reidsma. Towards a Reactive Virtual Trainer. Proceedings Intelligent Virtual Agents (IVA 2006), Marina del Rey, USA, LNAI 4133, Springer, 2006, 292-303.
[12] Z. Ruttkay & H. van Welbergen. On the timing of gestures of a Virtual Trainer. Central European Multimedia and Virtual Reality Conference 2006, 6-8 November 2006, Eger, Hungary.
[13] T. Shiratori, A. Nakazawa & K. Ikeuchi. Dancing-to-Music Character Animation. EUROGRAPHICS 2006, E. Gröller and L. Szirmay-Kalos (Guest Editors), Volume 25 (2006), Number 3, to appear.
[14] A. Witkin & Z. Popovic. Motion warping. Proceedings of the 22nd annual conference on Computer Graphics and Interactive Techniques (SIGGRAPH '95), ACM Press, 1995, 105-108.

Fundamentals of Verbal and Nonverbal Communication and the Biometric Issue
A. Esposito et al. (Eds.)
IOS Press, 2007

A POMDP approach to Affective Dialogue Modeling

Trung H. BUI[1], Mannes POEL, Anton NIJHOLT, and Job ZWIERS
Human Media Interaction, University of Twente, the Netherlands

Abstract. We propose a novel approach to developing a dialogue model that is able to take into account some aspects of the user's affective state and to act appropriately. Our dialogue model uses a Partially Observable Markov Decision Process approach with observations composed of the observed user's affective state and action. A simple example of route navigation is explained to clarify our approach. The preliminary results showed that: (1) the expected return of the optimal dialogue strategy depends on the correlation between the user's affective state & the user's action and (2) the POMDP dialogue strategy outperforms five other dialogue strategies (the random, three handcrafted and greedy action selection strategies).

Keywords. Dialogue modeling, dialogue management, affective computing, Partially Observable Markov Decision Processes, Dynamic Bayesian Networks

Introduction

We aim to develop dialogue management models which are able to act appropriately by taking into account some aspects of the user's affective state. These models are called *affective dialogue models*. Concretely, our affective dialogue manager processes two main inputs, namely the user's action (e.g., dialogue act) and the user's affective state, and selects the most appropriate system's action based on these inputs and the context. In human-computer dialogue, this work is difficult because the recognition results of the user's action and affective state are ambiguous and uncertain. Furthermore, the user's affective state can change over time. Therefore, an affective dialogue model should take into account both the basic dialogue principles (such as turn-taking and grounding) and the dynamic aspects of the user's affective state.

We found that Partially Observable Markov Decision Processes (POMDPs) are suitable for use in designing these affective dialogue models because of three main reasons. First, the POMDP allows for realistic modeling the user's affective state, user's intention, and other user's hidden states by incorporating them into the state space. Second, the recent dialogue management research for spoken dialogue systems [1,2,3,4] has shown that the POMDP-based dialogue model is able to cope well with the uncertainty that can occur at many levels inside a dialogue system from the speech recognition, natural language understanding to the dialogue management. Third, the transition model and observation model of a POMDP are usually represented by a set

---

[1] Corresponding Author: Trung H. Bui, Human Media Interaction, Department of Computer Science, University of Twente, Drienerlolaan 5, 7522 NB, Enschede, the Netherlands; Email: t.h.bui@ewi.utwente.nl

of Dynamic Bayesian Networks. These networks are suitable for modeling the user affect and for simulating the behavior of the user.

In this paper, we first introduce a short overview of POMDP and its application to the dialogue management problem. Second, a general affective dialogue model using POMDP is described. Then, we present a simple example to illustrate our ideas and discuss future work.

1. POMDP and Dialogue Management

A POMDP is defined by the tuple $<S, A, Z, T, O, R>$, where S is the set of states (of the environment), A is the set of the agent's actions, Z is the set of observations the agent can experience of its environment, T is the transition model, O is the observation model, and R is the reward model (Figure 2a).

In a dialogue management context (Figure 1), the agent is the system (i.e., the dialogue manager) and a part of the POMDP environment represents the user's state. The system uses a state estimator (SE) to compute its internal belief about the user's current state and a policy π to select actions. SE takes as its input the previous belief state, the most recent action and the most recent observation, and returns an updated belief state. The policy π selects actions based on the system's current belief state [5].

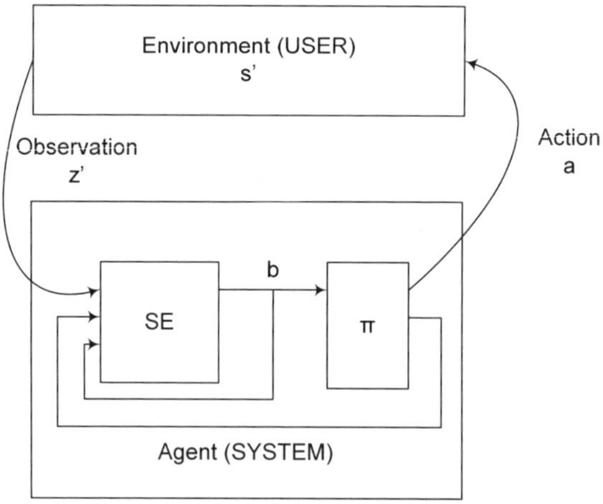

Figure 1. The interaction between the agent (the system) and its environment (the user) in a dialogue management context

Concretely, the system starts with an initial belief state b_0. At time t, the system belief is b, it selects action a and sends to the user. The user's state changes to s'. State s' is unobservable and the system only gets observation z'. At this moment the system needs to update its belief state b' given knowing b, a, z':

$$b'(s') = SE(b,a,z') = \alpha O(s',a,z') \sum_{s_i \in S} T(s_i,a,s')b(s_i),$$

where $\alpha = 1/P(z'|a,b)$ is the normalizing constant.

The system task is usually involved in finding the optimal policy (i.e. optimal dialogue strategy) $\pi^{*}=\mathrm{argmax}_{\pi}\mathrm{E}[V^{\pi}(b)]$, where $\mathrm{E}[V^{\pi}(b)]$ is the expected value function and

$$V^{\pi}(b) = E_{\pi}\left[\sum_{k=0}^{\infty}\gamma^{k}\sum_{s_i \in S} b_{t+k}(s_i)R(s_i,a_{t+k}) \mid b_t = b\right],$$

γ is a discount factor which ensures the sum is finite ($0\leq\gamma<1$) and the closer γ to 1 the more effect future rewards have on current system action selection.

The first work that applies POMDP for the dialogue management problem is proposed by Roy and his colleagues for the nursing home robot application [1]. In this application, a flat POMDP is used where the states represent the user's intentions; the observations are the speech utterances from the user; and the actions are the system responses. They show that the POMDP dialogue manager handles well with noisy speech utterances, for example their POMDP-based dialogue manager makes fewer mistakes than an MDP dialogue manager and it automatically adjusts the dialogue policy when the quality of the speech recognition degrades. Zhang's model [2] extends Roy's model in several dimensions: (1) a factored POMDP [6] is deployed for the state and observation sets, (2) the states are composed of the user's intensions and "hidden system states", (3) the observations are the user's utterances and other observation being inferred from lower-level information of speech recognizer, robust parser, and other input modalities. Williams's model [3,4] further extends Zhang's model by adding the state of the dialogue from the perspective of the user to the state set. All these approaches focus on spoken dialogue systems.

Our POMDP dialogue model extends the previous work by integrating the user's affective states and observed user's affective states into state and observation spaces. Furthermore, we propose to verify the performance of the POMDP-based dialogue strategy as well as to simulate the user behavior using a set of parameters (to generate the transition and observation models). The detail of the model is described in the next section.

2. A POMDP Approach to Affective Dialogue Modeling

We select the factored POMDP [6] for representing our affective dialogue model. The state set and observation set are composed of six features. The state set is composed of the user's goal (G_u), the user's affective state (E_u), the user's action (A_u), and the user's grounding state (D_u) (similar to the user's dialogue state described in [3,4]). The observation set is composed of the observed user's action (OA_u) and the observed user's affective state (OE_u). Depending on the complexity of the application's domain, these features can be represented by more specific features. For example, the user's affective state can be encoded by continuous variables such as valence and arousal, and can be represented using a continuous-state POMDP [7]. The observed affective state might be represented by a set of observable effects such as response speech, speech pitch, speech volume, posture, and gesture [8].

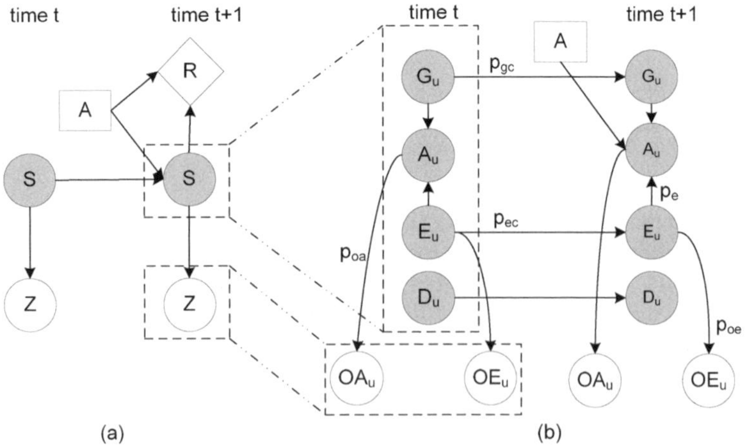

Figure 2. (a) Standard POMDP, (b) Two time-slice of factored POMDP for ADM, where state set S is factored into four features G_u, E_u, A_u, and D_u, the observation set Z is factored into two features OA_u and OE_u

At the moment we are focusing on finite-state discrete-time POMDPs. Figure 2b shows our affective dialogue model (ADM). The features of the state set, action set, observation set, and their correlations form a two time-slice Dynamic Bayesian Network (2TBN). The 2TBN in Figure 2b is built for our route navigation example that will be presented in Section 4. We can easily modify this 2TBN for representing other correlations, for example the correlation between the user's goal and affective state. Parameters p_{gc}, p_{ec}, p_e, p_{oa}, and p_{oe} are used to produce the transition and observation models in case no real data is available, where p_{gc} and p_{ec} are the probabilities that the user's goal and emotion change; p_e is the probability of the user's action error induced by emotion; p_{oa} and p_{oe} are the probabilities of the observed action and observed affective state errors.

The reward model depends on each specific application. Therefore, it is not specified in our general affective dialogue model.

3. Example: Route Navigation in an Unsafe Tunnel

We illustrate our affective dialogue model described in Section 2 by a simulated toy route navigation example. An accident happened in a tunnel. A rescue member (denoted by ``the user") is sent to the unsafe part of the tunnel to evacuate some injured victims. Suppose the user is in one of three locations (v_1, v_2, v_3). The user interacts with the system which is located at the operation center. The system is able to produce the route description when knowing the user's current location. Furthermore, the system can detect the user's stressful state (`nostress` or `stress`) and uses this information to act appropriately. In this simple example, the system can `ask` the user about his current location, `confirm` a location provided by the user, show route description (`ok`) of a given location, and `stop` the dialogue by connecting the user with the operator.

The POMDP for this problem is represented by $S=<G_u \times A_u \times E_u \times D_u>=$ $<\{v_1, v_2, v_3\} \times \{answer(v_1), answer(v_2), answer(v_3), yes, no\} \times \{stress, nostress\} \times \{notstated, stated\}>$, $A=\{ask, confirm(v_1), confirm$

(v2),confirm(v3),ok(v1),ok(v2),ok(v3),stop}>,O=<$OA_u \times OE_u$>=<{answer(v1),answer(v2),answer(v3),yes,no}×{stress,nostress}>. The full flat-POMDP model is composed of 61 states (including a special end state), eight actions, and ten observations.

The transition and observation models are generated from the 2TBN (Figure 2b). We assume that the observed user's action only depends on the true user's action (i.e. $P(oa_u|a_u)=(1-p_{oa})$ if $oa_u=a_u$, otherwise $P(oa_u|a_u)=p_{oa}/4$). The observed user's affective state is computed in a similar way. We use two criteria to specify the reward model, helping the user obtain the correct route description as soon as possible and maintaining the dialogue appropriateness [3]. Concretely, if the system confirms when the user's dialogue state is notstated, the reward is -2, the reward is -5 for action stop, the reward is 10 for action ok(v_i) if $g_u=v_i$, otherwise the reward is -10. The reward for any action taken in the absorbing end state is 0. The reward for any other action is -1. Whenever the system selects actions ok or stop, the current state changes to the end state and the dialogue episode (or dialogue session) is ended.

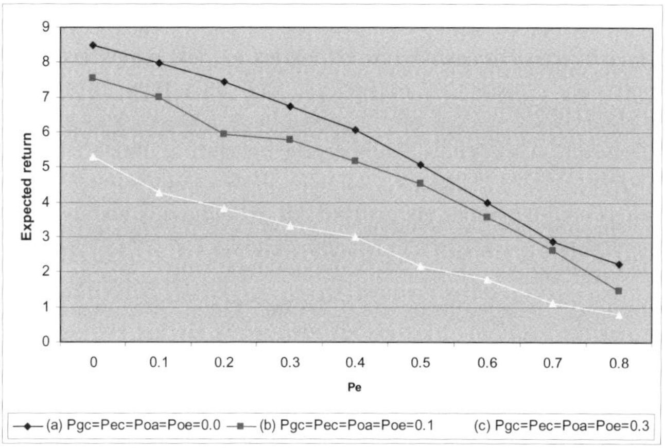

Figure 3. Expected return vs. the user's action error induced by stress p_e

The expected return (i.e. the expected amount of future discounted reward the agent can gather in a large number of steps) of the optimal policy (Figure 3) is computed using the Perseus [9] which is an approximate POMDP algorithm that requires two inputs, a number of belief points and a maximum runtime value. We found 1000 belief points and a runtime of 60 seconds be a good choice for testing our problem. The probability of the user's action error being induced by stress p_e changes from 0 (stress has no influence to the user's action selection) to 0.8 (the user is highly stressed and acts almost randomly). Three lines in Figure 3 are: no observation error ($p_{oa}=p_{oe}=0.0$); low observation error ($p_{oa}=p_{oe}=0.1$); and high observation error ($p_{oa}=p_{oe}=0.3$). All these lines show that the expected return of the optimal policy depends on p_e.

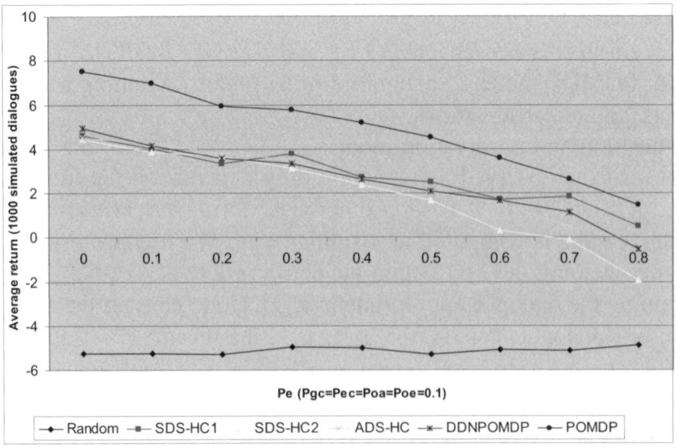

Figure 4. Comparison between POMDP and other dialogue strategies

Figure 4 shows a quantitative comparison between the POMDP dialogue strategy and a set of five other system dialogue strategies (1. select action randomly, 2. first ask and give route description (SDS-HC1), 3. first ask, then confirm, and give route description (SDS-HC2), 4. first ask, then confirm if stress, and give route description (ADS-HC), 5. select the greedy action using a set of 2 time-slice Dynamic Decision Networks (DDNPOMDP). Strategy 1 aims to show the difference between the random dialogue strategy and other dialogue strategies. Strategies 2 and 3 are considered as the non-affective dialogue strategies since they ignore the user's stress state. Strategy 4 uses commonsense rules to generate the system behavior. Strategy 5 is a special case of the POMDP-based dialogue with the discount factor $\gamma=0$ (this strategy is used in [10,11]). The result is obtained by letting each strategy interact with the simulated user (the simulated user model is constructed as a 2TBN described in Figure 2b). The average return is the average dialogue episode reward the agent receives (1000 dialogue episodes are carried out for each strategy). As expected, the POMDP dialogue strategy outperforms all other strategies, see Figure 4.

4. Conclusions and Future Work

We have presented a POMDP approach to affective dialogue modeling and illustrated our affective dialogue model by a simple example. The 2TBN representation allows integrating the features of states, actions, and observations in a flexible way. We have also shown that even if the observation is perfect, the expected return of the optimal dialogue strategy depends on the correlation between the user's affective state and the user's action. The POMDP dialogue strategy outperforms five other strategies (the random, three handcrafted and greedy action selection strategy). Furthermore, the POMDP dialogue strategy copes well with different types of errors such as speech recognition error [1,2,3,4] and the user's action error being induced by stress as showed in Section 3.

However, solving the POMDP problem (i.e. finding the optimal policy) is computationally expensive. Therefore, all currently developed POMDP dialogue management work is limited to toy frame-based dialogue problems with the size of several slots [4]. We are currently working with the scaling up issue; especially we focus on the online belief update for real-world dialogue systems.

Acknowledgements

This work is part of the ICIS program (http://www.icis.decis.nl). ICIS is sponsored by the Dutch government under contract BSIK 03024.

References

[1] N. Roy, J. Pineau, and S. Thrun. Spoken Dialogue Management using Probabilistic Reasoning. In Proceedings of the 38th Annual Meeting of the Association for Computational Linguistics, 2000.

[2] B. Zhang, Q. Cai, J. Mao, and B. Guo. Spoken Dialog Management as Planning and Acting under Uncertainty. In Proceedings of Eurospeech, 2001.

[3] J.D. Williams,P. Poupart, and S. Young. Factored Partially Observable Markov Decision Processes for Dialogue Management. In 4th Workshop on Knowledge and Reasoning in Practical Dialog Systems, 2005.

[4] J.D. Williams and S. Young. Scaling POMDPs for Dialog Management with Composite Summary Point-based Value Iteration (CSPBVI). In AAAI Workshop on Statistical and Empirical Approaches for Spoken Dialogue Systems, July 2006.

[5] A.R. Cassandra, L.P. Kaelbling, and M.L. Littman. Acting Optimally in Partially Observable Stochastic Domains. In Proceedings of the Twelfth National Conference on Artificial Intelligence (AAAI-94), volume 2, pages 1023–1028, Seattle, Washington, USA, 1994. AAAI Press/MIT Press.

[6] C. Boutilier and D. Poole. Computing Optimal Policies for Partially Observable Decision Processes using Compact Representations. In AAAI/IAAI,Vol.2, pages 1168–1175,1996.

[7] A. Brooks, A. Makarenko, S. Williams, and H. Durrant-Whyte. Planning in Continuous State Spaces with Parametric POMDPs. In IJCAI Workshop Reasoning with Uncertainty in Robotics, July 2005.

[8] E. Ball. A Bayesian Heart: Computer Recognition and Simulation of Emotion. In Paolo Petta, Robert Trappl, and Sabine Payr, editors, Emotions in Humans and Artifacts, chapter 11, pages 303–332. The MIT Press, 2003.

[9] M.T.J. Spaan and N. Vlassis. Perseus: Randomized Point-based Value Iteration for POMDPs. Journal of Artificial Intelligence Research, 24:195–220, 2005.

[10] T. Paek and E. Horvitz. Conversation as action under uncertainty. In UAI '00: Proceedings of the 16th Conference on Uncertainty in Artificial Intelligence, pages 455-464, San Francisco, CA, USA, 2000. Morgan Kaufmann Publishers Inc.

[11] S. Young, J.D. Williams, J. Schatzmann, M. Stuttle, and K. Weilhammer. The hidden information state approach to dialogue management. Technical Report CUED/F-INFENG/TR.544, University of Cambridge, 2005.

Author Index